NEURAL MODELING

Electrical Signal Processing in the Nervous System

NEURAL MODELING

Electrical Signal Processing in the Nervous System

Ronald J. MacGregor
University of Colorado, Boulder

and Edwin R. Lewis
University of California, Berkeley

PLENUM PRESS • NEW YORK AND LONDON

Library of Congress Cataloging in Publication Data

MacGregor, Ronald J
Neural modeling.

Bibliography: p.
Includes index.
1. Nervous system—Mathematical models. 2. Electrophysiology—Mathematical models. 3. Biomedical engineering. I. Lewis, Edwin R 1934- joint author. II. Title.
QP363.M18 612'.81'0184 77-8122
ISBN 0-306-30871-1

A Division of Plenum Publishing Corporation
227 West 17th Street, New York, N.Y. 10011

Printed in the United States of America

To our parents

Ronald E. and Vergene Y. MacGregor
and
Edwin M. and Sally R. Lewis

Preface

The purpose of this book is to introduce and survey the various quantitative methods which have been proposed for describing, simulating, embodying, or characterizing the processing of electrical signals in nervous systems. We believe that electrical signal processing is a vital determinant of the functional organization of the brain, and that in unraveling the inherent complexities of this processing it will be essential to utilize the methods of quantification and modeling which have led to crowning successes in the physical and engineering sciences. In comprehensive terms, we conceive neural modeling to be the attempt to relate, in nervous systems, function to structure on the basis of operation. Sufficient knowledge and appropriate tools are at hand to maintain a serious and thorough study in the area. However, work in the area has yet to be satisfactorily integrated within contemporary brain research. Moreover, there exists a good deal of inefficiency within the area resulting from an overall lack of direction, critical self-evaluation, and cohesion. Such theoretical and modeling studies as have appeared exist largely as fragmented islands in the literature or as sparsely attended sessions at neuroscience conferences.

In writing this book, we were guided by three main immediate objectives. Our first objective is to introduce the area to the upcoming generation of students of both the hard sciences and psychological and biological sciences in the hope that they might eventually help bring about the contributions it promises. In other words, we introduce the area as much for its potential value as for its accomplishments. Our second objective is to make the studies more easily available to our biologically and psychologically trained colleagues so as to integrate the area more effectively within the mainstream of ongoing brain research. Finally, our third objective is to introduce some cohesiveness (although not consolidation) in the area in the sense of encouraging more communication, more critical self-evaluation, and a sense of direction to higher levels of accomplishment than we have seen thus far.

Recently, several other admirable books which are related more or less directly to our area of interest have appeared. We have proceeded to write our book nonetheless, not because we think these books are inadequate, but simply because it seemed the natural and appropriate thing to do given our professional and research interests and activities. We hope only that this work might provide some contribution to our field along with those works of our colleagues who have been similarly motivated.

R. J. MacGregor
E. R. Lewis

Acknowledgments

Many people have contributed to the substance of this book or supported us in various critical ways during its creation. First, we express heartfelt thanks to our wives, Betsy and Millie, and our children, Ned, Sarah, Ronnie, Terri, and Felicia, for bearing so well the absenteeism required by this effort. We express thanks as well to many colleagues whose ideas and discussions have influenced our thinking, including most particularly P. S. Lykoudis, L. D. Harmon, S. K. Sharpless, T. H. Bullock, D. H. Perkel, R. F. Reiss, P. M. Groves, H. J. Hamilton, R. M. Oliver, T. E. Posch, and P. Vogelhut. Robert S. Zucker deserves particular thanks for his suggestions concerning Part II. Many other colleagues, too numerous to name, of the EECS, Biophysics, and Neurobiology faculties at Berkeley, and the EDEE, Chemical Engineering, and Psychology departments at Boulder, have contributed ideas to the material contained herein. We also thank various students of EECS 181 and IDS 201 at Berkeley and EDEE 571 and EDEE 572 at Boulder for many exciting discussions and suggestions concerning the material of the book. We are indebted to many of our graduate students who have worked often tirelessly and with great insight on various aspects of material described here, and welcome this opportunity to express admiration and thanks. In this regard we thank in particular M. J. Murray, M. Hassul, J. S. Charlton, E. Mayeri, Y. Y. Zeevi, R. Plantz, Y. Hazeyama, K. L. Lee, C. W. Li, I. Weeks, R. Prieto-Diaz, R. Spurlock, C. Radcliffe, R. Palasek, T. McMullen, B. Sherman, T. Bundy, R. Adams, D. Glover, Eric Berg, and T. Detman. In very large part the book has grown from these courses and theses work. We are also grateful to the National Science Foundation (Grants GG 31427, GK 42054, and GB 33687) and to the

National Institute of Neurological and Communicative Disorders and Stroke (Grants 5R01NS12359 and 1R01NS10781-01) for support of much of our own original work discussed here.

We express particular thanks to the secretaries of the EECS department at Berkeley and to Mrs. Virginia ("Ginny") Birkey and Mrs. Marie Hornbostel of the EDEE department at Boulder for painstaking care and limitless patience in the essential task of typing the several drafts of the manuscript. Finally, we express appreciation to Seymour Weingarten, Evelyn Grossberg, and Betty Bruhns, of Plenum Press, who have shown great patience in tolerating delay after delay in the preparation of the book, and whose expert editorial guidance has enhanced the readability of the manuscript.

Finally, we should like to thank the following individuals and publishing companies for permission to use quotations from previously published materials: Dr. Ter Andersen, Dr. Jack Cowan, Sir John Eccles, Dr. Karl Lashley (deceased), Dr. David Marr, Dr. Wilfrid Rall, The Association for Research in Nervous and Mental Disease, Inc., Cambridge University Press, The Royal Society of London, and Springer-Verlag Company.

The substantive deficiencies of the book remain our own responsibility.

Contents

PART I

INTRODUCTION

1

Signal Processing in Nervous Systems

The processing of electrical signals by the brain has long held a particular fascination for students and researchers. When DuBois Reymond made the first successful laboratory measurements of the nerve impulse in the mid-19th century, he wrote: "If I do not greatly deceive myself, I have succeeded in realizing in full actuality . . . the hundred years' dream of physicists and physiologists, to wit, the identification of the nervous principle with electricity." About half a century later Sherrington, in a classic work summarizing his research on the integration of spinal reflex activity, characterized the nervous system as " . . . an enchanted loom where millions of flashing shuttles weave a dissolving pattern, always a meaningful though never an abiding one."

1.1. Neurons

Two broad tenets underlie contemporary research on electrical signal processing in the brain. The most fundamental structural concept is that the nervous system is organized in terms of an immense number of elemental units called neurons which are arranged in functional constellations or assemblies according to the synaptic contacts they make with one another. Neurons come in an immense variety of shapes and sizes, most of which are extremely intriguing in their own right as examples of naturally occurring structures independently of their physiological and psychological functioning. Various schemes for classification of neurons on the basis of morphology have been devised, but none is entirely satisfactory. Neuron structures range from the relatively simple bipolar cell of primary sensory systems to

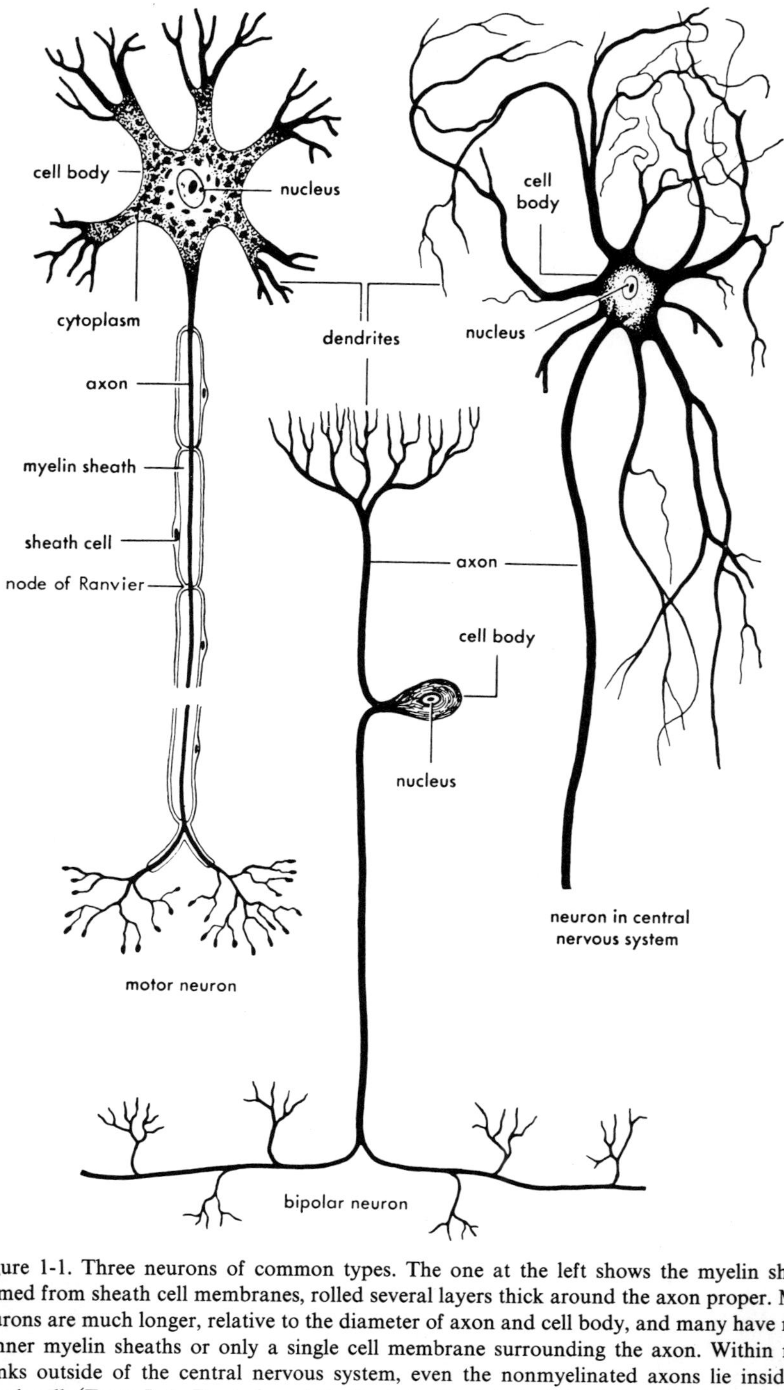

Figure 1-1. Three neurons of common types. The one at the left shows the myelin sheath, formed from sheath cell membranes, rolled several layers thick around the axon proper. Many neurons are much longer, relative to the diameter of axon and cell body, and many have much thinner myelin sheaths or only a single cell membrane surrounding the axon. Within nerve trunks outside of the central nervous system, even the nonmyelinated axons lie inside the sheath cell. (From J. A. Deutsch and D. Deutsch, *Physiological Psychology*, The Dorsey Press, Homewood, Illinois, 1973.)

the elaborate pyramidal cells of the cerebral cortex and hippocampus. The bipolar cell illustrated in Figure 1-1 illustrates the main anatomical features of neurons generally. As do almost all animal cells, neurons possess an expanded region that houses the nucleus and much protein-synthesis apparatus. This region is known as the *soma*, the *perikaryon*, or the *cell body*. Emanating from this region are long, often-branching fibers, the *dendrites*, and the *axon* and axon branches. There are structural distinctions between axons and dendrites, but questions still occasionally arise concerning functional distinctions. Generally, dendrites appear to provide receptive surfaces for input signals to the neuron and to conduct those signals passively (with no amplification) to the soma, which in turn conducts them passively over to the axon. In the axon, the signals generally are converted into trains of very brief pulses, called *nerve impulses*, *nerve action potentials*, or simply *spikes*, which are propagated with no attenuation along the axon and its branches to target cells (i.e., to other neurons, to muscles, to receptors, etc.).

There are many, many exceptions to this general pattern, however. Perhaps the most abundant neuron, in fact, is the unipolar interneuron of the typical invertebrate nervous system. This neuron has only one fiber emanating from its soma (hence the title *unipolar*); and this fiber, or its branches, may serve as both input and output channels. Figure 1-2 displays an array of neuronal silhouettes illustrating just a few of the many structural patterns that neurons assume. Some of these conform to the general pattern, others do not. In the upper left-hand side, we have a bipolar neuron, typical of the cells performing sensory transduction at the periphery of invertebrate nervous systems and a class of cell found in the vertebrate retina. This cell has a single, unbranched dendrite emerging toward the top of the figure and a single axon (which may branch as it reaches the central nervous system) emerging toward the bottom (but truncated in the drawing). The next two silhouettes to the right represent neurons with no apparent axons. The first is an interneuron from the optic lobe of the cuttlefish, and the second is an amacrine neuron from the eye of a fly (redrawn from Bullock and Horridge, 1965). The fibers emanating from these cells apparently are dendrites. The silhouette in the upper left and that in the lower right represent neurons with no apparent dendrites. That in the upper right represents an anastomosis of the axons from several somata to form a single segment of a giant fiber in the ventral nerve chord of an earthworm. That in the lower left represents the rather famous giant axon (shown truncated) of the squid. A large number of somata in the stellate ganglion project small fibers which merge to form the single giant. Immediately to the right of this giant axon is another giant cell. In this case it is the Mauthner's neuron of bony fish and amphibians with tails. As in the case of the other two giants in Figure 1-2, this cell mediates an

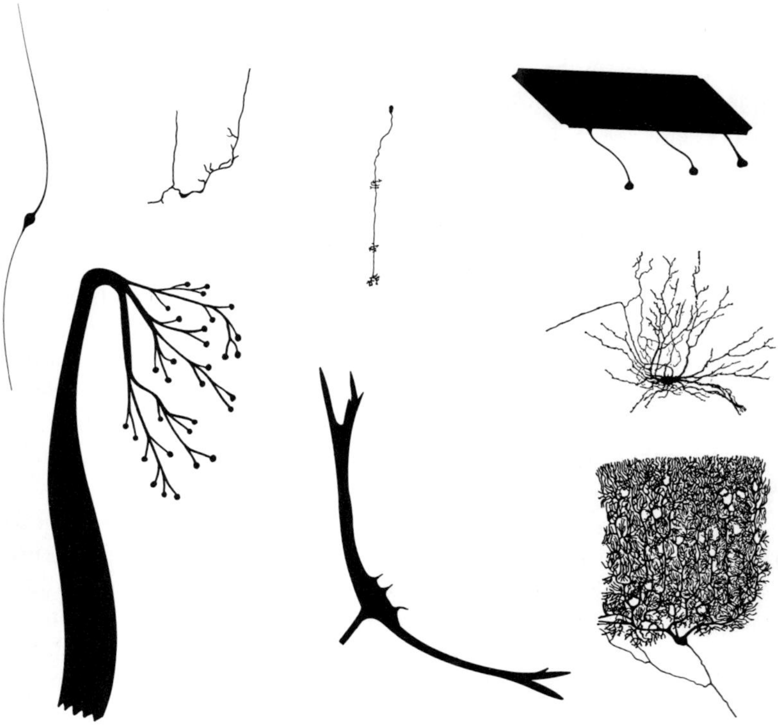

Figure 1-2. Silhouettes of a few neurons or portions of neurons as they might be seen through the light microscope after they have been stained with silver. See text for details.

emergency response, a reflex tail flip or series of tail flips that tend to carry the animal away from the source of any sudden sounds or vibrations in the water. The hornlike structures are dendrites and receive numerous inputs from auditory and vibratory receptors. The axon is shown truncated, emerging from the lower left. The last two silhouettes represent neurons with highly elaborate dendritic structures. In the middle of the right-hand side is a mammalian spinal motor neuron. Its axon is shown truncated, emerging to the upper left and bending downward. The remaining fibers are dendrites, which receive inputs from the brain, from position and tension receptors in the muscle and tendons, and from other neurons in the spinal chord. The axon projects to skeletal muscles. The last silhouette, in the lower right-hand corner, is the well-known Purkinje cell from the cerebellum. The axon emerges from the lower right-hand side of the soma and then branches. The fibers above the soma all are dendrites, which form an elaborate treelike structure that receives inputs from a huge number of other cells. From Figure 1-2, one can begin to see the enormous diversity of neuron shape and to sense that neurons indeed are very much *ad hoc* devices, apparently specifically adapted to a wide variety of niches in the structure and operation of the nervous system.

1.1.1. Synapses and Constellations of Neurons

Axons generally branch repeatedly and end, with swollen terminals called knobs or boutons, on the somata and dendrites of other cells. Synapses also exhibit a wide variety of morphological individuality and peculiarites. Presently, however, they usually are classified into two main types (Type I and Type II), as indicated in Figure 1-3. It is generally true that Type I synapses are excitatory and Type II are inhibitory, but this is an oversimplification, as we shall see later. The Type I synapse is characterized by a subsynaptic membrane which is everywhere thickened under the synapse. The Type II synapse has several small thickened patches in the subsynaptic membrane. Many synaptic contacts in the brain involve so-called postsynaptic "spines," as is also indicated in Figure 1-3. Typically, any given neuron makes outgoing synaptic contacts on hundreds of other neurons and, correspondingly, receives information from hundreds of still other neurons by way of input synapses distributed over its somadendritic

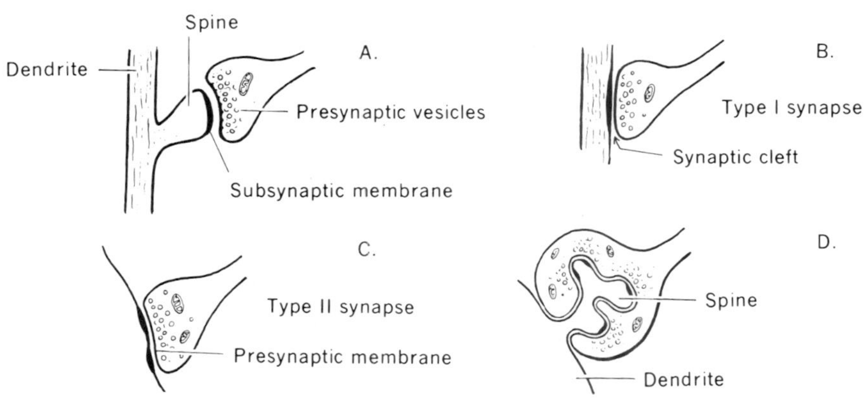

Figure 1-3. Synaptic structure within the central nervous system. Type I synapse has entire subsynaptic membrane thickened. Type II synapse has several small thickened patches on subsynaptic membrane. (A) Type I axodendritic synapse between an axon terminal and a dendritic spine; (B) Type I axodendritic synapse between an axon terminal and dendritic trunk; (C) Type II axodendritic synapse between axon terminal and cell body (axosomatic synapse); (D) a complex axodendritic synapse between an axon terminal and a dendritic spine on a pyramid cell of the hippocampus (archicortex), with several thickened subsynaptic patches. (Adapted from C. R. Noback and R. J. Demarest, *The Human Nervous System*, McGraw-Hill, New York, 1967.)

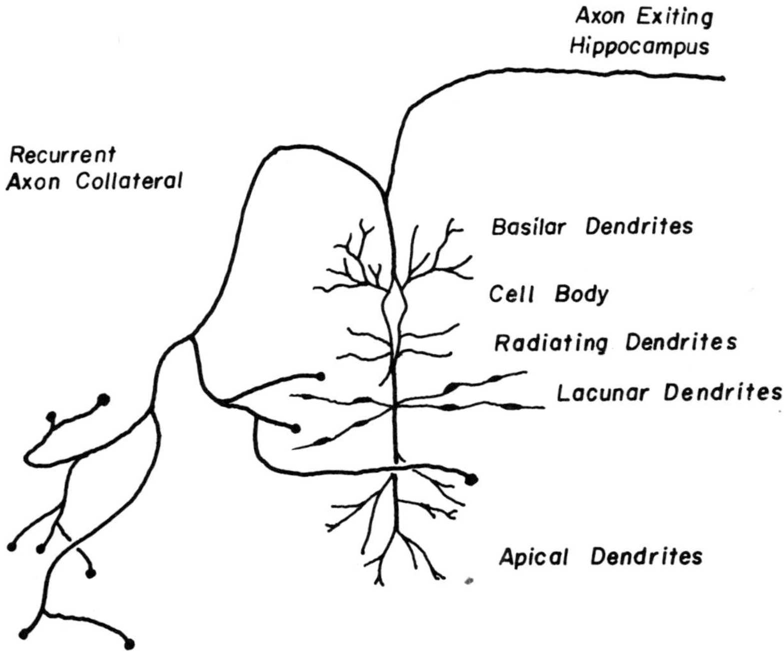

Figure 1-4. Schematic drawing of a pyramidal cell of the hippocampus.

surface. By this arrangement the nervous system is characterized at most junctions by a simultaneous divergence and convergence of connectivity channeling the flow of signals.

The two-dimensional drawing of the pyramidal cell of the hippocampus shown here as Figure 1-4 can only hint at the grandeur and complexity inherent in the basic information-processing networks of the central nervous system. This cell is typically about 1 mm in length; its cell body is approximately 20 μm in diameter. The cell exhibits two distinct dendritic trees each of which is highly elaborate in its own peculiar way, seems to serve separate input systems, and exhibits different physiological properties. Its single axon emanates from the cell body and traverses exceedingly long distances (several millimeters from the hippocampus through the fornix into the septum or hypothalamus). Moreover, before leaving the hippocampus, the axon gives rise to a collateral fiber which protrudes back into the receiving area of the parent pyramidal cell where it seems to synapse on both interneurons within the hippocampus and upon pyramidal cells including itself in some instances.

Vast numbers of such neurons are arranged into functional constellations or assemblies which somehow carry out the functioning of the central nervous system. Such organizing of assemblies can be determined by the channeling of patterns of electrical activity among different regions according to appropriate parallel and series grouping of synaptic contacts. Also, such groupings into assemblies can be determined by chemical specificity in that synapses of a single cell emit a single one of several chemical transmitters. The release into brain tissue of a given neurohumor, for example, can thereby selectively facilitate all synapses associated with a particular transmitter or family of transmitters and thereby selectively define a functional assembly on this basis. Thus, operational assemblies of neurons may be labile, changing from moment to moment in some presumably appropriate manner. On the other hand, certain assemblies may be more-or-less fixed (the term "hard-wired" often is used), thus constituting fixed neural circuits or networks.

1.1.2. Electrical Activity

The second main tenet in our current picture of neural coding refers to the dynamical properties of nervous systems. It is generally assumed that the information or meaning associated with experience, behavior, and the performance of vital internal processes is somehow contained and internally communicated within the chemical and electrical signals emitted more or less continually by the constituent neurons of the brain. In many, but not all neurons, signals apparently are communicated as temporal patterns of action potentials or spikes, as shown here in Figure 1-5. As we shall consider in more detail in later chapters, these signals represent fluctuations in the electrical potential across the membrane of neurons and are thought to reflect underlying patterns of slow, graded potentials in the neuron. These underlying graded potentials often are called *generator potentials*. Typically ongoing nervous activity apparently involves spike patterns being delivered more or less continually by each neuron to each of its synaptic knobs, and thereby, by means of a chemical transmitter, to each of many (perhaps several hundred or more) receiving neurons. Having passed through the synaptic junction, the spike pattern is reflected as a slower, graded potential, which is summed or integrated with perhaps several hundred or more similar responses to patterns received at synapses all over the soma and dendritic surfaces of the receiving neuron. Introducing

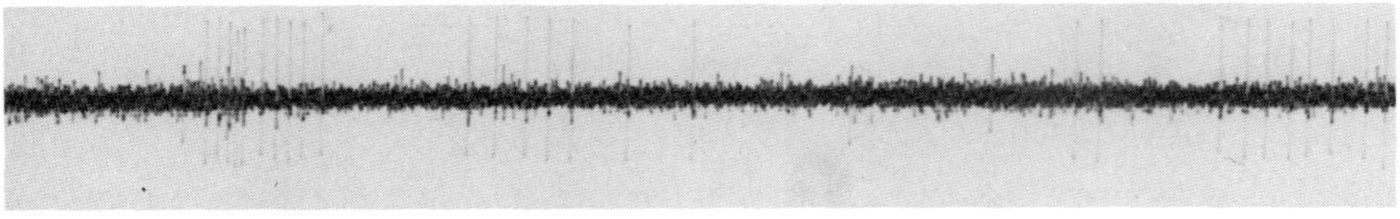

Figure 1-5. Representative time series of action potentials or "spike train" (recorded from midbrain reticular formation of the rat).

modifications based on its own metabolic state, its history of spike generation, and the like, the receiving neuron becomes the sender, translating the resultant slow potential into its own pattern of spikes, which is transmitted to many other neurons. Surrounding the neurons, and often outnumbering them, are the glial cells, whose functional significance is not clear and which exhibit slow, graded potentials of their own. These potentials probably reflect and influence the activity of nearby neurons. The entire system (neurons and glial cells) is immersed in a solution of electrolytes, which is a reasonably good conductor of electrical currents. Significant signals may be broadcast through this medium in the form of chemicals released from pools of neurons or in the form of slow, wavelike potentials reflecting the spike patterns of large assemblies of neurons and, in turn modulating to some extent the subsequent patterns in those and other assemblies.

With this in mind, it is easy to see that several types of recordings of electrical signals can be obtained from brain tissue and its constituent neurons. Relatively large electrodes (or macroelectrodes) can be placed on the surface of the skull or on the exposed surface of the brain, or can be inserted into brain tissue directly to record spontaneous slow-wave activity or fluctuations in that activity in response to certain prescribed input signals. Such records are called electroencephalograms (EEGs,) and the signals are called *brain waves* or *evoked responses*. Typical examples are shown in Figure 1-6. On the other hand, much smaller electrodes (or microelectrodes) with tip diameters of the order of a micrometer or less can be inserted into brain tissue. Such electrodes can be used to record spikes or wave activity from single neurons or small numbers of neurons when the electrode tip is external to but nearby the signaling cell or cells. A typical example is shown in Figure 1-6. Alternatively, such microelectrodes can be used to penetrate the surfaces of neurons and record electrical signals within particular cells. Microelectrodes are typically made with metal wire (tungsten is a popular material for this purpose), or with salt-filled glass pipettes. With modern laboratory methods, electrode tips can be placed with a relatively high degree of accuracy, and although neurons within the mammalian brain cannot be specified individually as target cells, definite brain regions and classes of cells within particular regions can be targeted and penetrated reliably.

1.2. Modeling

The goal of neural modeling is to relate, in nervous systems, function to structure on the basis of operation. The activities of neural modelers are thus largely concerned with the working out of various characteristics and

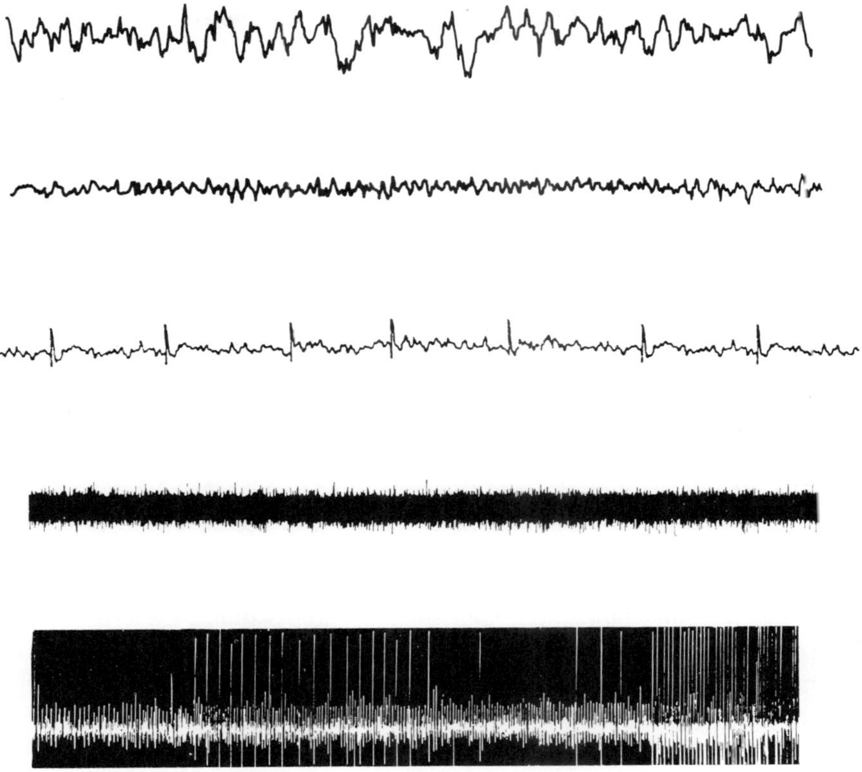

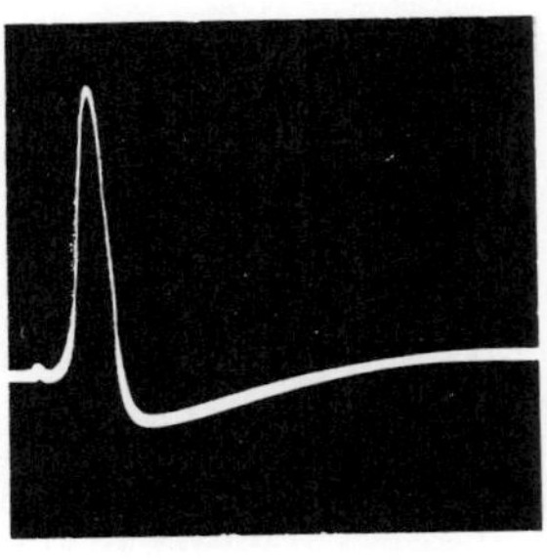

Figure 1-6. Electrophysiological signals which are commonly recorded from the nervous system. (A) EEG of a quiet subject; (B) EEG of an aroused subject; (C) evoked potentials superimposed on the cortical EEG; (D) multiple-unit activity; (E) single-unit activity recorded extracellularly; (F) single-unit activity recorded intracellularly. The EEG and evoked potentials are recorded from large electrodes, the single-unit activity from very small microelectrodes. (From S. P. Grossman, *Essentials of Physiological Psychology,* John Wiley and Sons, New York, 1973.)

peculiarities of the operational dynamics of the systems, subsystems, and cells of nervous tissue. In this light it is easy to see that the orientation of engineering is appropriate for the area, and to understand that modeling work has been performed in large part by engineers. This has been advantageous in that powerful quantitative and analytic techniques have been used and are available in the area. It also, however, has had the unfortunate side effect of producing much work characterized by a rather low level of biological understanding and perspective.

The main purpose of this book is to survey the state of the modeling art as it applies to the operations of electrical signal processing in nervous systems. We are particularly concerned with the actual and potential contributions of engineering and physical sciences to the area. The second main section of the book will deal with models of excitation and conduction in single neurons. The third main section deals with neural coding, and models for ongoing signal processing in single neurons and networks of neurons. Before undertaking these topics, however, we will complete this chapter with a discussion of some basic conceptions of modeling generally, and of some dangers of oversimplification in neural modeling specifically. The introductory section will be completed in Chapter 2 with the description of a highly oversimplified but basic concept of signal processing in single neurons preceded by a discussion of basic cell biology.

It is necessary to clear up some ambiquities concerning the terms "models" and "modeling," and their use and value. The term "model" is commonly used in at least three different senses: as a conceptual model (C-model), as a general-purpose computational model (GP-model), or as a special computational model (SC-model). A conceptual model is very much like a theory—a set of concepts and the interrelationships among them which seem to provide a comprehensive explanation for a given set of phenomena. A general-purpose computational model is a computational instrument for spelling out dynamical properties of certain conceptual models, for example, a general-purpose digital computer simulation scheme. The GP-models can be applied to various different conceptual models by alterations in internal configuration, choices of particular subroutines or parameter values, and the like. The SC-model is also a computational model; it differs from the GP-model in that it represents a realization of a particular conceptual model. In fact, the application of a GP-model to a particular C-model results in an SC-model. The construction of conceptual models is the domain of all researchers, whereas the construction of GP- and SC-models usually is restricted to researchers with special and particular backgrounds and training, particularly in mathematics, computers, or engineering.

The term "modeling" is also ambiguous. In a certain sense all of science is modeling; i.e., the main goal of all scientific research is to develop adequate C-models over the relevant problem area. Moreover, the collective research on a given area can be viewed as an eclectic process to attain progressively better C-models on the area in question. The whole process points toward a relatively final or "climax" conceptual model (or a group of them) which seems to adequately and comprehensively contain the phenomena of the area. At any instant in time in a developing research area, there exist one or more candidate C-models. These describe or contain with varying degrees of comprehensibility the data that may exist in the area up to that point of time. It makes no difference where the various conceptual candidate C-models come from; the only requirement for their validity is that they pass tests of refutability. Moreover, the key to their value is the extent to which they bring on experimental tests to evaluate and improve themselves. Thus, one arrives at the apparently paradoxical truth that the best candidate C-model is the one which brings on its own supercession most quickly. This, then, is the basic eclectic process of scientific research on a given area which results in a "climax" model over the area; it can indeed be conceived as a "modeling" process. It is also clearly the domain of all researchers.

Another, and more common, use of the term "modeling" refers specifically to the construction of computational models—GP-models and SC-models. Here, the modeler uses essentially mathematical or computer techniques to simulate some other system. The justification of this kind of modeling is that it can be applied to a given C-model to tell more about the C-model's behavior. It becomes a necessary part of the eclectic process when and only when the various candidate C-models reach a state of high complexity in terms of their elements and interrelationships such that their behavior is no longer immediately clear from the C-models themselves. When C-models are relatively simple, such as for example, $F = MA$, their implications are generally relatively clear, and experiments can be performed directly. In such a case, considerable progress can be made without the use of computational models. On the other hand, when elements and interrelationships within the domain of the C-model are very complicated, the collective process gets very sluggish. In this case the implications of the C-model are not immediately clear, and the step from candidate C-model to the next experiment is much delayed.

This then is the value of the computational models—the GP-models and SC-models: They can be used to spell out the implications of hypothesized C-models. Their primary purpose and indeed *raison d'etre* is to render candidate conceptual models more amenable to experimental test, i.e., to push candidate conceptual models to and through their necessary refutability exercise. It is important to note that the value of these models is totally dependent upon the existence of C-models; i.e., the existence of a meaningful C-model is a necessary prerequisite for any significant computational modeling. Moreover, it is clear that computational models cannot exist in isolation from experimental refutability testing any more than can conceptual models. Most of the modeling discussed in this book (and particularly that of Part III) makes use of the computational variety of models.

1.2.1. Minimal and Unique Models

Much of neural modeling has been directed toward the understanding of signal generation and processing by "hard-wired" networks. In general, such modeling efforts seem to have been most successful when they were based on reasonably good anatomical evidence combined with observations of signal-generating or signal-processing properties. When either of these bases is not present, the modeling effort is fraught with ambiguities and uncertainties. The doctrine of parsimony led some modelers of the 1950s and 1960s to the notion that a minimal modeling realization of an observed signal-generating or signal-processing property was very likely to be close to anatomical isomorphy with the neural networks in which that property was observed. Occasionally, in fact, one even finds the word "unique" applied to proposed realizations of that era. It is very clear that no modeling realization is unique in being able to duplicate an observed property; in fact, each property will have an infinite number of realizations. Furthermore, whether or not nature is parsimonious (e.g., for some teleological or thermodynamical reason), the development of neural networks in various nervous systems clearly has been limited by numerous constraints, including those imposed by phylogeny, ontogeny, and the basic structural and operational limitations of the neuron and its components.

For example, if one were designing a minimal network to fire simultaneously a large number of electric organs in an electric fish, he probably simply would connect them all to a single neuron through axonal branches that imposed identical delays on the single command to fire. This single neuron then could be driven by a firing-pattern generator (or, perhaps, it could generate the firing pattern itself). Apparently, however, electric organs are evolutionary adaptations of muscles; and the muscle cells already had central motor neurons associated with them. Thus, when a

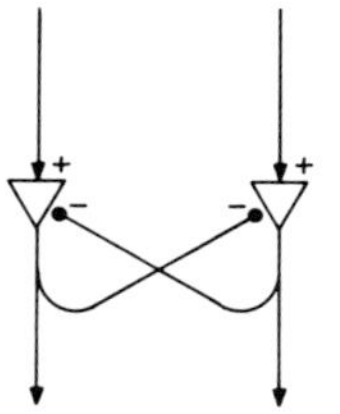

Figure 1-7. The reciprocally inhibiting pair, a simple neural-network model studied by many people, including Reiss and Landahl. The triangle represents the receiving and signal-processing parts of a neuron (usually the dendrites and soma), while the lines represent the signal-conducting parts (axons). Each neuron receives excitatory inputs (perhaps from a common source) and, in addition to sending corresponding outputs to other targets, sends an inhibitory signal to its counterpart.

large number of muscle cells became adapted to the function of electrical discharge, synchronous firing was assured not by rewiring of their inputs to a single cell, but by the development of very effective electrical connections among all of the existing, corresponding motor neurons. Thus, we find thousands of neurons coupled together to act as one. Without considerable knowledge of the phylogeny and ontogeny of the electric-organ system, a modeler might be hard-pressed to arrive at such an arrangement through considerations of minimal models.

As a second example, consider the model proposed by Landahl as a very simple pattern discriminator and, later, by Reiss as a pattern generator (Figure 1-7). Here we have two model neurons, each sending an axon branch to the other and inhibiting it. Thus, a train of spikes in either of the two neurons would suppress or tend to suppress the generation of spikes in the other. Each neuron is activated by an input channel whose signals are excitatory. Landahl demonstrated that if the synaptic parameters are adjusted appropriately, the cell receiving the stronger signal or the earlier signal could dominate completely, sending out a spike train along its own axon while completely suppressing the output signal from the other neuron. An elaboration of this scheme into a two-dimensional array of mutually inhibiting neurons was suggested (in slightly different terms) by Mach in the nineteenth century and adapted into more modern terms by Bekesy and by Hartline and Ratliff. Such an array could provide enchancement of edges and contrasts in sensory inputs, and anatomical evidence has confirmed its existence in several sensory systems.

In 1903, McDougall, expanding on an old Cartesian idea, suggested that this same network of two neurons could generate the alternating bursts of spikes that drive antagonistic muscles in repetitive activity. In 1962, Reiss confirmed this with neural models and studied the wide variety of patterns available from the network of Figure 1-7 when both neurons were driven by a single input source (i.e., the signals in the two input channels were identical). Reiss demonstrated that the network was capable of generating very stable patterns of alternating spike trains, and that these patterns were inherent in the network itself and tended to be quite independent of patterns in the input signal but depended instead on the average frequency of that signal.

Thus, a single, very simple network structure can exhibit two quite distinct functions. In fact, this same network is so rich in possibilities that several modelers in the years since Reiss's work have devoted considerable effort to cataloging its properties. Do such pattern discriminators and pattern generators exist in real nervous systems? It is clear now that they do. However, those that have been found generally are much more complicated than the simple network of Landahl and Reiss. This undoubtedly would bother neither Landahl or Reiss, since they were interested in operational principles rather than minimal realizations. It does point up once again, however, the danger of the minimal-modeling concept. For example, the present best estimate of the signal-discriminator/signal-generator network for emergency operation of antagonistic tail-flip muscles in the fish and tailed amphibia is shown in Figure 1-8. This network apparently

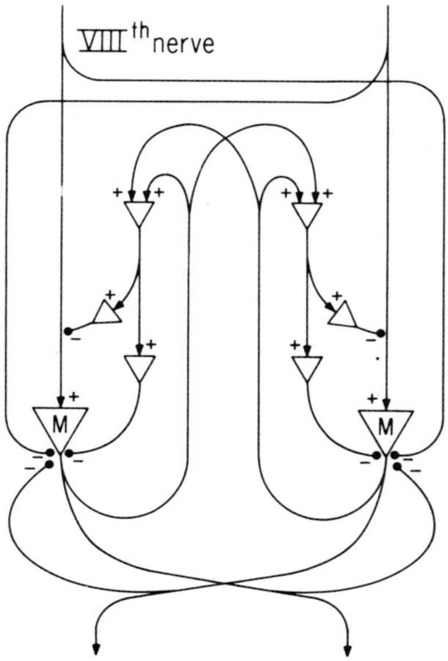

Figure 1-8. Network deduced to be driving the tail muscles of startled fish or tadpoles during escape behavior (from Horridge, 1968). Knobs denote inhibitory terminals, arrows denote excitatory terminals. The label "M" denotes a Mauthner's neuron. In effect, the two Mauthner's neurons are a reciprocally inhibiting pair; the pathways for inhibition, however, are more complicated than those in the simple model of Figure 1-7.

employs Landahl's principle for discrimination of the earlier or stronger input signal from vibrational or auditory receptors, and Reiss's principle for generating a pattern of alternating spikes that drive the tail in a strong swimming action. However, the network is much more elaborate than the simple pair of mutually inhibiting neurons proposed by them.

2

The Idealized "Standard Neuron"

In the previous chapter we identified the neuron as the presumed structural and functional unit in the organization of the nervous system, and the graded generator potentials and action potentials as its most characteristic signals. In this chapter we discuss in general and qualitative terms the dynamic signal-processing characteristics of an idealized neuron. In particular, we outline the basic mechanisms by which the large numbers of spike trains incident upon a neuron are integrated by the neuron, and by which the corresponding output spike train is continually produced. The concepts discussed here, although in themselves not sufficient to account for all neural operation, seem to be essential ingredients used by the brain in its ongoing signal processing and certainly comprise essential background material for understanding contemporary research in the area.

The material discussed in this chapter is also a keystone in the organization of this book. Part II is devoted very largely to a more detailed and penetrating study of the mechanisms described qualitatively in this chapter. Part III, on the other hand, is largely a discussion of attempts to describe signal-processing characteristics of pools or assemblies of interconnected neurons in terms of the properties of the idealized neuron described here. The presentation is preceded by a brief discussion of general cell biology.

According to the *cell doctrine*, as presently interpreted, a cell is the minimum biological unit capable of maintaining and propagating itself in an environment other than the interior of a cell. Regardless of whether or not this doctrine remains tenable in the light of future observations, it provides a useful conceptual basis for present-day biological thought. Cells appear to be strictly bounded systems, each being surrounded by the very thin (70 to 100 Å) plasma membrane which at least partially isolates the material inside the cell from the external environment and from the material inside other cells. Presently, the material inside the cell generally is viewed as comprising most of the machinery that makes life possible, all suspended or dissolved in an aqueous medium, the cytoplasm. Cytoplasm normally is 70 to 80% water (by weight), but much of the water may be bound in a gel-like matrix, since as little as 1% protein (by weight) dissolved in water is sufficient to produce a gel. Even if cytoplasm were gel-like, ions and small molecules very likely would be able to move about in it almost as freely as they do in liquid water, as illustrated by the fact that the mobilities of ions in agar gel are almost the same as they are in water. Larger molecules and various macroscopic structures, on the other hand, would be less mobile in the gel matrix.

Suspended in the cytoplasm are many of the important parts of the machinery of life. In the cells of animals and higher plants, the DNA, site of genetic memory, is condensed into a nucleus, the size of which apparently is more or less fixed for a given organism and seems to be the limiting factor with respect to the lower limit on overall cell size and thus to the upper limit on cell packing density. For example, Szentagothai has estimated that the human cerebellum contains close to 100 billion cells of a certain type (granule cells). The overall diameter of each of these cells is almost the same as the diameter of its nucleus, and thus it apparently exhibits maximum packing. Ribosomes, mitochondria, and numerous other structures complete the array of macroscopic components within the cytoplasm. Of course, the cytoplasm itself, with its organic and inorganic molecules, makes up an important part of the machinery. Also, important in many cells are surface structures that comprise projections of the plasma membrane and specialized internal structures. Among these are the cilia, which occur on the surfaces of many cells, including a large proportion of those specialized for reception of sensory inputs to the nervous system (see Figure 2-1), and microvilli, which occasionally have been implicated in secretory activity but which also occur in certain specialized sensory receptor cells.

Clearly, an important part of the machinery of the cell and one of paramount interest to neurobiologists is the plasma membrane itself. This structure apparently consists of a central bilayer of lipid molecules, covered on each side by a layer of protein molecules. Although it tends to be quite impermeable to ions and molecules whose diameters exceed a certain magnitude (approximately 5 or 6 Å), the plasma membrane is far more permeable to small ions than are pure lipid membranes manufactured by researchers. Thus, it appears that the protein surface layers somehow endow the membrane with special properties not inherent in the bilayer itself. Furthermore, there is now strong evidence to support the notion that different regions and subregions of the membrane are endowed with different properties either with respect to specific permeabilities themselves or with respect to specific stimuli (e.g., molecules) capable of modulating the local permeabilities. Thus, some regions of certain cell membranes appear to have permeation channels for sodium, while other regions appear to have permeation channels for calcium; and membrane permeability to a specific ion at one region may be sensitive to a specific pharmacological agent, while permeability in another region is not. If the membrane is endowed with these localized, *ad hoc* properties by virtue of the proteins on its surfaces, and if, as is generally believed, proteins are specified individually by the genetic memory, then the properties of the membrane are *direct* genetic expressions and presumably are subject to the same control and modification as other direct expressions (i.e., via control genes and modified protein synthesis). The neuron doctrine, discussed in the previous chapter, is a corollary of the general cell doctrine. It was debated rather strenuously 70 years ago but now is generally accepted. According to this doctrine, as we have seen, the functional and structural unit of a nervous system is the neuron, a cell with the same basic machinery as other cells and with a specialized capability of receiving, processing, and sending signals. The nature of those signals is a topic that has intrigued natural scientists for many centuries.

2.1. The Idealized "Standard Neuron"

The concepts discussed here collectively comprise an idealized or so-called "standard neuron." It is largely the result of the theoretical and experimental work performed by J. C. Eccles and various collaborators in the 1950s and early 1960s. Although neurons deviate in structure and operation from the idealization in various ways, many of the principles seem to be common to almost all cells and probably underlie the gamut of function in at least some.

In the idealized neuron, shown in Figure 2-2, the cell membrane is selectively permeable and thus can sustain a difference in composition between the fluid inside the cell and that external to it. The cell is immersed in a fluid characterized by a relatively low concentration of potassium ions and relatively high concentrations of sodium and chloride ions. The relative concentrations of these ions in the intracellular fluid, on the other hand, are just the reverse (high K^+, low Na^+ and Cl^-). These differing ion concentrations in the external and internal fluids result in a potential difference across the cell membrane.

During the course of cell operation, the permeabilities of the membrane to the different ions change, resulting in a flow of ions through the

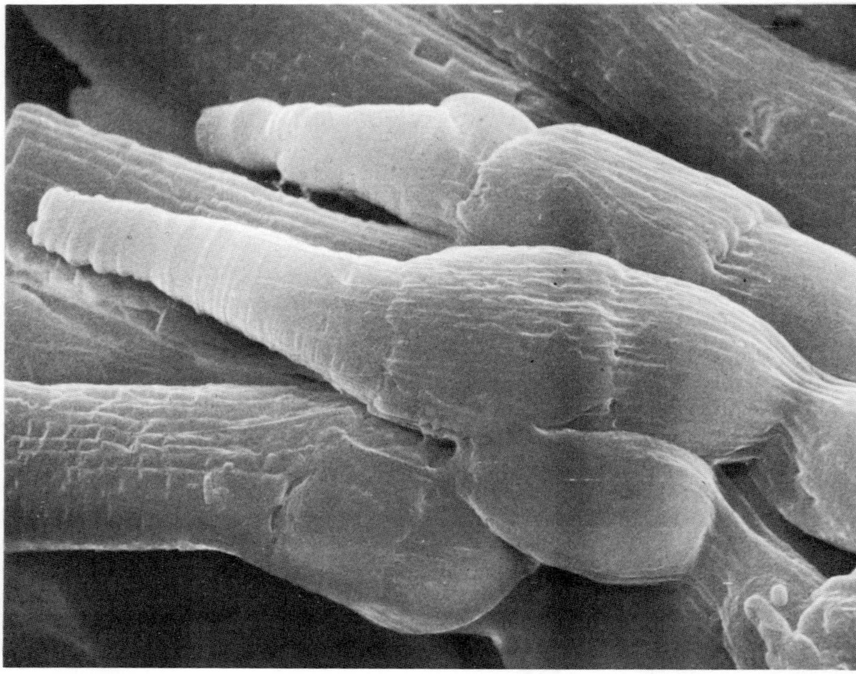

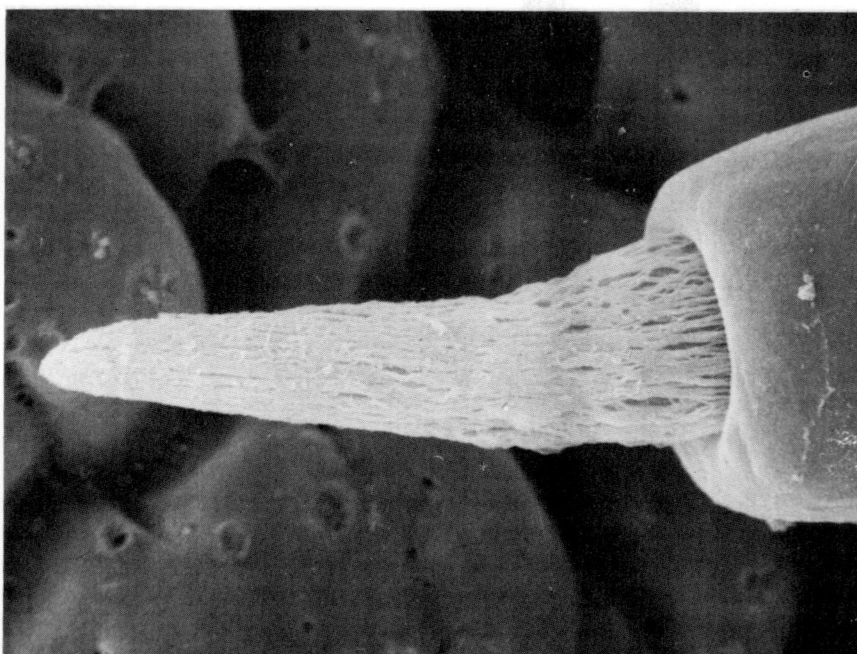

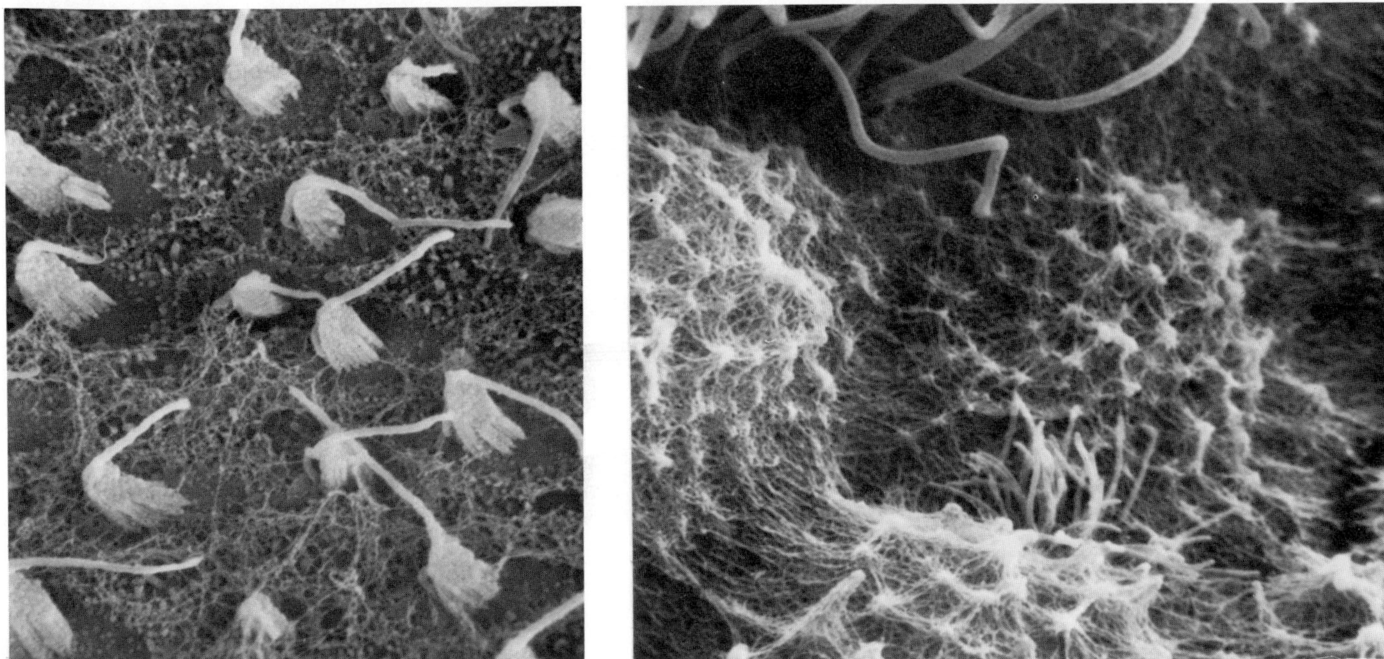

Figure 2-1. Some cilia specially adapted to sensory transduction. Upper left: a retractable tuft of celia used by a predatory sea slug (*Navanax inermis*) to taste its way along the mucus trails of the mollusks upon which it preys. Upper right: rods and cones from the retina of a salamander (*Necturus maculosus*); the conical outer segment of the cone and the cylindrical outer segment of the rod are cilia adapted to photoreception. Lower left: "hair cells" from the inner ear of a bullfrog (*Rana catesbeiana*); the one long "hair" on each cell is a cilium, the shorter "hairs" are modified microvilli; the entire array is adapted to detect mechanical stresses resulting from sound, vibration, movement, and the like, depending upon a filtering superstructure associated with the sensory organ involved (there are eight sensory organs in the bullfrog inner ear). Lower right: A tuft of small cilia peaking out from among mucus-covered microvilli is a structure adapted to detecting odors in the bullfrog; there are many such tufts in the lining of the frog's nasal cavity; above this tuft of olfactory cilia, one can see the motile cilia of the nasal epithelium, which are much larger in diameter. [From Murray and Lewis (1974); Lewis, Zeevi, and Werblin (1969); Lewis and Li (1975).]

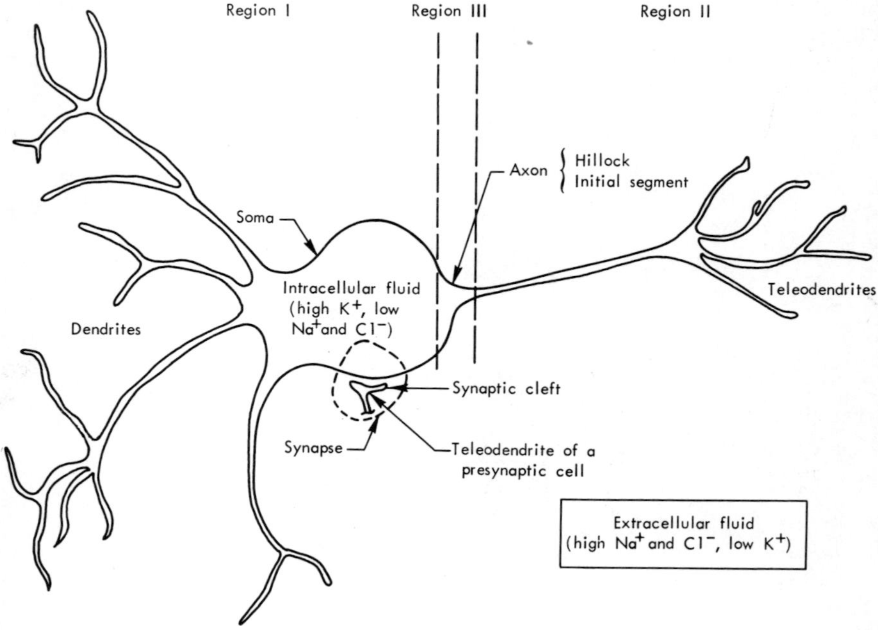

Figure 2-2. Idealized neuron.

membrane and consequent variation of the transmembrane potential. Recordings of these transmembrane-potential variations (and their correlations to physical parameters of sensory stimuli or muscle movement) are the basis of many investigations of signal processing.

The idealized neuron can be divided into three regions on the basis of different electrical characteristics (see Figure 2-2). Region I comprises the dendrites and soma (cell body) and is characterized by smooth, graded, low-level potential changes. Region II comprises the axon and termination fibers and is characterized by a train of discrete potential changes of large amplitude (spikes) which are essentially identical in form. Region III comprises the axon hillock (where the axon sprouts from the soma), on which slow potentials converge and outgoing spike trains are initiated.

A description of the potential changes occurring in each of these regions and an indication of the ionic bases of these changes follows.

2.1.1. The Dendrites and Soma: Incoming Information

The signals arriving at the soma and dendrites of a given cell are carried by the axonal terminations of preceding cells and are converted to a time series of postsynaptic potentials.

A synaptic connection between neurons is functionally either excitatory or inhibitory. Figure 2-3 shows a recording of the potential change at the receiving-cell soma following a presynaptic pulse at an excitatory synapse; such a response is called an excitatory postsynaptic potential (EPSP). It is postulated that the following ionic process underlies these EPSPs: A chemical transmitter substance is released during a spike from the synaptic knob of the sending cell into the adjacent synaptic cleft. This transmitter diffuses across the synaptic cleft and causes a brief, very large increase in the permeability of the subsynaptic membrane to several ions, including potassium and sodium.

Figure 2-3 also shows a recording of the postsynaptic soma potential (inhibitory postsynaptic potential or IPSP) following a presynaptic pulse at an inhibitory synapse. The ionic mechanisms underlying these potentials are similar to that for the EPSPs, but with permeability to K^+ and Cl^- increasing again greatly and briefly, while that to the larger Na^+ ions remains low.

The postsynaptic potentials (PSPs) shown in Figure 2-3 are associated with synapses upon the soma. The PSPs are modified if the synapse is on a dendrite some distance from the soma, necessitating conduction of the signal along the membrane. This conduction is associated with a passive membrane (i.e., the membrane does not provide energy to the signal). In this process, PSPs are spread out in time and decreased in amplitude. The effect of the location of the synapse, as computed by Rall, is shown in Figure 2-4. Another influence on PSPs at their site of initiation is the

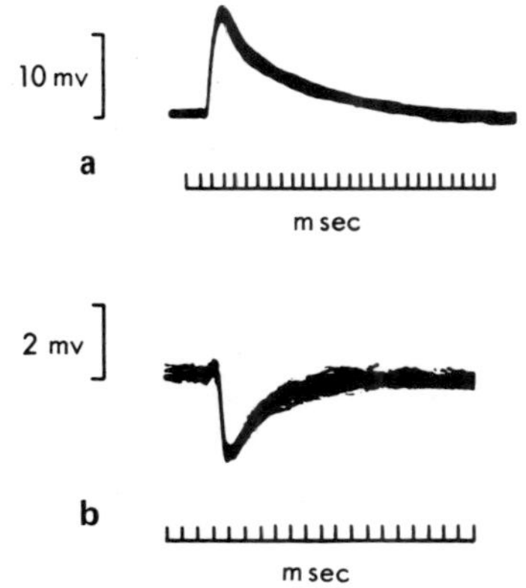

Figure 2-3. Postsynaptic potentials. (a) Excitatory postsynaptic potential (EPSP). (b) Inhibitory postsynaptic potential (IPSP). (From Eccles, 1964.)

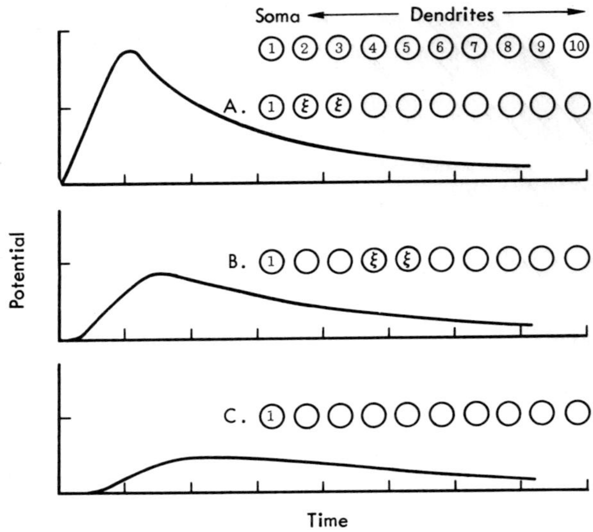

Figure 2-4. Effect of dendritic location of synapse on EPSP (marked circles indicate relative locations of synapses). (From Rall, 1964.)

transmembrane potential existing at the time of their arrival. The potentials are a reflection of ion concentrations and fluxes across the membrane. Excitatory (depolarizing) and inhibitory (hyperpolarizing) PSPs are associated with different combinations of membrane permeabilities and hence exhibit different dependences on the transmembrane potential. The observed effect of the membrane potential on PSP peak amplitude is shown in Figure 2-5. The equilibrium potential of the EPSP [i.e., the electrical potential at which the currents producing the synaptic potential are zero (See Chapter 5.)] is some seven to 10 times farther removed from the normal resting potential than that of the IPSP. Thus, this effect would appear to be more crucial in the latter.

The shapes of individual PSPs arriving at the soma, then, are determined by (1) the relative efficacy of the synapse at which the PSP originated (i.e., the width of the synaptic cleft, the surface areas of the membranes involved, and the mechanisms of transmitter release, action, and removal), (2) the length and geometry of the postsynaptic membrane over which the PSP must be transmitted, (3) the properties of the postsynaptic membrane (essentially, the permeabilities to the different ion types), and (4) the transmembrane potential existing at the time of PSP initiation.

2.1.2. The Axon and Its Terminations: Outgoing Signals

The signals transmitted down the cell axon to succeeding cells or muscles consist of a time series of essentially identical spikes. Such spike trains are shown in Figures 1-5 and 1-6.

The membrane-permeability changes usually associated with these trains of pulses are shown in Figure 2-6. The permeability to Na^+ increases before the permeability to K^+. The former decreases to its normal value quickly but is nonetheless responsible for the large increase in intracellular potential.* The subsequent increase in permeability to K^+ is responsible for the sharp return of the potential to its resting level, and its relatively long-lasting tail probably is the basis of the hyperpolarized "afterpotential" that occurs in many cells (see Figure 2-6).

Spike trains may differ greatly in the order and length of time intervals between spikes. Furthermore, other cells transmit similar spike trains to the same postsynaptic cell. The spike-elicitation process may be sensitive to small differences among temporal patterns of input, and this sensitivity may be important in normal function.

2.1.3. The Axon Hillock: Signal Processing

The axon hillock (or "initial segment") is the region of the idealized neuron at which the incoming PSP's converge and the outgoing spike trains are initiated. (Spike initiation actually occurs elsewhere in many cells; in fact, the site of spike initiation may move over a considerable distance as the neuron's activity varies.) The amplitude and time course of an individual PSP arriving at the axon hillock are determined essentially by the factors listed previously.

If the cell has not initiated a spike for a relatively long time, the hillock membrane may be more or less passive. In this case, the incoming PSPs may combine via an operation which deviates very little from a linear algebraic summation. (This operation is called "temporal summation" when all the PSPs are initiated at the same synapse, and "spatial summation" when they come from different synapses.) A dramatic example of such a summation process is shown in Figure 2-7. This process continues until the potential at the hillock exceeds a certain critical value, the "threshold" potential. If this potential is exceeded, the permeability changes, as shown in Figure 2-6, and the concomitant spike ensue.

* The transmembrane equilibrium-potential values (internal minus external) of Na^+, K^+, and Cl^+ are roughly + 65, − 95, and − 90 mV, respectively. A typical "resting potential" would be − 70 mV.

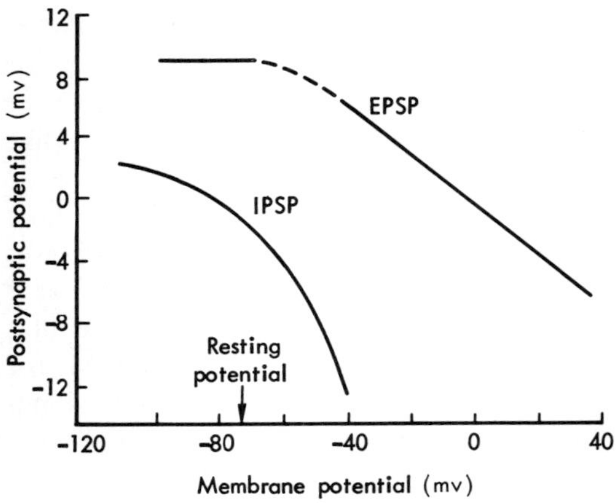

Figure 2-5. Dependence of PSP on potential. (From T. H. Bullock and C. A. Horridge, *Structure and Function in the Nervous Systems of Invertebrates*, Freeman Press, San Francisco, 1965.)

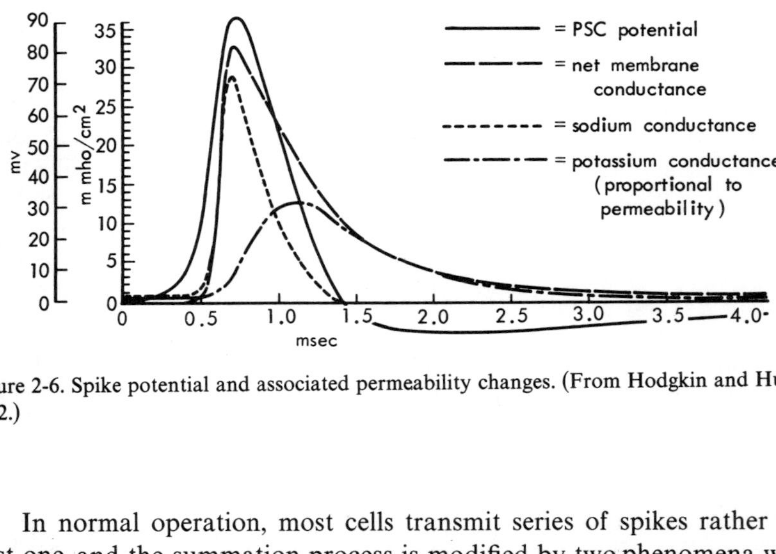

Figure 2-6. Spike potential and associated permeability changes. (From Hodgkin and Huxley, 1952.)

In normal operation, most cells transmit series of spikes rather than just one, and the summation process is modified by two phenomena which follow a spike: First, the threshold potential is very high immediately following a spike and declines essentially exponentially to its resting value. Second, the permeability changes associated with the spike have two manifestations: (1) they may bring about a spike afterpotential which is significantly different from the resting level (see Figure 2-6) and upon which incoming PSPs are superimposed, and (2) they affect the time courses (and indirectly the amplitudes) of any PSPs arriving at the hillock during their existence. The effect of any PSP arriving during the large Na^+ permeability

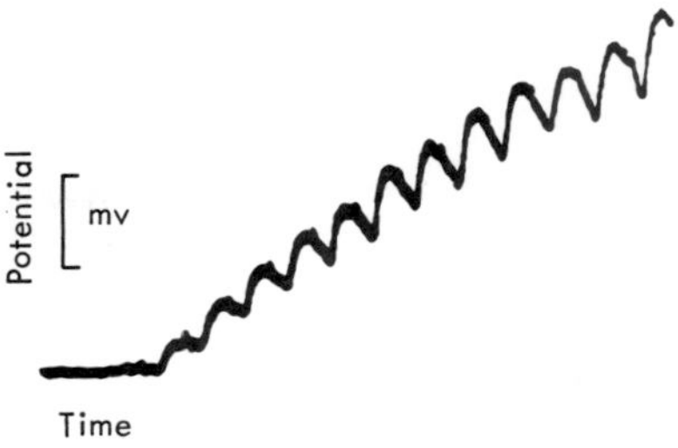

Figure 2-7. Subthreshold summation process. (From Eccles, 1964.)

change associated with the spike is very nearly obliterated. The permeability value drastically alters the shape and magnitude of the PSP and simply incorporates it into the spike. Also, any current associated with the PSP is quite small relative to that of the spike. However, as the spike potential rapidly diminishes, PSPs arriving at its later stages are again significant, and the marked influence of the "tail" of the K^+ permeability change on their courses is critical to the integrating mechanism. This increased K^+ permeability diminishes the EPSP and facilitates the IPSP. Figure 2-8, taken from Eccles' work, illustrates this effect on excitatory input. (The curves in Figure

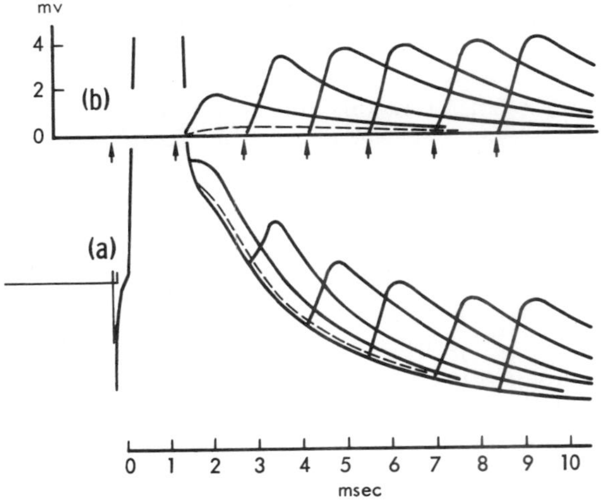

Figure 2-8. Postspike EPSP modification (arrows indicate initiation times of single EPSPs). (From Eccles, 1957.)

2-8b are obtained by subtracting the afterpotential from a superposition of the afterpotential and single EPSPs initiated at several different time intervals after the spike.) It should be noted that both of these postspike influences, the raised threshold and the increased permeabilities (with the possible consequent afterpotential and the influences on excitatory and inhibitory PSPs), tend to reduce the effectiveness of a given set of PSPs for releasing another spike. Hence, the set of effects collectively produces a "refractory period."

The idealized neuron provides a simplified, but highly useful means, for conceptualizing the main ingredients of signal processing in neurons. Even for the many types of neurons whose behavior deviates significantly from the idealization, it nonetheless provides a background conception against which those deviations are more readily comprehended and catalogued. Moreover, the concept lends itself to further simplification for modeling studies of neuronal operation in several ways.

The idea of a triggering section, bordered on one side by a passive input region and on the other side by an active region whose main function seems to reside simply in transmitting signals, allows one to consider "point neurons" as reasonable (although crude) approximate models for the dynamic activity of entire neurons. In its simplest form one may represent the dynamic activity of an entire cell by a quantitative model for its triggering section only. The input to the triggering section is specified as patterns of current waveforms which correspond to the graded passive currents conducted through the dendritic regions and soma; the output is a series of voltage pulses determined by the appropriate quantitative rules used to represent the activity of the triggering section. In the simplest form of such modeling, the influences of the passive dendritic trees on input wave forms is entirely neglected.

The natural extension of this approach is the consideration of "compartmentalized neurons" in which various morphological regions of a neuron are each represented by a compartment. In this approach, the dynamic activity of each compartment is represented by an appropriate set of dynamical equations or rules. The compartments can be interconnected to represent various morphologies with, for example, simple resistive or more complicated connecting elements as discussed in Part II. Obviously, model neurons of arbitrary complexity can be constructed by this approach. The inputs to such models can be simulated by conductance changes or current waveforms with temporal patterns specified and distributed among compartments as desired by the modeler. The outputs from such models are time series of action potentials generated at the triggering section which, in keeping with the idealized neuron concept, is assumed to be an essential component of the model neuron.

Certain other theoretical simplifications which are very useful for studies of signal processing in neuronal networks have been suggested by Eccles. The Dale chemical principle states that any single nerve cell produces a single synaptic transmitter, and that that transmitter is released at all the synaptic terminals of the neuron. In general, the effects of a given transmitter on a receiving cell depend on the detailed biochemical processes associated with the receiving cell and the various types of receptors that might be active on the surface of the receiving cell. A given transmitter may have an excitatory effect with one type of receptor, an inhibitory effect with another type of receptor, or may have another kind of effect which is not readily classifiable as excitatory or inhibitory—including, for example, a tendency to clamp membrane potential at a particular value. However, in the majority of known cases a given transmitter seems to have a single kind of effect at the vast majority of its terminals. Most transmitters seem to be either very largely excitatory or very largely inhibitory in their effects. Eccles has suggested the generalization that cells of the mammalian brain are typically either excitatory or inhibitory at all their terminals. This means that, from a dynamic point of view, it might be possible in many cases (at least as a first approximation and for certain purposes) to neglect the tremendous diversity exhibited by various neurons and classes of neurons and retain only the assumptions as to whether particular classes of cells are excitatory or inhibitory in their effects on other cells. Such a simplification is clearly tremendously powerful in terms of constructing dynamical models of networks or assemblies of neurons.

Another generalization suggested by Eccles which is very useful to the modeler concerns the distribution of excitatory and inhibitory synaptic terminals. Eccles has suggested that inhibitory terminals are for the most part restricted to the cell body or somatic regions of neurons; i.e., inhibitory synapses, in Eccles view, occur only very rarely on dendritic trees. Excitatory synapses, on the other hand, may occur any place on the soma or dendritic tree. One generally supposes that the proportion of excitatory synapses on the dendritic tree as opposed to the soma is approximately proportional to the surface area of the dendritic tree relative to that of the soma. A typical neuron of the mammalian central nervous system might have 80% of the surface area of its soma-dendritic region within the dendritic tree. For such a cell 80% of incident synapses might be expected to occur on the dendritic tree as opposed to the soma. These generalizations can be combined with those of the idealized standard neuron to construct relatively simple quantitative models for simulating the dynamic activity of networks or assemblies of interconnected neurons. Thus, for example, one may represent the dynamic activity of each constituent neuron according to the point-neuron or compartmentalized-neuron strategy already discussed. One then may suppose that the neuron axon corresponds functionally to merely a time delay to be added to the delay inherent in the slow-rising phase of the synaptic response and, moreover, suppose that the output spike train of a given cell initiates an excitatory or inhibitory current or a change in permeability in the appropriate receiving or triggering section of each receiving cell to which it is synaptically connected. Simulations can be performed of large networks simply by specifying the dynamic characteristics of each compartmental block by appropriate equations or rules, appropriate time delays for axonal and synaptic delays, and appropriate interconnection matrices to represent the paths over which the output trains lead to pulses of input current or conductance change in appropriate receiving cells. Examples of such simulations will be taken up in Part III.

The concept of an isolated section or compartment of the neuron thus emerges as a critical element in modeling electrical-signal processing in both single neurons and in neural networks. The unit compartment used in almost every neural model is some representation of a patch of neuronal membrane, with its associated ionic solutions (intracellular and extracellular). In Part II we shall discuss the foundations and elaborations of this unit of modeling.

PART II

NEURAL MODELING: MODELS OF EXCITATION AND CONDUCTION

In this section we discuss quantitative models of the development, propagation, and interaction of electrical signals in the membranes of single nerve cells. These models are best conceived as attempts to relate the electrical activity observed in single neurons to their underlying physiological mechanisms. It is the area of activity referred to as "neural modeling" when that term is used in its most narrow sense. Chapters 3 and 4 deal with the dynamic properties of single isolated patches of passive neural membrane. Chapter 5 considers the primary mechanisms of signal generation in neurons, i.e., the modulation of neural signals at receptors, at synapses, and spontaneously. Chapter 6 considers spatial as well as temporal features of behavior in the passive membrane comprising an extended region of neuron. Finally, the last chapter of this section, Chapter 7, addresses the mechanisms of the generation and conduction of action potentials in neurons. Note that the material discussed in Chapters 3, 4, and 6 refers to energetically passive neural operations, whereas the mechanisms of signal and spike generation discussed in Chapters 5 and 7 are active operations wherein both metabolic and potential energy are used and the very constituent properties of neural membrane in the active regions undergo alterations during the course of activity.

3

Models of Passive Membrane

3.1. The Nernst–Planck Model

From the point of view of irreversible thermodynamics, one of the most widely used models of passive membrane is the Nernst–Planck equation, which simply combines Ohm's law for the flow of charged particles in an electrical-potential gradient and Fick's first law for the diffusion of particles in a concentration gradient. In writing the Nernst–Planck equation, it is customary to deal with flow densities, i.e., the number of particles flowing through a unit of area in a unit of time. This quantity, usually labeled J, is particularly useful for membranes since one often is interested in the flow of particles through a given area of membrane.

Ohm's law is simply a model relating the "drift" of charged particles in an electric field to the strength of the field. For ion species B, for example, Ohm's law would be written

OHM'S LAW

$$\mathbf{J}_{\mathrm{B}_{\text{drift}}} = (\pm Zq)\mu_{\mathrm{B}}[\mathrm{B}]\mathbf{E} = (\pm Zq)\mu_{\mathrm{B}}[\mathrm{B}](-\text{grad } V) \tag{3-1}$$

where Zq is the charge on each individual ion, Z being the ionic valence, q being the magnitude of the electronic charge ($q = 1.602 \times 10^{-19}$C), and the sign associated with Zq being the ionic polarity (positive for cations, negative for anions); μ_B is the mobility of ion B in the system under consideration (e.g., in water, in cytoplasm, in membrane); [B] is the concentration of B ions (the number of B ions per unit volume); and $\mathbf{E} = -\text{grad } V$ is the strength of the electric field acting on the ions. Although Ohm's law generally is taken to be one of the basic empirical laws of irreversible thermodynamics, justified by its observed near validity in case after case, it can be derived from two more-primitive models: Coulomb's law, which states that the force exerted on a charged particle by an electric field is directly proportional to the product of the particle's charge Zq and the strength $\mathbf{E}$ of the electric field, and Stokes' law, which states that the frictional force on a moving particle is directly proportional to the velocity $\boldsymbol{v}$ of the particle. Thus, when these two laws are applicable, as they very often are, a charged particle will accelerate in an electric field until the force due to friction is just equal but opposite to the force due to the field itself, and its terminal velocity will be *directly* proportional to the strength of the electric field:

STOKES' LAW AND COULOMB'S LAW

$$\boldsymbol{v}_{\text{terminal}} \propto (Zq)\mathbf{E} \tag{3-2}$$

The sign associated with either of the expressions on the right in equation (3-1) simply indicates the direction of flow of the particles. By convention, a positive sign associated with either of the expressions implies flow in the same direction as the vector of that expression. Thus, positive ions flow in the same direction as the electric field vector, but in the direction opposite (or negative to) the electrical potential gradient; and negative ions flow in the same direction as the potential gradient, but in the direction opposite the electric field vector.

Fick's first law is a model relating the diffusional flow of particles to the concentration gradient of those same particles. Once again, it is considered to be one of the basic empirical laws of irreversible thermodynamics, but it too can be derived from more primitive models. For ion species B,

FICK'S LAW

$$J_{\text{B}_{\text{diffusion}}} = -D_{\text{B}}\,\text{grad}[\text{B}] \tag{3-3}$$

where D_B is the "diffusivity," the constant of proportionality between the rate of diffusional flow and the concentration gradient. Of course, the sign here is fixed, implying that the flow due to diffusion always is directed down (or opposite to) the concentration gradient.

The Nernst–Planck equation combines these two components of flow to give the total rate of flow due to drift and diffusion:

NERNST–PLANCK EQUATION

$$J_{\text{B}} = (\pm Zq)\mu_{\text{B}}[\text{B}](-\text{grad}\ V) - D_{\text{B}}\,\text{grad}[\text{B}] \tag{3-4}$$

In membrane systems, one very often is interested in ion flow in a single direction (e.g., radial flow across a spherical or cylindrical membrane surface), in which case the Nernst–Planck equation reduces to a more familiar form:

ONE-DIMENSIONAL NERNST–PLANCK EQUATION

$$J_B = (\pm Zq)\mu_B[B]\left(-\frac{\partial V}{\partial x}\right) - D_B\frac{\partial[B]}{\partial x} \tag{3-5}$$

Furthermore, in membrane systems one generally deals with specific monovalent or divalent ions such as chloride ions:

$$J_{Cl^-} = (-q)\mu_{Cl}[Cl]\left(-\frac{\partial V}{\partial x}\right) - D_{Cl}\frac{\partial[Cl]}{\partial x} = q\mu_{Cl}[Cl]\frac{\partial V}{\partial x} - D_{Cl}\frac{\partial[Cl]}{\partial x}$$

or potassium ions

$$J_{K^+} = -q\mu_K[K]\frac{\partial V}{\partial x} - D_K\frac{\partial[K]}{\partial x}$$

or calcium ions

$$J_{Ca^{+2}} = -2q\mu_{Ca}[Ca]\frac{\partial V}{\partial x} - D_{Ca}\frac{\partial[Ca]}{\partial x}$$

Combined with two additional constraints, space-charge neutrality and the Einstein relation, the Nernst–Planck equation leads to a number of very interesting conclusions and provides the basis of classical membrane theory.

3.2. The Einstein Relation

The Einstein relation is merely a formal statement of the proposition that the resistance to the movement of a given particle is independent of the mechanism causing the particle to move. In the cases of ions drifting and diffusing in a medium of some sort, the Einstein relation implies that the resistance presented by that medium is the same for drift as it is for diffusion. As one might expect, the derivation of the relation usually invokes Stokes' law (e.g., see Kittel, 1958, pp. 154–155). The relation itself usually is given in the form

EINSTEIN RELATION

$$D_B = kT\mu_B \tag{3-6}$$

where k is the Boltzmann constant and T is the absolute temperature. It is valid under conditions of thermal equilibrium.

3.3. The Nernst Equation and Ionic Reversal Potentials

The Nernst equation is a direct corollary of the Einstein relation and applies to situations in which the net flow of an ion species due to drift and diffusion is zero. In other words, any diffusional flow is precisely balanced by an equal but opposite drift flow. Extending the Nernst–Planck equation to include nonlinear effects, one has

$$j_B = \pm(Zq)\mu_B\left(x, \frac{\partial V}{\partial x}, [B], \ldots\right)\left(-\frac{\partial V}{\partial x}\right)[B] - D_B\left(x, \frac{\partial V}{\partial x}, [B], \ldots\right)\frac{\partial[B]}{\partial x}$$

When the drift flow is just equal but opposite to the diffusion, one can set the right side equal to zero, which leads to

$$\frac{\partial V}{\partial x} = \pm\frac{D_B(x, \frac{\partial V}{\partial x}, [B], \ldots)}{\mu_B(x, \frac{\partial V}{\partial x}, [B], \ldots)}\frac{1}{Zq}\frac{1}{[B]}\frac{\partial[B]}{\partial x}$$

According to the Einstein relation and regardless of their dependence on other variables, $D_B = kT\mu_B$; therfore, we have the differential form of the Nernst equation,

$$\frac{\partial V}{\partial x} = \pm\frac{kT}{Zq}\frac{1}{[B]}\frac{\partial[B]}{\partial x} \tag{3-7}$$

and the integral form of the Nernst equation,

$$V_2 - V_1 = \pm\frac{kT}{Zq}\ln\frac{[B]_2}{[B]_1} \tag{3-8}$$

where V_i is the voltage at point i; $[B]_i$ is the concentration of B ions at that same point; ln is the logarithm to the Napierian base (e); and the *positive* sign on the right side applies to anions (A), and the negative sign to cations (C). If the solutions on either side of a membrane are essentially equipotential, then the voltage difference $V_2 - V_1$ would be that across the membrane itself. In that case, the subscripts would denote side 2 and side 1, respectively. If each of several ion species present has equal but opposite drift and diffusion, then (3-8) leads to

$$\frac{[A_1^-]_1}{[A_1^-]_2} = \frac{[C_1^+]_2}{[C_1^+]_1} = \left(\frac{[C_2^{+2}]_2}{[C_2^{+2}]_1}\right)^{1/2} = \left(\frac{[A_2^{-2}]_1}{[A_2^{-2}]_2}\right)^{1/2} = \left(\frac{[A_3^{-n}]_1}{[A_3^{-n}]_2}\right)^{1/n}$$

$$= \left(\frac{[C_3^{+n}]_2}{[C_3^{+n}]_1}\right)^{1/n} \tag{3-9}$$

The coefficient (kT/Zq) in equations (3-7) and (3-8) often is written (RT/ZF), where F = the Faraday = Avogadro's number × the protonic charge and R = the gas constant = Avogadro's number × the Boltzmann constant.

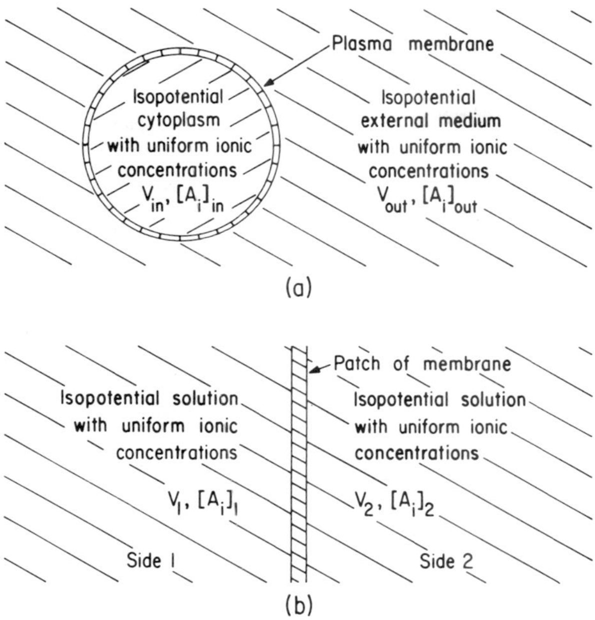

Figure 3-1. Conventional membrane system models with idealized geometries.

The abilities of ions to move about in cytoplasm and in the aqueous media surrounding cells apparently are very much greater than the abilities of those same ions to move into and through the plasma membrane. Thus, when ionic currents pass through the membrane, or when electrical potentials appear between the cytoplasm and the extracellular solutions, the concentration gradients and electrical potential gradients in the cytoplasm at large and the extracellular solutions at large should be very small compared to those in and close to the membrane itself. For this reason, modelers often neglect the gradients in the cytoplasm at large and extracellular solutions at large, treating those media as well-mixed, equipotential solutions. The resulting model is depicted in Figure 3-1.

Regardless of whether or not Ohm's law or Fick's law is valid for the membranes of Figure 3-1, as long as the Einstein relation holds (the principal requirement for which is thermal equilibrium) the Nernst equation will apply. Thus, for example, if the solutions bathing the membrane contain potassium ions, then the diffusion of potassium due to concentration differences across the membrane in Figure 3-1a will be offset precisely by an equal but opposite drift when

$$V_{in} - V_{out} = -\frac{kT}{q} \ln \frac{[K^+]_{in}}{[K^+]_{out}}$$

Clearly, in this case, the direction of electrical potential increase is opposite to the direction of concentration increase (i.e., the side of the membrane with positive electrical potential also has the lower potassium concentration). If the magnitude of the potential difference were increased slightly, the drift would become slightly larger than the diffusion, and there would be a net flow of potassium ions in the direction of the electrical potential gradient. On the other hand, if the magnitude of the potential were reduced slightly, the drift would become slightly smaller than the diffusion, and there would be a net flow of potassium ions down their concentration gradient. Thus, as the potential passes through the value given by the Nernst equation, the flow of potassium ions across the membrane should reverse its direction. However, it is quite possible that the Nernst equation might not be applicable for a particular membrane. Therefore, the potential given by the Nernst equation generally is called the *Nernst potential,* while the observed potential at which the flow of a given ionic species reverses is called the *reversal potential.* Comparison of these two potentials is one test of validity of the Nernst equation for a particular ion in a particular membrane system. Although the example here was limited to potassium ions, the conclusions are equally valid for all ions. In the case of negative ions, the polarity of the Nernst potential is such that the side of the membrane with higher concentration is positive. The effects of increasing and decreasing the magnitude of the potential will be qualitatively the same as they are for positive ions.

3.4. Space-Charge Neutrality

When the Nernst–Planck model is applied to membranes, an assumption that virtually always accompanies it is *space-charge neutrality.* This assumption can be stated as follows: "For the purpose of counting the charged particles present in a given volume, one can assume that the total number of exposed protonic charges is just equal to the total number of excess electronic charges." In bulk materials, this approximation usually is extremely accurate, simply because *very few* uncompensated charges generally will produce *very large* electrical potential gradients.

For example, if we wished to apply space-charge neutrality to the solutions at large on either side of the plasma membrane, we probably would derive very accurate ion counts, even if all of the uncompensated charges in the system were located in those solutions. The electrical potentials between the two solutions seldom exceed 100 mV, and the capacitance between the two solutions (i.e., the capacitance associated with the plasma membrane) is very close to 1 μF per square centimeter of membrane (i.e., 10^{-6} uncompensated coulombs on each side of the membrane is sufficient to produce a voltage difference of 1 V between the two solutions). Therefore, to produce 100 mV, the uncompensated charge on each side of the membrane must be 10^{-7} C for each square-centimeter patch of membrane. Since each singly charged ion contributes 1.6×10^{-19} C, the total number of uncompensated singly charged ions required to produce 100 mV is approximately 6×10^{11} per square centimeter of membrane. The concentration of ions of each polarity in the cytoplasm at large and in the extracellular solutions at large is at least 0.1 M, or 10^{-4} times Avagadro's number per cubic centimeter. This product is approximately 6×10^{19} ions per cubic centimeter. Therefore, the 6×10^{11} ions required for 100 mV across a 1 cm^2 patch of membrane are all expected to be available in the first 1 Å of solution on each side of the membrane. Actually, the effective (hydrated) radii of ions in aqueous solution is greater than 1 Å, so one can conclude that the necessary charge is available in the first layer of ions on each side of the membrane. As one moves beyond that layer, space-charge neutrality should be a thoroughly valid principle. In the layer itself, the principle undoubtedly is violated.

The layer in which space-charge neutrality is violated often is called simply a *space-charge layer.* As described in the previous paragraph, it is quite compact. There is evidence supporting the hypothesis that more diffuse space-charge layers also exist on each side of the plasma membrane (see Gilbert, 1971). In these more diffuse layers, which are postulated to result from fixed charge attached to the surface of the membrane, the violation of space-charge neutrality will not be as severe as it is in compact space-charge layers. The diffuse layers apparently extend as far as 10 Å or more into the solutions; but the greater their extent, the less the charge density is within them, and the more applicable is the principle of space-charge neutrality as far as the counting of ions is concerned.

Within the membrane itself, the validity of space-charge neutrality depends upon many unknown factors, chief among which is the concentration of ions inside the membrane. The average dielectric constant of the membrane is apparently approximately one-eighth that of water, so salts are much less likely to be dissociated in the membrane than they are in water, and much less soluble in the membrane. Thus, the concentration of dissociated ions in the membrane may be considerably smaller than it is in the immediately adjacent aqueous solutions. The ratio of the ion concentration immediately inside the surface of the membrane to that in the aqueous solution immediately outside the surface is often called the partition coefficient of the membrane for the particular ion species under consideration. If the partition coefficient is small, then the principle of space-charge neutrality is in danger of being violated. At this point, however, the best we can do is establish a criterion of validity and then apply it later to specific cases. The criterion comes directly from Poisson's equation for one dimension:

$$\frac{d^2V}{dx^2} = -\frac{\rho}{\epsilon}$$

where ρ is the local charge density and ϵ is the local dielectric constant. What we shall ask is whether or not the local charge density is comparable to the total local available charge density (i.e., to the local ion density times the ionic charge). If it is, then space-charge neutrality is violated. In order for space-charge neutrality to be valid for a system of monovalent ions, the following inequality must apply:

$$\left|\frac{2\rho}{qn_i}\right| = \left|\frac{2\epsilon}{qn_i}\frac{d^2V}{dx^2}\right| \ll 1 \tag{3-10}$$

3.5. Nernst–Planck Membranes with Space-Charge Neutrality and Obeying the Einstein Relation

Time and again, throughout the biophysical literature, one finds membrane models employing the Nernst–Planck equation, the assumption of space-charge neutrality, and the Einstein relation. Thus, implicitly, these models are also based on the assumption of thermal equilibrium and the validity of Ohm's law and Fick's first law. Within the framework of these assumptions, the models themselves take a wide variety of forms, including models of homogeneous, electrically neutral membranes, models of membranes with fixed internal charges (e.g., see Carnay and Tasaki, 1971), and models of membranes with fixed surface charges (e.g., see Gilbert, 1971). In this section, a few of these models are presented, along with their consequences. Since neurons and other cells generally are rich in internnl potassium ions and are bathed in solutions rich in sodium and chloride ions, the model membrane systems presented here include these ionic species.

3.5.1. Homogeneous, Neutral Membranes

The homogeneous, neutral membrane undoubtedly is the most common ideal employed for biophysical membrane considerations, and it has been used to generate or justify a large number of common membrane equations. For example, suppose that the solutions on either side of the membrane in Figure 3-1b occupied very large volumes and contained different concentrations of potassium chloride. Initially, the flow of one ion species across the membrane would exceed that of the other, depending on which species had the greater mobility in the membrane. Almost instantly, however, a potential would appear which brought the flows of the two species to essentially equal levels (retarding the flow of the more mobile ion and enhancing the flow of the other ion). At this point, the net rate of charge transport across the membrane would be essentially zero (the flow of ions being essentially equal), and the system would be governed by the equation

ION FLOW BETWEEN TWO LARGE RESERVOIRS CONTAINING KCl SOLUTIONS

$$-q\mu_K[\mathrm{K}]\frac{\partial V}{\partial x} - kT\mu_K\frac{\partial[\mathrm{K}]}{\partial x} = q\mu_{Cl}[\mathrm{Cl}]\frac{\partial V}{\partial x} - kT\mu_{Cl}\frac{\partial[\mathrm{Cl}]}{\partial x}$$

but according to space-charge neutrality, [K] = [Cl] everywhere. Letting the salt concentration be [KCl] and assuming complete dissociation,

$$-q\mu_K[\mathrm{KCl}]\frac{\partial V}{\partial x} - kT\mu_K\frac{\partial[\mathrm{KCl}]}{\partial x} = q\mu_{Cl}[\mathrm{KCl}]\frac{\partial V}{\partial x} - kT\mu_{Cl}\frac{\partial[\mathrm{KCl}]}{\partial x}$$

which leads to the familiar equation

$$V_2 - V_1 = \frac{\mu_{Cl} - \mu_K}{\mu_{Cl} + \mu_K}\frac{kT}{q}\ln\frac{[\mathrm{KCl}]_2}{[\mathrm{KCl}]_1} \tag{3-11}$$

If the two reservoirs were in fact infinite in volume, this equation would describe a permanent potential across the membrane. However, since the reservoirs will be finite, the flow of KCl will cause the concentration ratio to approach unity. Equation (3-11) in that case provides an excellent approximation of the electrical potential across the membrane as the concentration difference declines. The key factor in this approximation is space-charge neutrality. Changes in $V_2 - V_1$ will require an essentially negligible amount of charge and thus negligible imbalance between the chloride and potassium currents.

Equation (3-11) applies to another situation where the medium is essentially homogeneous, namely, an aqueous KCl solution with gradients in the KCl concentration. The potential between any two points in the solution should be very close to those predicted by the equation, where the mobilities employed are those for the two ion species in water. Actually, the mobility of potassium in water is very nearly equal to that of chloride, so the observed and predicted potentials (often called "diffusion potentials") resulting from concentration differences are quite small, even when the differences themselves are rather large.

When more than two ions are present in the system, matters become considerably more complicated within the framework of the Nernst–Planck equation. With the presence of sodium, potassium, chloride, and a homogeneous, neutral, Nernst–Planck membrane, the system of Figure 3-1b is governed by the three Nernst–Planck equations

$$\frac{J_K}{\mu_K} = -q[K]\frac{\partial V}{\partial x} - kT\frac{\partial[K]}{\partial x}$$

$$\frac{J_{Na}}{\mu_{Na}} = -q[Na]\frac{\partial V}{\partial x} - kT\frac{\partial[Na]}{\partial x}$$

and

$$\frac{J_{Cl}}{\mu_{Cl}} = q[Cl]\frac{\partial V}{\partial x} - kT\frac{\partial[Cl]}{\partial x}$$

along with the condition of space-charge neutrality

$$[K^+] + [Na^+] = [Cl^-]$$

Summing the Nernst–Planck equations for the anions and cations separately leads to

$$\frac{J_K}{\mu_K} + \frac{J_{Na}}{\mu_{Na}} = -q([Na] + [K])\frac{\partial V}{\partial x} - kT\frac{\partial([Na] + [K])}{\partial x}$$

$$\frac{J_{Cl}}{\mu_{Cl}} = q([Na] + [K])\frac{\partial V}{\partial x} - kT\frac{\partial([Na] + [K])}{\partial x}$$

Adding these two equations, one obtains

$$\frac{J_K}{\mu_K} + \frac{J_{Na}}{\mu_{Na}} + \frac{J_{Cl}}{\mu_{Cl}} = -2kT\frac{\partial([Na] + [K])}{\partial x} \quad (3\text{-}12)$$

while subtracting one from the other leads to

$$\frac{J_{Cl}}{\mu_{Cl}} - \frac{J_K}{\mu_K} - \frac{J_{Na}}{\mu_{Na}} = 2q([Na] + [K])\frac{\partial V}{\partial x} \quad (3\text{-}13)$$

According to the assumption of homogeneity throughout the membrane, the various ionic mobilities will not be dependent on x. Furthermore, if the system is reasonably close to being in steady state, then the rate at which ionic concentrations are changing within the membrane will be negligible compared to the rate at which ions are flowing through the membrane, so J_{Cl}, J_K, and J_{Na} all will be essentially constant with respect to x. Combining these two assumptions, we can immediately integrate equation (3-12) to find that the weighted sum of ionic flows is directly proportional to the total anion or cation concentration difference across the membrane:

$$J_{Cl}/\mu_{Cl} + J_K/\mu_K + J_{Na}/\mu_{Na} = 2kT([Cl]_{2m} - [Cl]_{1m})/w \quad (3\text{-}14)$$

where w is the thickness of the membrane; $[Cl]_{2m}$ is the chloride concentration just inside the membrane on side 2; and $[\mathrm{Cl}]_{1\mathrm{m}}$ is the chloride concentration just inside the membrane on side 1.

Equation (3-12) now implies that the steady-state (or near steady-state) gradient of the *total* ionic concentration within the homogeneous Nernst–Planck membrane is constant with respect to x. Note that this does not imply that the concentration gradient of an individual ion species is constant; in general, it will not be. Equation (3-13) implies that the electrical potential gradient (i.e., the electrical field strength) will not be constant within the membrane either, unless the sum $[\mathrm{Na}] + [\mathrm{K}]$ just happens to be constant through the membrane. This condition is equivalent to a constant total ion concentration through the membrane, a situation usually described as *isotonicity*.

Both of these results are easily extended to systems containing many ion species. The form and implications of (3-12) will not be altered by inclusion of additional species. The form of (3-13) will be altered, however, unless all of the ions of each polarity are of the same valence (e.g., all anions are monovalent, all cations are divalent). If more than one valence is represented among the ions of either polarity, then the condition for constant potential gradient becomes complicated and is no longer equivalent to isotonicity. For example, if calcium ions were present in addition to potassium, sodium, and chloride, then the condition for constant electrical field strength becomes

$$2[\mathrm{Ca}^{+2}] + [\mathrm{K}^{+}] + [\mathrm{Na}^{+}] = \text{constant with respect to } x$$

There is no particular reason that this weighted sum or others like it should be independent of x.

3.5.2. The Goldman Model

A homogeneous, neutral Nernst–Planck membrane in an isotonic system of ions of one valence clearly is a rather restricted idealization. However, under many circumstances, the solutions on the two sides of a neuronal membrane are in fact very nearly isotonic with respect to one another, and the predominant ionic flows through the membrane often comprise monovalent ions (on the other hand, it should be noted that calcium often is an important permeant). Under these circumstances, the restricted ideal is just as reasonable as any homogeneous, neutral Nernst–Planck membrane model could be for a plasma membrane. According to equation (3-13), when this ideal system is in steady state or near steady state (i.e., not undergoing rapid transient phenomena), then the electric field strength within the membrane is constant. The assumption of isotonicity throughout also leads directly to the conclusion that the right side of equation (3-12) is zero, leading to the following statement:

$$\sum_i J_i/\mu_i = 0 \tag{3-15}$$

The presence of a constant field strength facilitates calculation of steady-state membrane potentials, leading to the well-known *constant field equations of Goldman.*

Consider, for example, the system of Figure 3-1b. Let the surface of the membrane on side 1 be located at $x = 0$ and the surface on side 2 be located at $x = w$, where w is the membrane thickness. If the voltage difference $V_2 - V_1$ is equal to $-V_m$, then the electric field strength everywhere within the membrane is V_m/w. With this in mind, we can rewrite the Nernst–Planck equation as follows for monovalent anion species A:

$$\frac{d[\mathrm{A}]}{dx} = -\frac{J_\mathrm{A}}{kT\mu_\mathrm{A}} + \frac{qV_\mathrm{m}}{kTw}[\mathrm{A}]$$

If the system is in steady state, then J_A will be independent of x, and we have a simple differential equation for [A] as a function of x. The solution simply is

$$[\mathrm{A}] = a + be^{(qV_\mathrm{m}/kTw)x} \tag{3-16}$$

where a and b are constants that can be evaluated in terms of the concentrations of A ions just inside the membrane on each of the two sides,

$$[\mathrm{A}]|_{x=0} = [\mathrm{A}]_1 = a + b$$

$$[\mathrm{A}]|_{x=w} = [\mathrm{A}]_2 = a + be^{qV_\mathrm{m}/kT}$$

from which

$$b = ([\mathrm{A}]_1 - [\mathrm{A}]_2)/(1 - e^{qV_\mathrm{m}/kT})$$

$$a = ([\mathrm{A}]_2 - [\mathrm{A}]_1 e^{qV_\mathrm{m}/kT})/(1 - e^{qV_\mathrm{m}/kT})$$

Returning to the usual form of the Nernst–Planck equation for A, we have

$$J_\mathrm{A} = q\mu_\mathrm{A}(a + be^{(qV_\mathrm{m}/kTw)x})\frac{V_\mathrm{m}}{w} - kT\mu_\mathrm{A}\frac{d}{dx}(a + be^{(qV_\mathrm{m}/kTw)x})$$

which reduces immediately to

$$J_\mathrm{A} = q\mu_\mathrm{A} a\frac{V_\mathrm{m}}{w} = \left(\frac{q\mu_\mathrm{A}}{w}\frac{[\mathrm{A}]_2 - [\mathrm{A}]_1 e^{qV_\mathrm{m}/kT}}{1 - e^{qV_\mathrm{m}/kT}}\right)V_\mathrm{m} \tag{3-17}$$

The corresponding equation for monovalent cation species C is

$$J_\mathrm{C} = \left(\frac{q\mu_\mathrm{C}}{w}\frac{[\mathrm{C}]_2 - [\mathrm{C}]_1 e^{-qV_\mathrm{m}/kT}}{1 - e^{-qV_\mathrm{m}/kT}}\right)V_\mathrm{m} \tag{3-18}$$

Now, if we let side 2 be the inside of a cell and side 1 be the outside, then V_m becomes the electrical potential difference across the cell membrane, with its polarity defined so that V_m is positive when the inside of the cell is positive with respect to the outside. We also can use partition coefficients to relate the concentrations just inside the membrane to those just outside:

$$[\mathrm{Cl}]_2 = \zeta_{\mathrm{Cl}}[\mathrm{Cl}]_{\mathrm{in}}; \qquad [\mathrm{Cl}]_1 = \zeta_{\mathrm{Cl}}[\mathrm{Cl}]_{\mathrm{out}}$$
$$[\mathrm{K}]_2 = \zeta_{\mathrm{K}}[\mathrm{K}]_{\mathrm{in}}; \qquad [\mathrm{K}]_1 = \zeta_{\mathrm{K}}[\mathrm{K}]_{\mathrm{out}}$$
$$[\mathrm{Na}]_2 = \zeta_{\mathrm{Na}}[\mathrm{Na}]_{\mathrm{in}}; \qquad [\mathrm{Na}]_2 = \zeta_{\mathrm{Na}}[\mathrm{Na}]_{\mathrm{out}}$$

where the quantities on the left sides of the equations are concentrations just within the membrane itself, next to its surface; ζ_i is the partition coefficient of the ith ion species; and $[\ ]_{\mathrm{out}}$ and $[\ ]_{\mathrm{in}}$ are ion concentrations just adjacent to the membrane on the outside and the inside of the cell, respectively. Considering the system of Figure 3-1a with potassium, sodium, and chloride ions as an example, we now can arrange equations (3-17) and (3-18) as follows:

$$J_{\mathrm{Cl}} = \frac{q\zeta_{\mathrm{Cl}}\mu_{\mathrm{Cl}}}{w}\frac{[\mathrm{Cl}]_{\mathrm{in}} - [\mathrm{Cl}]_{\mathrm{out}}e^{qV_m/kT}}{1 - e^{qV_m/kT}}V_m \tag{3-19a}$$

$$J_{\mathrm{K}} = \frac{q\zeta_{\mathrm{K}}\mu_{\mathrm{K}}}{w}\frac{[\mathrm{K}]_{\mathrm{out}} - [\mathrm{K}]_{\mathrm{in}}e^{qV_m/kT}}{1 - e^{qV_m KT}}V_m \tag{3-19b}$$

$$J_{\mathrm{Na}} = \frac{q\zeta_{\mathrm{Na}}\mu_{\mathrm{Na}}}{w}\frac{[\mathrm{Na}]_{\mathrm{out}} - [\mathrm{Na}]_{\mathrm{in}}e^{qV_m/kT}}{1 - e^{qV_m/kT}}V_m \tag{3-19c}$$

From these equations, we can calculate the voltage–current characteristics of our ideal system, which, at first impression, would appear to be nonlinear. Before we do so, however, we shall determine the steady-state voltage that one should expect in the absence of currents applied from external sources, and then we shall consider the phenomenon of bulk storage of ions, which may well affect the voltage–current characteristics.

In the absence of applied currents, the net flow of charge across the membrane as a result of drift and diffusion will be zero in the steady state. Therefore,

$$qJ_{\mathrm{Na}} + qJ_{\mathrm{K}} - qJ_{\mathrm{Cl}} = 0$$

Applying this relationship to equations (3-19) and solving for V_m, one obtains the familiar Goldman equation for the steady-state potential of an unstimulated (resting) membrane:

THE GOLDMAN EQUATION

$$V_G = \frac{kT}{q} \ln \frac{P_K[K]_{out} + P_{Na}[Na]_{out} + P_{Cl}[Cl]_{in}}{P_K[K]_{in} + P_{Na}[Na]_{in} + P_{Cl}[Cl]_{out}} \tag{3-20}$$

where $P_i = \zeta_i \mu_i$ and is defined as the *permeability* of the membrane to the ith ion species. We can call V_G the resting membrane potential according to the Goldman equation, or simply the *Goldman potential.* It is given by equation (3-20) in terms of the steady-state ion concentrations on the two sides of the membrane.

3.5.3. Bulk Storage of Ions in a Uniform Neutral Nernst–Planck Membrane

Although the situation depicted in the previous section, with isotonicity throughout and with ions of one valence only, is rather restricted, it nevertheless provides a convenient basis for considering several interesting phenomena. One such phenomenon is the shift in individual ion distribution profiles in the membrane as a result of a shift from one steady-state value of V_m to another. This shift represents a change in the total number of the ions under consideration that must be added to or taken from the membrane. Thus, we might think of the uniform membrane as storing ions throughout its volume. By analogy with similar storage of charge carriers in solids (e.g., in semiconductors), and to distinguish it from charge storage at surfaces, we can label this phenomenon *bulk storage.*

To estimate the importance of this phenomenon, consider the potassium ions stored in the membrane under two distinct conditions: V_m equal to zero and V_m equal to the Nernst Potential (V'_K) for potassium ions. Modifying equation (3-16) so that it applies to potassium ions, we can write

$$[K] = \zeta_K \frac{[K]_{in} - [K]_{out} e^{qV_m/kT}}{1 - e^{qV_m/kT}} + \zeta_K \frac{[K]_{out} - [K]_{in}}{1 - e^{qV_m/kT}} e^{(qV_m/kTw)x} \tag{3-21}$$

Taking the limit of this function as V_m approaches zero, one obtains

$$[K] = \zeta_K [K]_{out} + \zeta_K ([K]_{in} - [K]_{out})(x/w) \tag{3-22a}$$

which is a linear function of x. This is just as one would expect, regardless of whether or not the specific constraints of the Goldman equation were met. When V_m equals zero, there is no electrical potential gradient within the membrane; so the flow of potassium ions (or any other ions) is due to diffusion alone. If the system has attained steady state, then there should be no local accumulation of any ion species taking place within the membrane, which means that there must be no divergence of the flow of any ion species (i.e., J_K must be independent of x). If the membrane is uniform in the sense that μ_K is independent of x, then J_K will be independent of x only if $d[K]/dx$ is independent of x (i.e., if [K] varies linearly through the membrane). Moving on to the situation in which $V_m = V'_K$, we have another simple version of equation (3-21):

$$[K] = \zeta_K [K]_{out}([K]_{in}/[K]_{out})^{x/w} \tag{3-22b}$$

In other words, in this case the potassium ion concentration varies exponentially through the membrane.

The total number of potassium ions stored in the membrane will be quite different in the two cases; so if one suddenly changed the potential across the membrane from one of these levels to the other, a considerable transfer of potassium ions could occur between the membrane and the surrounding solutions. As an example, consider the situation in which $[K]_{in}/[K]_{out} = 20$ (a fairly typical ratio for cells). Integration of equations (3-22a) and (3-22b) in this case leads to

$$N_K \simeq 10.5\zeta_K [K]_{out} w \qquad \text{for} \qquad V_m = 0 \tag{3-23}$$

$$N_K \simeq 6.3\zeta_K [K]_{out} w \qquad \text{for} \qquad V_m = V'_K \tag{3-24}$$

If the membrane were 100 A thick and $[K]_{out}$ were 10 μmoles/cm^3 (see Katz, 1966), then the number of potassium ions in the homogeneous membrane would be

$$N_K \simeq \zeta_K \times 10^{-10} \text{ moles/cm}^2 \qquad \text{for} \qquad V_m = 0$$

$$N_K \simeq \zeta_K \times 6 \times 10^{-11} \text{ moles/cm}^2 \qquad \text{for} \qquad V_m = V'_K$$

The magnitude of the potassium Nernst potential in this case can be evaluated by noting that $\ln 20 \simeq 3$ and

$$kT/q \simeq 25 \text{ mV} \qquad \text{at} \qquad 300\,^\circ\text{K}$$

Therefore, $V'_K \simeq -75$ mV in this case. The number of univalent ions that must be transferred across the membrane in order to charge or discharge its capacitance by this amount is approximately

$$N_i = \frac{\Delta V \times C}{q} \simeq \frac{75 \times 10^{-3} \times 10^{-6}}{1.6 \times 10^{-19}} \text{ ions/cm}^2 \simeq 5 \times 10^{11} \text{ ions/cm}^2$$

$$N_i \simeq 10^{-12} \text{ moles/cm}^2$$

which, if ζ_K were not small, could be comparable to the number of ions that must be transferred into or out of the membrane in order to move from one steady-state situation to the other (i.e., from $V_m = 0$ to $V_m = -75$, or vice versa). In fact, if ζ_K were close to unity, the number of ions required to charge the capacitance would be a small fraction of the number moving into or out of bulk storage in the membrane.

When significant, bulk storage of ions in the membrane very likely would endow it with interesting and complicated electrical properties. In response to a stepwise change in V_m, for example, the drift term would dominate initially in the Nernst–Planck equation, since concentrations and concentration gradients would not change instantaneously. Gradually, however, the concentrations and concentration gradients would assume their steady-state levels appropriate to the new value of V_m. During this gradual change, the ionic flows through the membrane and the apparent electrical conductance of the membrane also would change. It should be noted here, however, that the conductance changes expected as a result of this phenomenon are not nearly as large as those observed in the voltage-clamped squid axon (see Chapter 7), and that bulk storage of ions, by itself, cannot be postulated as the mechanism underlying the time- and voltage-dependent conductances of active membrane.

Furthermore, considerable evidence exists supporting the proposition that plasma membranes are not uniformly permeable to ions, but that permeation is limited to scattered, localized sites (e.g., see Narahashi, 1971). In that case, the total number of ions stored in the membrane would be considerably lower than that expected if the membrane were uniformly permeable. Such limited permeation also would imply that ionic flow densities at the permeation sites themselves are proportionately larger than they would be in the case of uniform permeation of a homogeneous membrane. The significance of this increased flow density with respect to bulk storage of ions can be estimated by considering the computational results of Cohen and Cooley (1965), who studied numerically the dynamics of a Nernst–Planck system. They found that the computed concentration profile within the membrane underwent the transition from one steady-state situation to another in times comparable to the diffusion time T_D across the membrane:

$$T_D = w^2/D \tag{3-25}$$

where w is the membrane thickness and D is the diffusivity of the ion in question. Considering the steady-state flow of potassium ions through a Nernst–Planck membrane when $V_m = 0$ (i.e., when there are no electric fields), we have

$$J_K(0) = -D_K \frac{d[K]}{dx} = -D_K \frac{[K]_{out} - [K]_{in}}{w} \zeta_K$$

Thus,

$$D_K = \frac{wJ_K(0)}{([K]_{in} - [K]_{out})} \zeta_K \tag{3-26}$$

and

$$T_{D_K} = \frac{w([K]_{in} - [K]_{out})}{J(0)}\zeta_K \tag{3-27}$$

In a passive neuronal membrane, the current density due to potassium ions when $V_m = 0$ is approximately 25 $\mu A/cm^2$, which corresponds to approximately 16×10^{13} ions/sec-cm^2. A typical value for $[K]_{in} - [K]_{out}$ is 2.4×10^{20} ions/cm^3 (see Katz, 1966, p. 43), and a typical value for w is 10^{-6} cm. Thus, we could estimate

$$T_{D_K} \simeq \frac{10^{-6} \times 2.4 \times 10^{20}}{16 \times 10^{13}}\zeta_K = 1.5\zeta_K \text{ sec} \tag{3-28}$$

if the membrane were permeated uniformly. On the other hand, if only 1% of the membrane were permeated by potassium ions, then the value of J_K in the permeation channels would be 100 times our previous estimate, and T_{D_K} would be decreased by a factor of 100. In such a case, however, it is quite conceivable that the value of ζ_K in the channels themselves would be reasonably close to unity. In active membranes (see Chapter 7), $J_K(0)$ will be increased by a factor lying somewhere between 10 and 100, so T_{D_K} in such membranes is likely to be very small (i.e., probably considerably less than 1 msec).

3.6. Space-Charge Neutrality in a Homogeneous Nernst–Planck Membrane

Throughout the preceding sections, we have based all of our calculations on the principle of space-charge neutrality throughout the membrane, namely, that

SPACE-CHARGE NEUTRALITY

$$\sum_i Z_i[A_i] = \sum_j Z_j[C_j] \tag{3-29}$$

where A_i is the ith anion species and Z_i is its valence; and C_j is the jth cation species and Z_j is its valence. At this point, we are in a position to test that assumption for the homogeneous, Nernst–Planck membrane with monovalent cations and anions. Let $n_i(x)$ be the total ion density at position x in the membrane. Equations (3-12) and (3-13) can be generalized to

$$-kT\frac{\partial n_i}{\partial x} = \sum_i \frac{J_{A_i}}{\mu_{A_i}} + \sum_j \frac{J_{C_j}}{\mu_{C_j}}$$

$$qn_i\frac{\partial V}{\partial x} = \sum_i \frac{J_{A_i}}{\mu_{A_i}} - \sum_j \frac{J_{C_j}}{\mu_{C_j}} \tag{3-30}$$

Noting that under steady-state conditions the terms on the right side of equation (3-30) are independent of x, one can differentiate both sides to obtain

$$q\frac{\partial n_i}{\partial x}\frac{\partial V}{\partial x} + qn_i\frac{\partial^2 V}{\partial x^2} = 0$$

which can be rearranged to yield

$$\frac{\partial^2 V}{\partial x^2} = \frac{-1}{n_i}\frac{\partial n_i}{\partial x}\frac{\partial V}{\partial x} \tag{3-31}$$

Substituting the right side of this equation into the term on the left side of inequality (3-10), we have

$$\left|\frac{2\rho}{qn_i}\right| = \left|\frac{2\epsilon\frac{\partial n_i}{\partial x}\frac{\partial V}{\partial x}}{qn_i^2}\right| \tag{3-32}$$

which must be very much less than unity if space-charge neutrality is to be valid. We can establish a *rough* estimate of the magnitude of this term as follows: The average value of ϵ in the membrane is approximately $10^{-14}\,\mathrm{F/cm^2}$; the product of the two partial derivatives might be of the same order of magnitude as $2\langle n_i\rangle V_m/w^2$, where $\langle n_i\rangle$ is the average total ionic concentration in the membrane and can be written as $\zeta n_{i_{out}}$, where ζ is an appropriately defined average partition coefficient and $n_{i_{out}}$ is the total ionic concentration outside of the membrane; similarly, n_i should be of the same order of magnitude as $\zeta n_{i_{out}}$; and $n_{i_{out}}$ is approximately 3×10^{20} ions/cm^3, w is approximately 10^{-6} cm, and V_m is approximately 0.1 V. All of this leads to

$$|2\rho/qn_i| \simeq 10^{-4}/\zeta$$

which must be very much less than unity. Of course, if isotonicity prevails, the partial derivative of n_i with respect to x will be zero, and the right side of equation (3-32) therefore also will be equal to zero; so space charge neutrality will be valid regardless of any other factors. On the other hand, if isotonicity does not prevail, then the principle of space-charge neutrality may be in jeopardy, especially at high current densities where the partial derivatives of n_i and V become large.

3.7. Sources of Permanent Potential Differences across Membranes

If all the ions in the systems of Figure 3-1 can permeate the membrane, then the conditions set forth in equations (3-29) (space-charge neutrality) and (3-7) (the Nernst equation) are met only if there is no concentration difference across the membrane for any ion species, implying [equation (3-8)] that no permanent electrical potential difference will exist in the system if the only modes of flow are drift and diffusion and all charged particles are free to move into and out of the membrane. A permanent potential difference will exist, on the other hand, if any of these conditions is not met.

3.7.1. Donnan Equilibrium

Donnan's model applies to the situation in which some of the charged particles in the system have strictly limited mobility (e.g., they may be permanently attached to the structural molecules of the membrane itself, or they may be ions in solution that are very large and thus unable to permeate the membrane). According to the Donnan model, the permanent potentials in such cases can be calculated from the Nernst equation and the condition of space-charge neutrality. Consider, for example, the system of Figure 3-1a with Na, K, and Cl ions free to permeate the membrane, but with the addition of large monovalent protein anions (A^-) to the cytoplasm. Under permanent (equilibrium) conditions, the drift of each permeating ion will be balanced by an equal but opposite diffusion, and the net flux of each ion species across the membrane will be zero:

$$\frac{[\mathrm{K}]_{\mathrm{in}}}{[\mathrm{K}]_{\mathrm{out}}} = \frac{[\mathrm{Na}]_{\mathrm{in}}}{[\mathrm{Na}]_{\mathrm{out}}} = \frac{[\mathrm{Cl}]_{\mathrm{out}}}{[\mathrm{Cl}]_{\mathrm{in}}} \tag{3-33}$$

According to the condition of space-charge neutrality in the cytoplasm,

$$[\mathrm{K}]_{\mathrm{in}} + [\mathrm{Na}]_{\mathrm{in}} = [\mathrm{Cl}]_{\mathrm{in}} + [\mathrm{A}]_{\mathrm{in}} \tag{3-34}$$

while, in the membrane,

$$[\mathrm{K}] + [\mathrm{Na}] = [\mathrm{Cl}] \tag{3-35}$$

and, in the external medium,

$$[\mathrm{K}]_{\mathrm{out}} + [\mathrm{Na}]_{\mathrm{out}} = [\mathrm{Cl}]_{\mathrm{out}} \tag{3-36}$$

Equations (3-33), (3-34), and (3-36) can be combined as follows: From equation (3-34),

$$\frac{[\mathrm{K}]_{\mathrm{out}}}{[\mathrm{K}]_{\mathrm{out}}}[\mathrm{K}]_{\mathrm{in}} + \frac{[\mathrm{Na}]_{\mathrm{out}}}{[\mathrm{Na}]_{\mathrm{out}}}[\mathrm{Na}]_{\mathrm{in}} = [\mathrm{Cl}]_{\mathrm{in}} + [\mathrm{A}]_{\mathrm{in}}$$

Combining this result with equations (3-33) and (3-36), we have

$$\frac{[\mathrm{Cl}]_{\mathrm{in}} + [\mathrm{A}]_{\mathrm{in}}}{[\mathrm{Cl}]_{\mathrm{out}}} = \frac{[\mathrm{Cl}]_{\mathrm{out}}}{[\mathrm{Cl}]_{\mathrm{in}}} \tag{3-37}$$

Equation (3-37) can be transformed into a simple quadratic equation if the volumes of the intracellular and extracellular spaces along with the total number of chloride ions occupying those volumes are known. Once the quadratic equation is solved, one can apply the Nernst equation to determine the corresponding permanent electrical potential.

Note that isotonicity cannot exist permanently in this system as it stands. Isotonicity requires that $[\mathrm{Cl}]_{\mathrm{in}} + [\mathrm{A}]_{\mathrm{in}} = [\mathrm{Cl}]_{\mathrm{out}}$, which contradicts equation (3-37) except in the uninteresting case of $[\mathrm{A}]_{\mathrm{in}} = 0$. Thus, the Goldman equation is not valid for this system.

3.7.2. Donnan Jumps

A *space-charge layer* very likely will exist at the inner surface of the membrane in this system. The limits of integration of the differential form of the Nernst equation [equation (3-7)] are completely arbitrary. If one integrates from a point in the external medium to any point within the membrane, he obtains equation (3-8) for the included region. Applying also the principle of space-charge neutrality, and noting that all ions in the included region are unrestrained, one immediately is led to the conclusion that the concentration of each ion species within the membrane (or within those parts of the membrane that are permeable to the ions) is the same as it is in the external medium. On the other hand, if one integrates from a point within the membrane, but just adjacent to its internal surface, to a point in the cytoplasm, but just adjacent to the membrane, the A ions within the region are constrained and one has

$$\frac{[\mathrm{Cl}]_{\mathrm{m}}^{*}}{[\mathrm{Cl}]_{\mathrm{in}}} = \frac{[\mathrm{Cl}]_{\mathrm{out}}}{[\mathrm{Cl}]_{\mathrm{in}}} = \frac{[\mathrm{Cl}]_{\mathrm{in}} + [\mathrm{A}]_{\mathrm{in}}}{[\mathrm{Cl}]_{\mathrm{out}}} \tag{3-38}$$

where $[\mathrm{Cl}]_{\mathrm{m}}^{*}$ is the chloride concentration within the membrane, just adjacent to its inner surface. Thus, according to the Donnan model, there is an abrupt discontinuity of chloride concentration and an abrupt change in the electrical potential, occurring together at the inner surface of the membrane. These abrupt shifts in concentration and potential often are called *Donnan jumps*.

Actually, the charge separation required to produce the predicted potential difference cannot occur across an infinitesimal boundary but must occupy a finite distance, which may or may not be small compared to the thickness of the membrane. In the regions immediately adjacent to the inner surface of the membrane a space-charge layer will exist, and the principle of space-charge neutrality will be invalid. The geometry of the space-charge layer could be estimated by means of the Debye–Hückel model. As long as the Einstein relation remains valid, however, the integration of equation (3-8) can be taken between points on opposite sides of the space-charge layer and well beyond it, where space-charge neutrality once again applies. Thus, equation (3-38) remains valid, and the Donnan model can be used to predict the total potential drop across the Donnan jump.

Similar considerations can be applied to membranes with *fixed internal* charges. For example, if the A ions were locked into position in the internal structure of the membrane rather than being free to move about in the cytoplasm, then the Donnan model would predict abrupt changes in concentrations and potentials at both surfaces of the membrane, and there very likely would be significant space-charge layers associated with both surfaces. The presence of two abrupt interfaces with associated space-charge layers suggests an analogy with the geometry of transistors. In fact, most of the equations presented so far in this section are commonly applied to models of transistors. The analogy has been pointed out by Wei and others, but to date it has not led to any deeper understanding of either neural membranes or transistors. In this respect, it is very much like the analogy between nerve fibers and iron wires, promoted so vigorously in years past (see Chapter 7).

One very significant consequence of fixed charges within a membrane is the resulting relative exclusion of ions of the same polarity as the fixed charge. Thus, if large concentrations of fixed negative sites were located within the membrane, equation (3-29) would imply that the cations within the membrane would outnumber the anions by a considerable margin. In fact, if all the ions and the fixed charges were univalent, then the margin would be simply the number of fixed charges. The membrane then would exhibit a marked tendency to pass cations in preference to anions.

3.7.3. Ion Pumps

Active transport of ions provides another means by which permanent potential differences can be maintained across a cell membrane. Whereas the potentials in the Donnan model are maintained without expenditure of energy, maintenance of potentials by the continual pumping of ions requires energy expenditure. For that reason, the transport mechanisms are labeled *active*. Electrophysiological evidence indicates that ions are often pumped without concomitant transport of charge. This apparently is accomplished by ion exchange on a one-for-one basis (e.g., one-for-one exchange of internal sodium ions for external potassium ions). The steady-state concentration difference generated by such pumps across homoge-

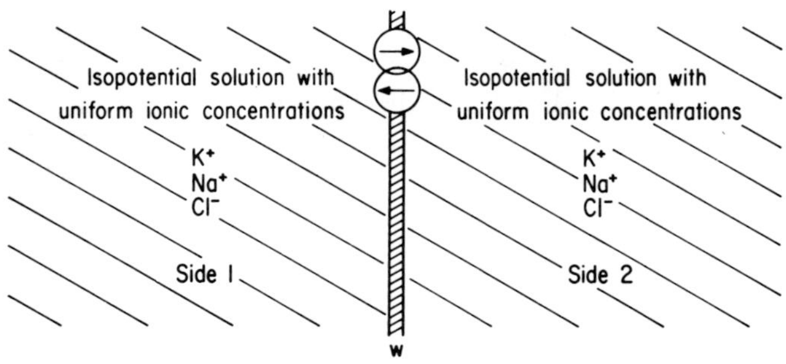

Figure 3-2. An idealized membrane system with three ion species and a pump (represented by the two arrows enclosed in circles) that passes one potassium ion in (from side 2 to side 1) for every sodium ion that it passes out (from side 1 to side 2).

neous, neutral Nernst–Planck membranes can be calculated rather easily. Consider, for example, the system of Figure 3-2, in which the membrane is permeable to potassium, chloride, and sodium, and in which potassium and sodium ions are exchanged on a one-for-one basis by the ion pump.

The steady-state equations for the system are

$$J_{\mathrm{Cl}} = 0 = q\mu_{\mathrm{Cl}}[\mathrm{Cl}]\frac{dV}{dx} - kT\mu_{\mathrm{Cl}}\frac{d[\mathrm{Cl}]}{dx} \tag{3-39a}$$

$$J_{\mathrm{K}} = 0 = -q\mu_{\mathrm{K}}[\mathrm{K}]\frac{dV}{dx} - kT\mu_{\mathrm{K}}\frac{d[\mathrm{K}]}{dx} + J_{\mathrm{p}} \tag{3-39b}$$

$$J_{\mathrm{Na}} = 0 = -q\mu_{\mathrm{Na}}[\mathrm{Na}]\frac{dV}{dx} - kT\mu_{\mathrm{Na}}\frac{d[\mathrm{Na}]}{dx} - J_{\mathrm{p}} \tag{3-39c}$$

where J_{p} is the pumped flow density of sodium ions outward and potassium ions inward. In steady state, the pumped flow density for each ion species must be just equal but opposite to the net passive flow density (the combination of drift and diffusion). Equations (3-39) can be combined with the constraint of space-charge neutrality to yield

$$[\mathrm{Cl}]_2 - [\mathrm{Cl}]_1 = \frac{(\mu_{\mathrm{Na}} - \mu_{\mathrm{K}})}{2kT\mu_{\mathrm{Na}}\mu_{\mathrm{K}}} J_{\mathrm{p}}w \tag{3-40}$$

Since equation (3-39a) is equivalent to the differential form of the Nernst equation for chloride, equation (3-8) can be used to estimate the permanent membrane potential in terms of the ratio of the chloride concentrations on the two sides of the membrane. In order to derive that ratio from equation (3-40), however, we must have some additional information (e.g., the concentration of chloride ions on one side of the membrane, or the total number of chloride ions in the system and the volumes of the spaces on the two sides of the membrane). The computations become slightly more complicated when additional, nonpumped ions are present.

A second type of pump, implicated in several membrane systems, is the "electrogenic" pump, which produces net charge transfer as well as ion transport. If such a pump is specified, the resulting concentration differences can be calculated rather easily for the uniform, neutral Nernst–Planck membrane. Consider, for example, the situation depicted in Figure 3-2, but with only the sodium ions being pumped. The steady-state chloride concentration difference across the membrane now becomes

$$[\mathrm{Cl}]_2 - [\mathrm{Cl}]_1 = \frac{J_\mathrm{p} w}{2kT\mu_{\mathrm{Na}}} \tag{3-41}$$

which, with a little additional information, can be used to calculate the steady-state potential difference.

Notice that a permanent potential in the system of Figure 3-2 is inconsistent with isotonicity (i.e., if $[\mathrm{Cl}]_1 = [\mathrm{Cl}]_2$, the potential difference is zero). On the other hand, near isotonicity is consistent with a permanent potential if *properly matched* pumps and Donnan jumps are present.

3.8. Impedance to Ion Flow at Donnan Jumps

Up to this point, we have based all our calculations on the Nernst–Planck model, which tacitly assumes that the principal impediment to the flow of ions is viscous drag, as described by Stokes' law. Viscous drag comes into play as a result of collisions between ions and other particles or structures. According to Stokes' law, the retarding force on a moving ion will be directly proportional to the radius of the ion, the viscosity of the medium in which it is moving, and the velocity with which it is moving. If the ions were permeating the membrane uniformly, directly through its lipid phase, then one would expect their mobility within the membrane to be considerably lower than it is in water, simply because the viscosity of lipids usually is several orders of magnitude greater than the viscosity of water (and the ionic mobility in a given medium should be inversely proportional to the viscosity of that medium). However, if the ions are permeating at special sites, then the viscosity of the lipid phase of the membrane at large is not relevant. The mobility of the ion through the permeation channel will be a special property of the channel itself.

One region of special interest is that of the Donnan jumps. Here, the electrical potential gradients and ionic concentration gradients are extremely large; yet the distances across the jumps are comparable to the mean free

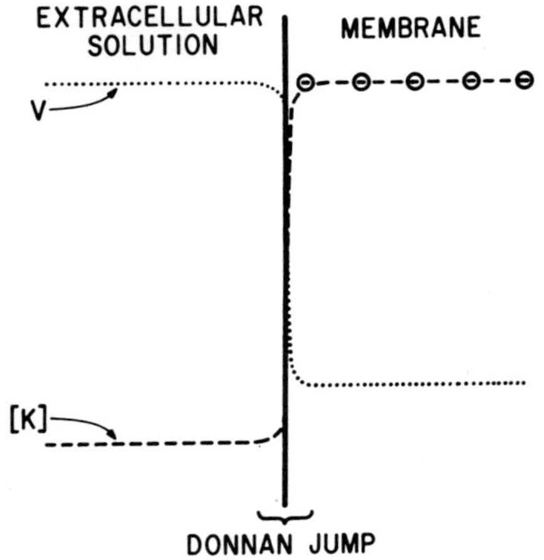

Figure 3-3. Idealization of a Donnan jump. Fixed negative charges within the membrane result in an abrupt change in potassium concentration (dashed line) at the membrane–water interface. The abrupt concentration shift is matched by an abrupt change in electrical potential (dotted line) in such a way that Nernst equilibrium conditions are approximated closely through the interface.

paths of the ions, so the importance of collisions and viscous drag becomes problematical. Consider, as a limiting case, the idealized Donnan jump, in which the space-charge layer is so compact that ions are likely to proceed across it without any collisions at all. In such situations, the Nernst–Planck model clearly is not appropriate, since flow is not limited by viscous drag. As an alternative, we can picture the ideal Donnan jump as an energy barrier for ions on one side of it, as depicted in Figure 3-3. Here we have a high concentration of potassium ions in the membrane, compensating for a large concentration of fixed charge there, and a lower concentration of potassium ions outside the membrane. The corresponding Donnan jump generates an electrical potential difference that is positive just outside the surface of the membrane and negative just inside the surface.

Potassium ions approaching the jump from the outside of the membrane are swept across the jump by enormous electric fields in the jump itself. Those approaching from inside, on the other hand, face the same fields, but in a direction that tends to retard their movement and fling them back into the membrane. A few of those approaching from this side, however, will possess enough momentum in the right direction (i.e., normal to the jump) to carry them through the jump to the solution on the other side. According to the Boltzmann distribution, the number of ions with sufficient kinetic energy to pass through the energy barrier will be directly proportional to

$$e^{-qV/kT}$$

where V is the voltage across the jump. The actual number with velocities in the correct direction will depend on the geometry of the situation, but the rate at which the ions approach the barrier from within the membrane should be directly proportional to the concentration of ions within the membrane and to their average velocity in the direction normal to the ideal Donnan jump. The concentration as depicted in Figure 3-3 should be of the order of 10^{20} ions/cm^3, and for most ion species (including potassium) at room temperature, the average velocity in a given direction is close to 10^4 cm/sec. Thus, ions should flow toward the barrier at a rate of approximately 10^{24} ions/cm^2-sec. This is a truly enormous flow density, and it underscores the significance of viscous drag in diffusional and drift processes. For our idealized Donnan jump with its presumed absence of drag, the rate of flow across it from within the membrane should be given by

$$J_{\mathrm{K}} = A_{\mathrm{K}} e^{-qV/kT}$$

where the magnitude of A_K is close to 10^{24} ions/cm^2-sec. The flow in the opposite direction, being unretarded by the fields in the jump, is limited only by the rate at which ions approach the surface of the membrane from the outside. When the system is at equilibrium, the voltage across the Donnan jump will equal the Nernst potential corresponding to the potassium-ion concentration ratio across the jump, and the flows of potassium ions from the two sides will be equal (but in opposite directions). Therefore, as long as the polarity of the voltage across the jump is that shown in Figure 3-3, the net flow density of potassium ions across the jump should be

$$J_{\mathrm{K}_{\mathrm{net}}} = A_{\mathrm{K}}(e^{-q\Delta V/kT} - 1) \tag{3-42}$$

where ΔV is the difference between the voltage across the jump and the Nernst Potential for the jump.

Because of the enormous size of A_K, large net flow densities will be generated by vanishingly small values of V. In other words, the ideal Donnan jump should offer no impedance to ion flow and should not affect the membrane voltage-current characteristics. Limitations to flow therefore must reside in the interior of the membrane, where viscous drag is effective.

The situation is not so clear in the case of more diffuse space-charge layers, where viscous drag undoubtedly comes into play. The width of the diffuse space-charge layer at the membrane surface has been estimated by several authors employing the Gouy–Chapman theory (e.g., see Gilbert, 1971). Theoretically, the electrical potential gradient should decline exponentially with distance from the center of a space-charge layer, and the distance required for each reduction by the factor $1/e$ is called the *Debye shielding distance* or *ion atmosphere radius*. This distance is proportional to the square root of the dielectric constant of the medium divided by the total ionic concentration in that medium. Typical values are of the order of 5 Å for physiological solutions; for membranes, they are not known, since the dielectric constants and ion concentrations associated with permeation channels are not known. However, the dielectric constant of water is exceptionally high, and that in the membrane is not likely to be higher. Therefore, in the model of Figure 3-3, one would expect the Debye shielding distance in the membrane to be less than 5 Å, perhaps considerably less. The overall space-charge layer, then, might have an effective width of between 5 and 10 Å. Over such distances, the average electric field strength undoubtedly is very large, and small changes in the electrical potential difference across the layer will produce large changes in field strength, which in turn will result in large changes in ionic flow. Thus, even the more diffuse space-charge layers probably present very low impedances to ion flow.

In general, the steady-state ion flows through space-charge layers, whether compact or diffuse, should follow expressions of the same form as the right side of equation (3-42). Noble (1972*c*) has used such expressions to model a membrane with two space-charge layers separated by a central Nernst–Planck region. By varying the relative values of A_K in his models, he has been able to demonstrate a wide range of behavior that one might expect in membranes and which, indeed, has been observed.

3.9. Ion-Concentration and Electrical Potential Profiles for Various Membrane Models

As a summary to the discussion so far in this chapter, Figure 3-4 illustrates qualitatively the profiles one might expect in various membrane situations under steady-state conditions.

The main principles employed in the construction of these graphs were space-charge neutrality, the linearity of total ion concentration [equation (3-12)], the omnipresence of near-equilibrium conditions across Donnan jumps [which follows from the discussion concerning equation (3-42)], and continuity of flow in steady-state situations. The last principle was mentioned earlier but bears reemphasis here since it is so important in profile considerations. Basically, continuity of flow simply means that the flow of

Figure 3-4. Idealized ionic and electrical potential profiles under various circumstances. (a) Donnan equilibrium with nonpermeating (A^-) ions in the solution on the left. Nernst equilibrium conditions prevail across the Donnan jump, space-charge neutrality prevails everywhere else. (b) Donnan equilibrium with fixed (B^-) charge uniformly distributed within the membrane. (c) Steady state with a simple electrogenic pump, homogeneous membrane. According to equation (3-12), the total ion concentration gradient (i.e., $d[\mathrm{KCl}]/dx$) is constant throughout the membrane. Since the diffusional flow of each ion species in this case is

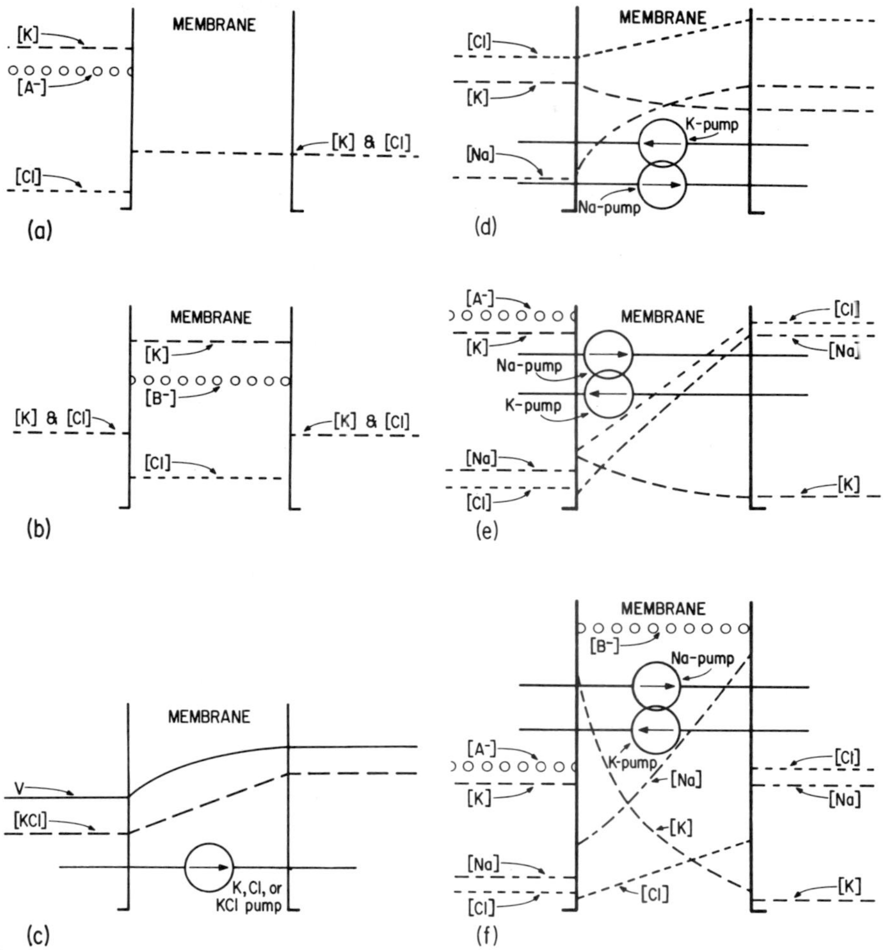

therefore constant throughout the membrane, the drift flow also must be constant (otherwise local accumulation of ions would occur and steady state would not prevail). Constant drift flow in the face of varying concentrations requires a varying potential gradient (steeper where the concentrations are lower). (d) Steady state in a three-ion system with a nonelectrogenic potassium/sodium pump, homogeneous membrane. Potassium mobility throughout the membrane is assumed to be considerably greater than sodium mobility, leading to a potassium concentration gradient considerably smaller than the sodium concentration gradient. The curvatures of the potassium and sodium concentrations should be equal but opposite to provide a constant total-ionic-concentration gradient. (e) The system of (d) with nonpermeating anions added to the left side. A large potassium concentration change occurs at the Donnan jump. The chloride concentration is locked to the potassium and sodium concentrations through space-charge neutrality. (f) The same system, with uniformly distributed, fixed charge added to the membrane. Marked Donnan jumps now occur at each surface of the membrane; anions tend to be excluded from the membrane, reducing their permeabilities relative to those of the cations.

a given ion species into a region (of any size) is precisely equal to the flow of that same ion species out of that region, and it is nothing more than a statement of conservation for steady state. To apply it to a Nernst-Planck membrane, one simply must keep in mind that any divergence in diffusional flow must be offset by an equal but opposite divergence of drift flow. Thus, if the concentration of a given ion species varies linearly with distance through a homogeneous membrane, then the diffusional flow through that membrane does not diverge, and the drift flow must not diverge either. If there is in fact a drift flow in this situation, then the electrical potential gradient *must* vary with distance through the membrane in such a manner as to keep the drift flow constant. Similarly, if the electrical potential gradient is constant, then any ion gradients present must not be constant.

3.10. Summary: A Qualitative Model of Neuronal Membrane

In developing a qualitative model of neuronal membrane, one must consider a fairly large number of factors, many of which have been mentioned previously in this section:

1. What are the major permeating ion species?
2. Are there important populations of nonpermeating ions on either side of the membrane?
3. Is the membrane homogeneous across its width with respect to ionic mobilities?
4. Is the membrane uniform over its surface with respect to permeation?
5. Are fixed charges present within the membrane or on its surface?
6. Is bulk ion storage within the membrane large enough compared with ionic flow rates to affect the dynamics of flow?
7. Are significant space-charge layers present within the membrane?
8. Which ionic species are actively transported across the membrane?

The answers apparently are:

1. Potassium and sodium are the major permeants, accompanied by chloride in some systems. Under certain circumstances, calcium also becomes an important permeant.
2. Magnesium and sulfate are also present, but their degree of permeation is not well known. The two divalent cations, magnesium and calcium, apparently significantly affect the permeability of the membrane to potassium and sodium. Most of the counterions for potassium in the cytoplasm apparently are nonpermeating organic anions.
3. It is not known whether or not the permeation channels in the membrane are homogeneous.

4. The membrane apparently is not uniform over its surface with respect to permeation.

5. In some cases, the plasma membrane appears to be somewhat selective for cations, implying some fixed negative charges in the permeation channels. Furthermore, titration experiments (e.g., see Gilbert, 1971) indicate the presence of fixed charges at the membrane surfaces.

6. Although bulk storage of ions generally has been ignored by membrane modelers, it may be important, especially in passive membrane.

7. It is quite possible that the space-charge layers associated with Donnan jumps or fixed surface charges are sufficiently diffuse to be important with respect to the overall voltage–current characteristics of the plasma membrane.

8. Any ion species whose Nernst potential is not the same as the resting potential of the membrane presumably is actively transported. Such ions include sodium (pumped outward), calcium (pumped outward), occasionally chloride, and potassium (pumped inward). Occasionally, the resting potential appears to be very close to the potassium-ion Nernst potential, in spite of the presence of a potassium pump (simply because potassium permeability is high).

Putting some of these bits of information together, we have the qualitative model depicted in Figure 3-5.

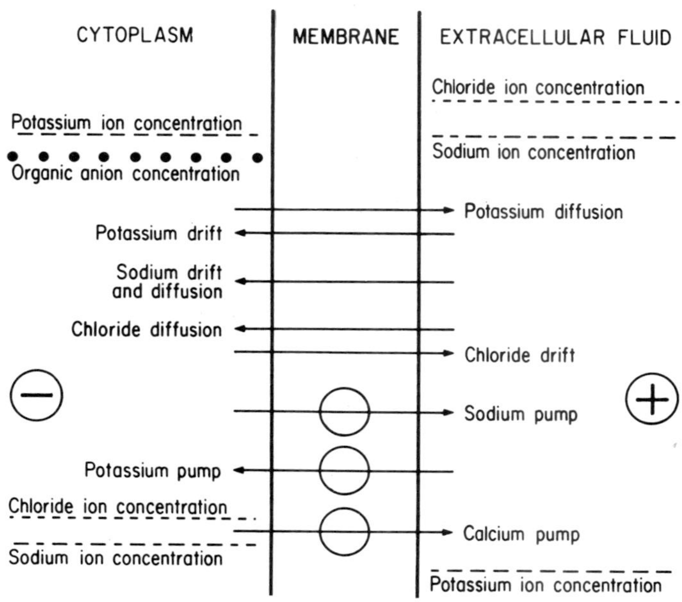

Figure 3-5. Qualitative model of the plasma membrane. Here the major permeants are taken to be potassium, chloride, and sodium. In muscle cells, certain receptors, synapses, and the somata of some neurons, calcium ions are also thought to be important permeants.

Suggested Further Readings

For further readings on this chapter, we refer the reader to Adelman (1971); Aidley (1971); Barr (1965); Bennett (1967); Cohen and Cooley (1965); Cole (1968); Eigen and DeMaeyer (1971); Katz (1966); Kittel (1958); Noble (1972); Plonsey (1969); and Wallach (1972).

4

Equivalent Circuits for Passive Membrane

Although the Nernst–Planck equations and the various constraints that might be placed upon them are interesting from the point of view of membrane physics, they become awkward and unrevealing when one is interested in signal flow over distributed neuronal structures. In such cases, equivalent electrical circuits are both convenient and revealing. However, it also is important to reconcile the circuit models with the more primitive models in order to delineate the assumptions and constraints inherent in the former. On the other hand, the circuit models themselves have a strong basis in empirical observation, and one is perfectly justified in allowing them to stand on their own merits.

In this chapter, we shall discuss four basic types of equivalent circuits for membranes, and we shall develop their parameter values and properties from the Goldman model, namely, a membrane with homogeneous permeation channels across which the Nernst–Planck equations apply, with the permeating ions being univalent and isotonic throughout each channel. The channels may or may not have homogeneously distributed fixed charges (see Finkelstein and Mauro, 1963). Taken together, these assumptions allow us to use a constant electric field in our calculations, which greatly simplifies the analysis.

4.1. The Basic Equivalent Circuit for a Patch of Membrane

Perhaps the most commonly invoked equivalent circuit for a local patch of plasma membrane is that shown in Figure 4-1. Here we have three elements—an electrical conductance, an electrical capacitance, and a

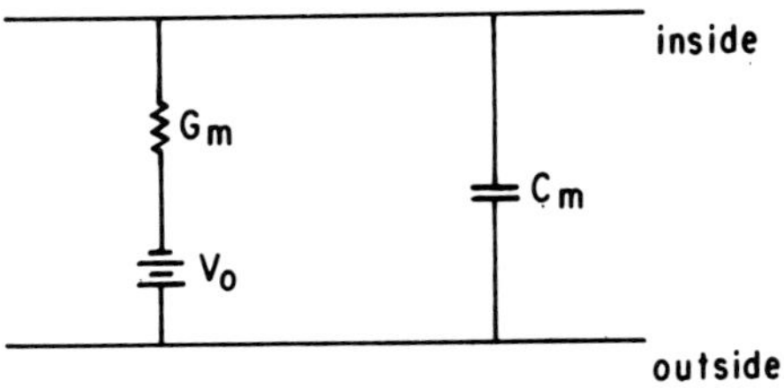

Figure 4-1. Common equivalent circuit for a patch of plasma membrane.

source of electrical potential difference. The local intracellular solutions and extracellular solutions are taken to be isopotential and are represented by perfect conductors on each side of the equivalent circuit.

Apparently, the capacitance in Figure 4-1 represents a property of the membrane as a whole and is independent of local variations in permeation channels. Although it varies to some extent from one cell type to another (apparently being higher for muscle-cell membranes), the electrical capacitance per square centimeter of membrane seems to remain quite constant in a given cell. For neurons, it is very close to 1 μF/cm^2, which corresponds to an average (or effective) membrane dielectric constant of 10^{-14} F/cm for membranes 100 Å thick.

While the capacitance in Figure 4-1 can be represented quite reasonably as being constant, the parameters of the other two elements may very well depend on V_m and time. The importance of these dependences hinges upon the signals of interest and upon whether or not bulk storage of ions is significant. For very small excursions in the membrane potential, an equivalent circuit with constant parameter values is probably quite reasonable. For large excursions, on the other hand, such an equivalent circuit could be inadequate unless the diffusion time across the membrane [see equation (3-28)] is long compared to the durations of the excursions. These points are discussed in more detail below.

4.1.1. The Finkelstein–Mauro Model

The electrical conductance of a Nernst–Planck membrane clearly will depend upon the ion profiles within the membrane, increasing as the number of high-mobility ions in the membrane increases. Finkelstein and Mauro (1963) considered the conductance of the system depicted in Figure 4-2, with a homogeneous Nernst–Planck membrane separating two solu-

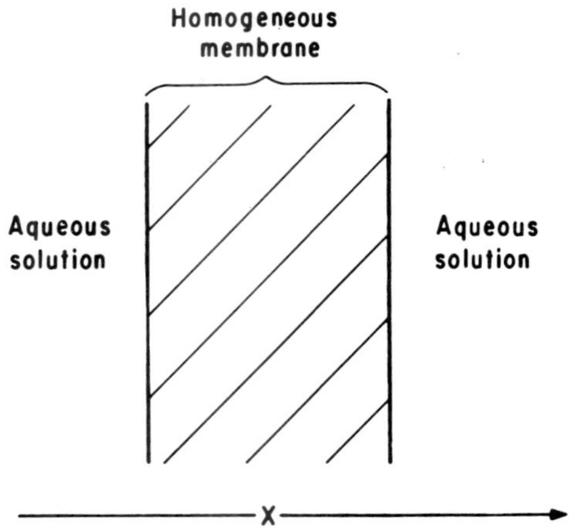

Figure 4-2. The basic geometry of the Finkelstein–Mauro model. A homogeneous membrane separating two well-stirred aqueous solutions.

tions of monovalent ions. The net electrical current density through the membrane is

$$I = q \sum_j J_{C_j} - q \sum_i J_{A_i} \tag{4-1}$$

where C_j is the jth monovalent cation species and A_i is the ith monovalent anion species.

According to the principle of space-charge neutrality, the local rate of net charge accumulation will be negligible (compared with local flows of ions); so the rate at which charge flows into a given volume of membrane or solution will be very nearly equal to the rate at which charge flows out of that same volume. Thus, the divergence of current density across the membrane will be essentially zero. This is Kirchhoff's current law for bulk materials. For the membrane system of Figure 4-2, it simply means that the net electrical current density will not vary significantly with position through the membrane (i.e., I is independent of x). If the rates of ion transfers as a result of shifts in concentration profiles are significant, then the individual ion flows themselves will have significant divergences (dependence on x) through the membrane until new steady-state profiles are attained.

Substituting the Nernst–Planck expressions for the ion flow densities in Equation (4-1), one obtains

$$I = -q^2 \frac{dV}{dx}(A^* + C^*) + kTq\frac{d}{dx}(A^* - C^*) \tag{4-2}$$

where

$$A^* = \sum_i \mu_i[A_i] \qquad \text{and} \qquad C^* = \sum_j \mu_j[C_j]$$

A_i being the ith anion species and C_j being the jth cation species. Since I is independent of x, we can divide all terms in equation (4-2) by $(A^* + C^*)$ and integrate from one surface of the membrane to the other to obtain the voltage across the membrane:

$$V_m = -I\left(\frac{1}{q^2}\right)\int_0^w \frac{dx}{A^* + C^*} + \frac{kT}{q}\int_0^w \frac{d(A^* - C^*)}{A^* + C^*} \tag{4-3}$$

Identifying the first term on the right side as a voltage due to current passing through a resistance and the second term as an offset potential, Finkelstein and Mauro established the following parameters for the equivalent circuit of Figure 4-1:

FINKELSTEIN–MAURO MODEL

$$G_m = \frac{q^2}{\displaystyle\int_0^w \frac{dx}{A^* + C^*}} \tag{4-4a}$$

$$V_0 = \frac{kT}{q}\int_0^w \frac{d(A^* - C^*)}{A^* + C^*} \tag{4-4b}$$

which are valid in general for systems with monovalent ions, and could rather easily be generalized to systems with ions of different valences. In general, since the ionic concentration profiles in the membrane vary from one steady-state membrane potential to another, A^* and C^* will be time and voltage dependent, leading to time and voltage dependence of both G_m and V_0 in the circuit of Figure 4-1. For purposes of further discussion, we shall call this representation the Finkelstein–Mauro model. It applies to both steady-state and transient conditions and does not depend on the assumptions of the Goldman model.

4.1.2. The Steady-State Conductance According to the Goldman Model

Employing the Goldman model to facilitate calculations, one can simplify the Finkelstein–Mauro model in two ways. First of all, the Goldman potential provides a useful value at which to fix V_0, in which case the membrane conductance would embody all of the voltage and time dependence owing to shifting ion concentration profiles in the membrane (exclusive of those resulting from changes in mobilities or from changes in ionic concentrations outside the membrane, both of which we shall ignore at this time). In the second place, we can use the assumption of a constant electric field in the membrane to reduce the description of G_m to relatively simple terms.

The steady-state conductance of the membrane can be defined as follows:

$$G_{\mathrm{m_{ss}}} = \frac{I_{\mathrm{ss}}}{V_{\mathrm{m_{ss}}} - V_{\mathrm{G}}} \tag{4-5}$$

where $V_{\mathrm{m_{ss}}}$ is the steady-state membrane potential; I_{ss} is the steady-state current density through the membrane, which is equal to zero when the membrane potential equals the Goldman potential V_{G}. Thus, when it is at rest, our ideal membrane will exhibit no current and will reside at the Goldman potential. If one applies a current to the membrane but allows the internal and external ion concentrations to remain unchanged, then equation (4-5) will give the resulting change in potential. Similarly, if one applies a voltage, equation (4-5) will give the resulting current. In order to use the equation, however, we must find the steady-state conductance G_{ss}. This can be done simply by combining equations (3-19) and (4-1) to yield I_{ss} and then dividing both sides by $V_{\mathrm{m_{ss}}} - V_{\mathrm{G}}$:

$$G_{\mathrm{m_{ss}}} = \frac{V_{\mathrm{m_{ss}}}}{V_{\mathrm{m_{ss}}} - V_{\mathrm{G}}} \frac{q^2}{w} \sum P_i a_i(V_{\mathrm{m_{ss}}}) \tag{4-6}$$

where

$$a_i = \frac{[\mathrm{A}_i]_{\mathrm{in}} - [\mathrm{A}_i]_{\mathrm{out}} e^{qV_{\mathrm{m_{ss}}}/kT}}{1 - e^{qV_{\mathrm{m_{ss}}}/kT}} \quad \text{for the } i\text{th anion} \tag{4-7a}$$

$$a_i = \frac{[\mathrm{C}_i]_{\mathrm{out}} - [\mathrm{C}_i]_{\mathrm{in}} e^{qV_{\mathrm{m_{ss}}}/kT}}{1 - e^{qV_{\mathrm{m_{ss}}}/kT}} \quad \text{for the } i\text{th cation} \tag{4-7b}$$

w is the width of the membrane, and $P_i = \zeta_i \mu_i$ is the permeability of the ith ion. Equation (4-6) applies when

$$V_0 \equiv V_{\mathrm{G}}$$

Clearly, the steady-state conductance as given by (4-6) is nonohmic. The function $a_i(V_{\mathrm{m_{ss}}})$ represents the concentration profile of the ith ion species through the membrane. The profile will be an instantaneous function of $V_{\mathrm{m_{ss}}}$ only if the ionic concentration profiles can change instantaneously [i.e., if T_{D} in equation (3-25) is effectively zero]. Otherwise, the conductance will not assume its steady-state value immediately in response to a new membrane potential, but will lag the potential to some extent. Thus, G_{m} will depend not only upon membrane potential but also upon time.

4.1.3. The Instantaneous Conductance according to the Goldman Model

Suppose that the electrical potential across the membrane suddenly were shifted from one steady-state value to another and then held at the new level. If the bulk storage of ions in the membrane were not negligible and the various concentration profiles required time to change, then in the instant following the abrupt shift of V_m, the only variable on the right side of the Nernst–Planck equation that would have changed is dV/dx (or grad V). Therefore, the instantaneous shift in membrane current density, ΔI, is given by

$$\Delta I = -q^2(\mathrm{A}^* + \mathrm{C}^*)\Delta \frac{dV}{dx} \tag{4-8}$$

where A* and C* are the terms defined with equation (4-2) and can be given specific forms with the aid of equation (3-16):

$$\mathrm{A}^* = \sum_i \mu_i[\mathrm{A}_i] = \sum_i P_i(a_i + b_i e^{(qV_{m_{ss}}/kT)x}) \tag{4-9a}$$

$$\mathrm{C}^* = \sum_j \mu_j[\mathrm{C}_j] = \sum_j P_j(a_j + b_j e^{-(qV_{m_{ss}}/kT)x}) \tag{4-9b}$$

where a_i and a_j are given by equations (4-7a) and (4-7b), respectively, and

$$b_i = \frac{[\mathrm{A}_i]_{\mathrm{out}} - [\mathrm{A}_i]_{\mathrm{in}}}{1 - e^{qV_{m_{ss}}/kT}} \tag{4-10a}$$

$$b_j = \frac{[\mathrm{C}_j]_{\mathrm{in}} - [\mathrm{C}_j]_{\mathrm{out}}}{1 - e^{qV_{m_{ss}}/kT}} \tag{4-10b}$$

Equation (4-8) can be integrated in exactly the same manner as equation (4-2) to yield the instantaneous conductance. This conductance will be precisely the same as the Finkelstein–Mauro conductance for the membrane at the steady-state potential from which the abrupt change was made. In other words, the instantaneous conductance is independent of the magnitude of the abrupt shift in membrane voltage but depends solely on the voltage from which that shift was made (or, to be more precise, upon the recent history of voltages just prior to the shift). Thus, the instantaneous conductance is ohmic.

4.1.4. The Reversal Potential Associated with the Instantaneous Conductance

Suppose that we have imposed a steady-state potential across the membrane of Figure 4-2 and produced as a result a steady-state current density I_{ss}. If we now suddenly shifted the voltage to a new value, the resulting change in current density would be directly proportional to the magnitude of the voltage shift; and for some value of voltage shift, the current would go to zero. The magnitude of the shift required to drive the current to zero is simply $-I_{ss}/G_{inst}$, where G_{inst} is the instantaneous ohmic conductance of the membrane. The membrane voltage corresponding to zero current is the instantaneous reversal potential for membrane current and is given by

$$V_{r_{inst}} = V_{m_{ss}} - I_{ss}/G_{inst} \tag{4-11}$$

and must be precisely equal to the Finkelstein–Mauro potential [equation (4-4b)].

4.1.5. Instantaneous Conductance and Reversal Potential in a Three-Ion Goldman System

If we restrict the permeating ions in our system to potassium, sodium, and chloride, then the expressions for the instantaneous conductance and its reversal potential become very simple indeed. This is due to the constraint of space-charge neutrality:

$$[\mathrm{K}] + [\mathrm{Na}] = [\mathrm{Cl}] \tag{4-12}$$

which simplifies the integrations involved in the Finkelstein–Mauro model:

$$\begin{aligned} \mathrm{C}^* &= \mu_{\mathrm{K}}[\mathrm{K}] + \mu_{\mathrm{Na}}[\mathrm{Na}] \\ \mathrm{A}^* &= \mu_{\mathrm{Cl}}[\mathrm{Cl}] = \mu_{\mathrm{Cl}}([\mathrm{K}] + [\mathrm{Na}]) \\ \mathrm{C}^* + \mathrm{A}^* &= (\mathrm{P_K} + \mathrm{P_{Cl}})(a_{\mathrm{K}} + b_{\mathrm{K}}\, e^{-qV_{m_{ss}}x/kTw}) \\ &\quad + (P_{\mathrm{Na}} + P_{\mathrm{Cl}})(a_{\mathrm{Na}} + b_{\mathrm{Na}}\, e^{-qV_{m_{ss}}x/kTw}) \\ &= a' + b'e^{-c'x} \end{aligned}$$

$$\begin{aligned} \int \frac{dx}{a' + b'e^{-c'x}} &= \frac{w}{a'} - \frac{1}{a'c'}\ln(a' + b') + \frac{1}{a'c'}\ln(a' + b'e^{-c'w}) \\ &= \frac{w}{a'}\left(1 - \frac{kT}{qV_{m_{ss}}}\ln\frac{P_{\mathrm{K}}[\mathrm{K}]_{out} + P_{\mathrm{Na}}[\mathrm{Na}]_{out} + P_{\mathrm{Cl}}([\mathrm{Na}]_{out} + [\mathrm{K}]_{out})}{P_{\mathrm{K}}[\mathrm{K}]_{in} + P_{\mathrm{Na}}[\mathrm{Na}]_{in} + P_{\mathrm{Cl}}([\mathrm{Na}]_{in} + [\mathrm{K}]_{in})}\right) \end{aligned} \tag{4-13}$$

but, owing to the constraint of isotonicity implicit in the Goldman model,

$$[\mathrm{Na}]_{out} + [\mathrm{K}]_{out} = [\mathrm{Na}]_{in} + [\mathrm{K}]_{in}$$

so

$$\int \frac{dx}{a' + b'e^{-c'x}} = \frac{w}{a'}\left(1 - \frac{V_G}{V_{m_{ss}}}\right) \tag{4-14}$$

and

$$G_{m_{inst}} = \frac{q^2[P_{Na}a_{Na}(V_{m_{ss}}) + P_K a_K(V_{m_{ss}}) + P_{Cl}a_{Cl}(V_{m_{ss}})]}{w} \frac{V_{m_{ss}}}{V_{m_{ss}} - V_G} \tag{4-15}$$

This is exactly the same conductance as that given by equation (4-6) for $G_{m_{ss}}$. Therefore, for our three-ion Goldman system, the instantaneous conductance, the steady-state conductance, and the Finkelstein–Mauro conductance all are one and the same. Furthermore, it is quite easy to show that the potential sources associated with each of these conductances also are one and the same and equal to the Goldman potential.

From equation (4-12) combined with equations (4-5), (4-6), and (4-15), we have

$$V_{r_{inst}} = V_{m_{ss}} - \frac{G_{m_{ss}}(V_{m_{ss}} - V_G)}{G_{m_{inst}}} = V_G \tag{4-16}$$

As a result of isotonicity

$$\frac{d([\mathrm{Na}] + [\mathrm{K}])}{dx} = (b_K + b_{Na})(-c)e^{-cx} = 0$$

Therefore

$$b_K = -b_{Na} \tag{4-17}$$

and

$$dA^* = 0 \tag{4-18a}$$

$$dC^* = (P_{Na}b_{Na} + P_K b_K)(-c)e^{-cx}\,dx \tag{4-18b}$$

$$A^* + C^* = P_K a_K + P_{Na}a_{Na} + (P_K b_K + P_{Na}b_{Na})e^{-cx}$$

which leads to the following result from equation (4-4b):

$$V_0 = V_G \tag{4-19}$$

Therefore,

$$V_0 = V_{r_{inst}} = V_G$$

4.1.6. Current–Voltage Characteristics of a Three-Ion Goldman Membrane

In a Goldman membrane permeated by potassium, sodium, and chloride ions, the steady-state current density can be considered as comprising three components:

the current density due to potassium-ion flow

$$I_K = -\frac{q^2}{w} P_K a_K V_{m_{ss}} = \frac{q^2}{w} P_K \frac{[K]_{in} e^{qV_{m_{ss}}/kT} - [K]_{out}}{1 - e^{qV_{m_{ss}}/kT}} V_{m_{ss}} \tag{4-20a}$$

the current density due to sodium-ion flow

$$I_{Na} = -\frac{q^2}{w} P_{Na} a_{Na} V_{m_{ss}} = \frac{q^2}{w} P_{Na} \frac{[Na]_{in} e^{qV_{m_{ss}}/kT} - [Na]_{out}}{1 - e^{qV_{m_{ss}}/kT}} V_{m_{ss}} \tag{4-20b}$$

the current density due to chloride-ion flow

$$I_{Cl} = -\frac{q^2}{w} P_{Cl} a_{Cl} V_{m_{ss}} = \frac{q^2}{w} P_{Cl} \frac{[Cl]_{out} e^{qV_{m_{ss}}/kT} - [Cl]_{in}}{1 - e^{qV_{m_{ss}}/kT}} V_{m_{ss}} \tag{4-20c}$$

The shapes of these three current–voltage relationships depend principally upon the ratios of the ion concentrations on the two sides of the membrane. For example, if the ratios have the values

$$[K]_{in}/[K]_{out} = 20$$
$$[Na]_{in}/[Na]_{out} = 0.1$$
$$[Cl]_{in}/[Cl]_{out} = 0.067$$

then the relationships plotted in Figure 4-3 would apply.

To obtain the overall current–voltage relationship for the membrane, one simply would take the sum of the ordinates, each weighted appropriately (i.e., by $P_B[B]_{in}$). Figure 4-4 shows one possible outcome, with potassium and chloride dominating to make the overall characteristics convex rather than concave.

The current–voltage curve in Figure 4-4 intercepts the voltage axis at the Goldman potential. Furthermore, according to equation (4-16) the instantaneous current–voltage characteristics also intercept the voltage at the Goldman potential. Therefore, if a steady-state situation were established at V_1, I_1, then any instantaneous shift in current density or voltage would carry the system along the straight line intercepting both that point and the point V_G, 0, as depicted in the figure.

Although the current–voltage characteristics shown in Figure 4-4 were derived from a model representing a highly idealized system, they nonetheless are illustrative of several important general points concerning equivalent circuits; so we shall employ them as the basis for discussion in the following sections.

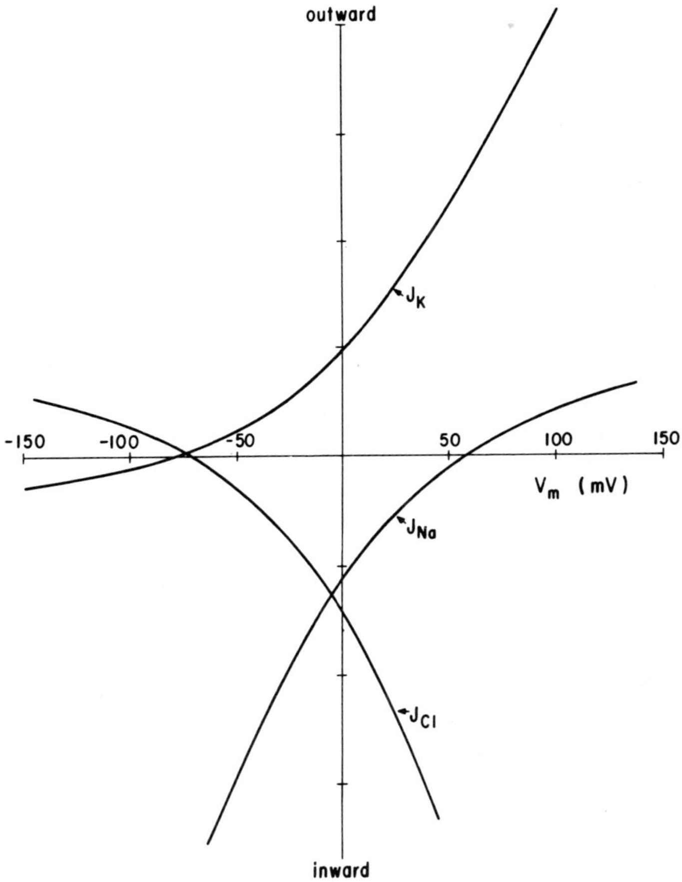

Figure 4-3. Ionic flux densities for a three-ion Goldman membrane with a 20:1 potassium concentration ratio, a 1:10 sodium concentration ratio, and a 1:16 chloride concentration ratio. The flux density (ordinate) in each case can be converted to current density by a multiplication by the appropriate ionic charge (q for Na and K, $-q$ for Cl).

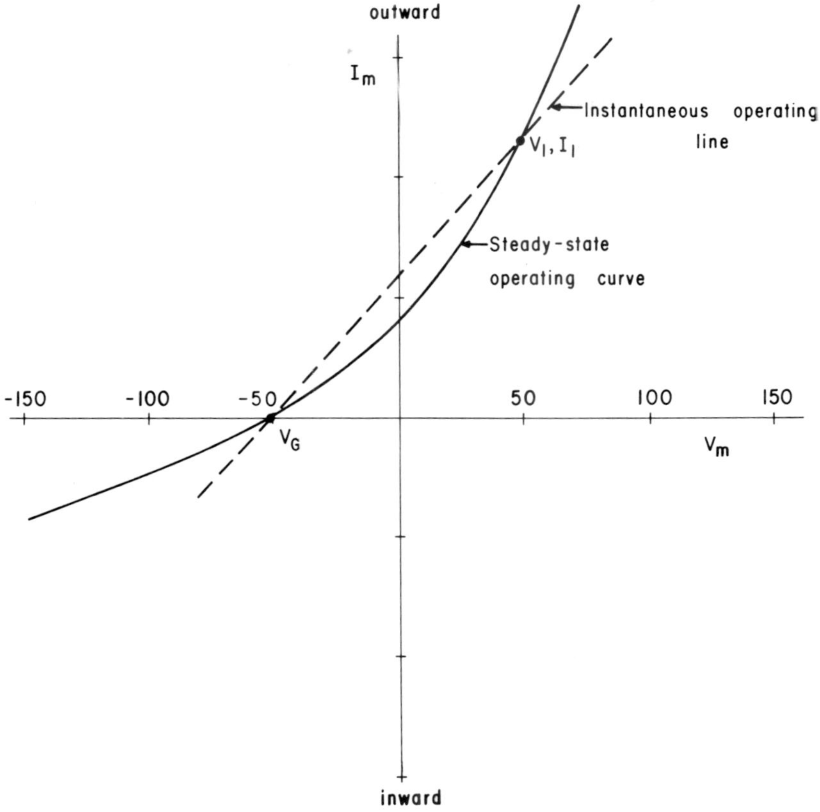

Figure 4-4. One possible outcome of the superposition of the current-density components in Figure 4-3, yielding one possible voltage–current relationship for the membrane. V_G is the Golman potential, and the dashed line shows the instantaneous voltage–current characteristics for the system operating from steady state at V_1, I_1.

4.2. Small-Signal Equivalent Circuits

For small excursions of voltage or current about a steady-state operating point, such as V_1, I_1 in Figure 4-4, the system could move along the steady-state operating curve, the instantaneous operating line, or some other operating curve, depending upon whether the excursions are fast or slow relative to the rate at which ion concentration profiles within the membrane can change. The latter is inversely proportional to the diffusion time T_D for ions in the membrane. Thus, if the duration of an excursion is short compared to T_D, then the system will operate along or close to the instantaneous operating line. If the duration of the excursion is long compared to T_D, then the system will operate along the steady-state operating curve. If the duration of the excursion is comparable to T_D, then the system will operate along some other curve. If the system is isotonic throughout, then the total concentration of cations in the membrane will be constant, and concentration-profile shifts will involve exchange of one cation for the other. Thus, the appropriate value of T_D with which to make the comparison is that for the slowest cation (i.e., the cation with the lowest mobility in the membrane).

For such excursions, one can define an effective operating conductance as follows:

$$G_m = \Delta I / \Delta V_m \tag{4-21}$$

which can be evaluated by taking a Taylor's series expansion of the operating characteristics about the operating point. Let $I(V_m)$ be the function describing the operating curve,

$$\Delta I = I(V_m + \Delta V_m) - I(V_m) = \frac{dI(V_m)}{dV_m}\Delta V_m + \frac{d^2 I(V_m)}{dV_m^2}\frac{(\Delta V_m)^2}{2} + \frac{d^3 I(V_m)}{dV_m^3}\frac{(\Delta V_m)^3}{6} + \cdots \tag{4-22}$$

$$G_m = \frac{dI(V_m)}{dV_m} + \frac{d^2 I(V_m)}{dV_m^2}\frac{\Delta V_m}{2} + \frac{d^3 I(V_m)}{dV_m^3}\frac{(\Delta V_m)^2}{6} + \cdots \tag{4-23}$$

If the excursions are sufficiently fast to allow the system to operate along the instantaneous operating line, then only the first term on the right side of equation (4-23) is nonzero. On the other hand, if the excursions are not that fast, then the higher-order terms very likely will be present to some extent. However, if the excursions are very small (i.e., involve very small values of ΔV_m), then the effects of the higher-order terms can become negligible, in which case the effective membrane conductance is simply the slope of the operating characteristics. For excursions that are very slow compared to T_D^{-1}, the effective conductance is the slope of the steady-state operating curve.

The value of G_m appropriate for the speed of a particular, very small excursion will be called the *small-signal conductance* of the membrane. The slope of the steady-state operating curve at a particular operating point is generally called the *slope conductance* at that point. The slope of the line connecting the voltage-axis intercept of the steady-state operating curve and any particular operating point along that curve is generally called the *chord conductance* for that point. The conductance defined by equation (4-4) we shall call the *Finkelstein–Mauro* conductance. For membranes in general,

FOR SLOW EXCURSIONS

$$G_{\mathrm{m}_{\text{small-signal}}} = G_{\mathrm{m}_{\text{slope}}} \tag{4-24}$$

FOR FAST EXCURSIONS

$$G_{\mathrm{m}_{\text{small-signal}}} = G_{\mathrm{m}_{\text{Finkelstein–Mauro}}} \tag{4-25}$$

and for our ideal, three-ion Goldman membrane,

$$G_{\mathrm{m}_{\text{Finkelstein–Mauro}}} = G_{\mathrm{m}_{\text{chord}}} \tag{4-26}$$

so

FOR FAST EXCURSIONS

$$G_{\mathrm{m}_{\text{small-signal}}} = G_{\mathrm{m}_{\text{chord}}} \tag{4-27}$$

In general, one employs a small-signal equivalent circuit to deduce the fate of small signals; so the various dc potentials (such as the Goldman potential and the potential of the operating point) are ignored, which leads to the configuration of Figure 4-5.

The slope conductance for our ideal three-ion system can be determined from the steady-state current

$$I_{\mathrm{ss}} = -\frac{q^2}{w}[P_{\mathrm{K}} a_{\mathrm{K}} + P_{\mathrm{Na}} a_{\mathrm{Na}} + P_{\mathrm{Cl}} a_{\mathrm{Cl}}] V_{\mathrm{m_{ss}}} \tag{4-28}$$

$$\begin{aligned} G_{\mathrm{m}_{\text{slope}}} &= -\frac{q^2}{w}[P_{\mathrm{K}} a_{\mathrm{K}} + P_{\mathrm{Na}} a_{\mathrm{Na}} + P_{\mathrm{Cl}} a_{\mathrm{Cl}}] \\ &\quad + \frac{q^2 V_{\mathrm{m_{ss}}}}{w}\left[P_{\mathrm{K}} \frac{da_{\mathrm{K}}}{dV_{\mathrm{m_{ss}}}} + P_{\mathrm{Na}} \frac{da_{\mathrm{Na}}}{dV_{\mathrm{m_{ss}}}} + P_{\mathrm{Cl}} \frac{da_{\mathrm{Cl}}}{dV_{\mathrm{m_{ss}}}}\right] \end{aligned} \tag{4-29}$$

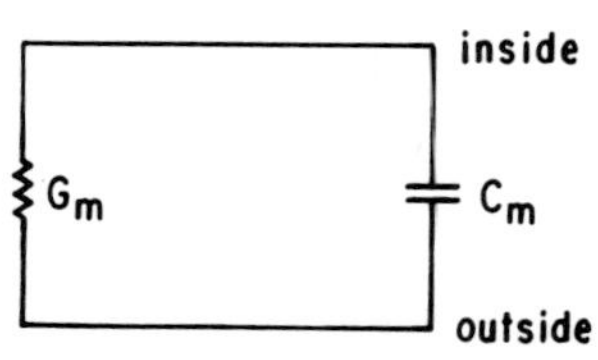

Figure 4-5. Small-signal equivalent circuit for a membrane.

which can be expanded by means of equations (4-7a) and (4-7b). Clearly, by virtue of the definitions of slope conductance and chord conductance, the two become one and the same when $V_{\mathrm{m_{ss}}} = V_{\mathrm{G}}$.

4.3. Equivalent Circuits for Large Signals

As in the case of small excursions about an operating point, the membrane system will respond differently to large excursions, depending upon whether they are fast or slow. For very fast excursions the system will be confined to the instantaneous operating line, while for very slow excursions it will be confined to the steady-state operating curve. Thus, for very fast, large signals, the membrane will appear to be ohmic; and for very slow, large signals, it will appear to be distinctly nonohmic. Once again, the speed of a signal is measured against the speed at which the ion-concentration profile in the membrane can relax from one steady-state condition to another, which is given roughly by the reciprocal of the diffusion time across the membrane of the slowest ion being exchanged.

4.3.1. Mixed Equivalent Circuit for Slow Signals

The equivalent circuit for a patch of membrane often is drawn with a separate path for each of the ion species considered to be of importance in the system. Thus, for our three-ion system, we might have the equivalent circuit of Figure 4-6. Finkelstein and Mauro labeled such representations as "mixed equivalent circuits," since they provided information about the individual ionic components of membrane current as well as the overall electrical properties of the membrane. We shall use their nomenclature in the discussions that follow.

For steady-state or for signals that are very slow compared to T_{D}^{-1}, each ion flow path in the mixed equivalent circuit would represent the steady-state current--voltage operating curve of the ion species in question.

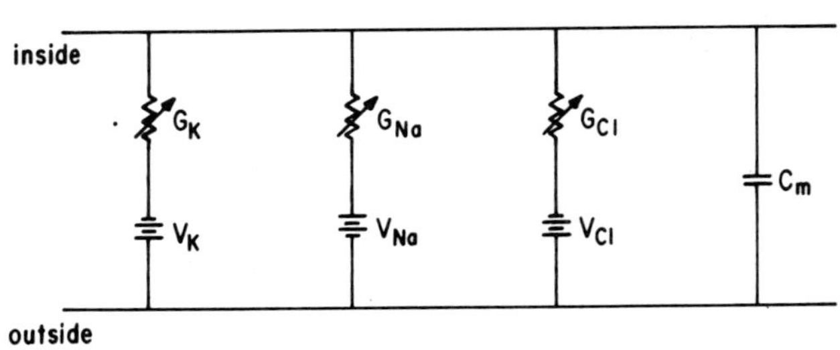

Figure 4-6. Mixed equivalent circuit for a membrane system with three ions.

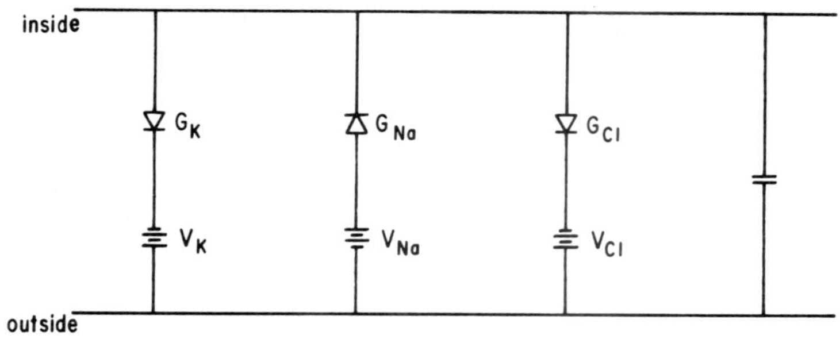

Figure 4-7. Mixed equivalent circuit for a three-ion system. Here the nonohmic conductors are represented as imperfect diodes, which have essentially the same types of current–voltage curves as those in Figure 4-3.

Thus, for our three-ion Goldman system, each of the three curves in Figure 4-3 would be represented by a separate path. The potential at which the current reverses direction in each case is the Nernst potential of the ion species. Thus, the voltage source in each branch of the mixed equivalent circuit simply is the Nernst potential for the ion species represented by that branch. The conductance in each branch clearly is nonlinear. It is often depicted by means of the symbol for a variable resistor. However, the current–voltage characteristics shown in Figure 4-3 are very much like those of an imperfect semiconductor diode. This is not especially surprising, since the underlying physical phenomena apparently are basically the same for the two systems. Thus, as an alternative to the variable-resistor notation, we might use the equivalent circuit of Figure 4-7, which would serve to remind us that the conductances associated with the individual branches are indeed nonlinear.

Taking the limits of equations (4-20) as V_m approaches plus and minus infinity, one finds that the asymptotic slope of the current–voltage relationship in each case is directly proportional to the ion concentration on one side of the membrane or the other. The ratio of the asymptotic slopes for a given ion species, therefore, simply is the ratio of the concentrations of that ion species on the two sides of the membrane. Thus, in our three-ion system, the ratio of the asymptotic slopes for potassium is greatest, being 20 to 1, and in this way the potassium current–voltage characteristics most closely approach those of an ideal diode.

4.3.2. Mixed Equivalent Circuit for Signals That Are Not Very Slow

If excursions of membrane current or voltage are not sufficiently slow to allow steady-state conditions to prevail, then ion exchange will be occurring within the membrane during the excursions, and the flow of a given ion species into the membrane from one side will not necessarily be equal to the flow of that same species out the other side. In our three-ion Goldman system, for example, sodium ions and potassium ions will be exchanged for one another to maintain space-charge neutrality as the ion-concentration profiles are shifting. Thus, the current contributed by a given ion species does not have a single value but diverges markedly through the membrane, completely contrary to Kirchhoff's current law. Under such circumstances, the mixed equivalent circuit, whose structure is based on Kirchhoff's current law, simply is of no value. This deficiency in the mixed equivalent circuit was pointed out by Finkelstein and Mauro.

Conceivably, if membrane excursions were sufficiently fast that they were completed before any significant changes in ion-concentration profiles had occurred, divergence of individual ionic flows could be avoided. However, the local change in voltage gradient during such an excursion will be inversely proportional to the total local conductivity in the membrane [see equation (4-8)], which has contributions from all ion species. The new voltage gradient thus is not forced by constraints of space-charge neutrality or steady state to provide nondiverging flow of an individual ion species. A given concentration profile requires a specific voltage-gradient profile to offset precisely any divergence in diffusional flow with an equal but opposite divergence in drift flow. In a system of three or more permeating ions, the steady-state voltage-gradient profile and the ion-concentration profiles will have adjusted to one another according to the two constraints, and no two ion-concentration profiles will be exactly the same. Therefore, it is essentially impossible for a new voltage-gradient profile (resulting from the membrane excursion) to provide the compensation required for all of the different ion-concentration profiles.

However, there is one situation in which nondivergence is guaranteed, regardless of the speed of the excursions, and another situation in which nondivergence is approximated for very fast excursions. The first situation was pointed out by Finkelstein and Mauro, and is one in which only two ion species (one anion and one cation) are allowed to enter each permeation channel. In other words, for a given permeation channel, the partition coefficient is zero for any ion species other than the two allowed. In such a situation, the ion concentrations will vary linearly from one side of the membrane to the other [see equation (3-12)] and will be invariant under changes in membrane current or voltage. Thus, not only will the flows of individual ion species be nondivergent, even under transient membrane excursions, but the equivalent conductances associated with the individual ion species will be ohmic, and all of the equivalent conductances (slope, chord, instantaneous, steady-state) of the overall membrane will be one and the same and ohmic. Unfortunately, this very simple model does not apply to plasma membranes that have been characterized electrically, since all of them exhibit nonohmic steady-state current–voltage relationships.

The second situation is slightly more complicated. In this case, the partition coefficients for the various ions can be comparable for a given permeation channel, but the mobilities within the channel are very low for all ion species but one. Thus, the channel is selective, but through a different mechanism. In such a situation, the rate of ion-concentration profile change during membrane excursions will be limited by the movement of the various low-mobility ions, and, for excursions that are fast compared to those movements, the constraint of space-charge neutrality will force the flow of the high-mobility ion species to be essentially nondivergent. The appropriate equivalent conductance for such fast excursions is the Finkelstein–Mauro conductance [equation (4-4a)], and the appropriate voltage source is the Finkelstein–Mauro potential [equation (4-4b)]. In each case, the high-mobility ion would dominate. However, the presence of the low-mobility ions within the channel would markedly affect the concentration distribution of the high-mobility ion species and thus could indirectly but very strongly affect both the conductance and the voltage source. However, according to equation (4-4b), which is quite general for Nernst–Planck systems, if there is only one ion with high mobility, then the reversal potential for the flow of that ion species should be very close to its Nernst potential [e.g., consider equation (4-13) for a system in which only one cation species has high mobility]. The steady-state current–voltage relationship of the high-mobility ion species would be nonohmic, being given by an expression similar to that of equation (4-20).

4.4. The Frankenhaeuser–Hodgkin Space

Frankenhaeuser and Hodgkin (1956) observed that the magnitude of the apparent reversal potential for potassium ions decreased slightly after each spike in the squid giant axon. This decrease could be accounted for completely by the assumption that each spike caused a fixed increase in the potassium concentration outside the membrane, and that the potassium concentration subsequently returned in a rather slow, exponential manner to its original magnitude. In response to a train of spikes, with a constant interspike interval, one would expect a situation such as that shown in Figure 4-8. If the increase in external potassium in response to each spike were ΔN, and if the total excess potassium (ΔN_T) were removed by a linear process, then at the time of a spike ΔN_T would increase suddenly,

$$\Delta N_T(t_0) = \Delta N_T(t_0^-) + \Delta N \qquad \text{at the time of a spike}$$

and between spikes, ΔN_T would fall exponentially,

$$d\Delta N_T/dt = -b\Delta N_T$$

$$\Delta N_T(t) = \Delta N_T(t_0)e^{-b(t-t_0)} \qquad \text{following a spike at } t_0$$

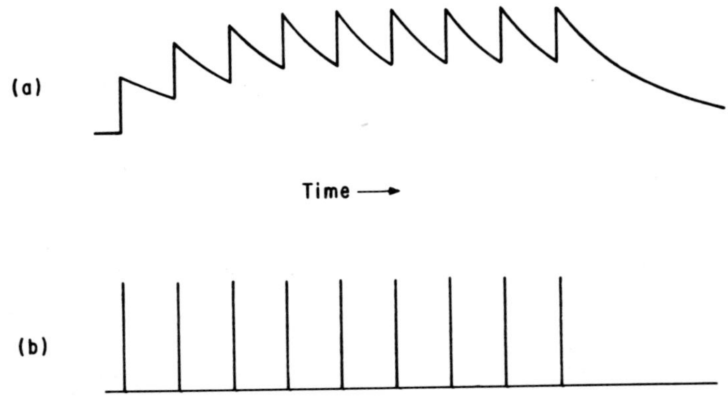

Figure 4-8. Hypothetical linear accumulation of extracellular potassium (a) in response to a train of evenly spaced spikes (b).

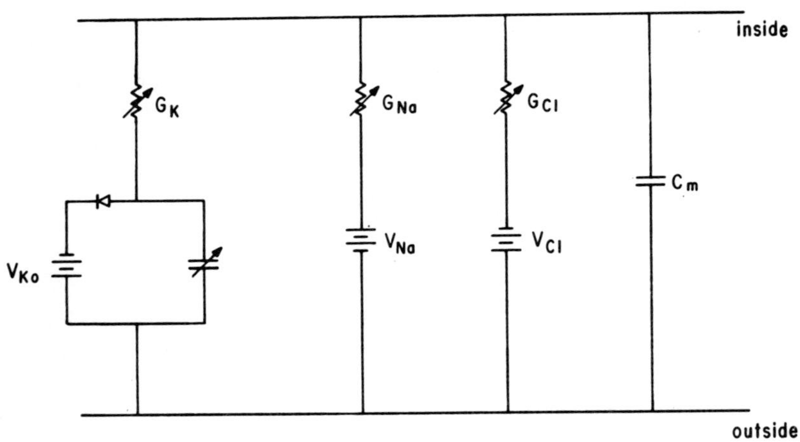

Figure 4-9. Three-ion membrane model incorporating a representation of the Frankenhaeuser–Hodgkin space. The diode is imperfect, representing the nonlinear relationship between voltage and ion flow; the capacitor in the G_K circuit is nonlinear, representing the log dependence of voltage on potassium concentration.

If the axon is in a steady-state resting condition at the time ($t = 0$) that a train of spikes is initiated, then the excess potassium should accumulate according to the following equation:

$$\Delta N_{\mathrm{T}}(n\tau) = \Delta N \sum_{m=0}^{n-1} e^{-mb\tau} \tag{4-30}$$

where τ is the interval between the spikes. This equation should be valid as long as the interval between spikes is large compared to the duration of the individual spikes themselves. When the interval becomes too short, successive spikes will interfere with one another and ΔN will gradually diminish. In the Frankenhaeuser–Hodgkin experiments, the interval remained rather long. Equation (4-30) can be written in more concise form through the application of a very common generating function

$$\Delta N_{\mathrm{T}}(n\tau) = \Delta N \frac{e^{-b(n\tau)} - 1}{e^{-b\tau} - 1} \tag{4-31}$$

from which one can easily derive the envelope of ΔN_{T} (see Figure 4-8):

UPPER BOUNDARY OF ΔN_{T} IN THE FRANKENHAEUSER–HODGKIN SPACE

$$\Delta N_{\mathrm{T}}(t) = \frac{\Delta N}{1 - e^{-b\tau}}(1 - e^{-bt}) \tag{4-32}$$

Indeed, in response to a train of spikes, Frankenhaeuser and Hodgkin observed a gradual change in the potassium-ion reversal potential that could be accounted for by just such an exponential increase in external potassium ions. The concentration increment after each spike appeared to be approximately 1.6 mM (16% of the assumed steady-state potassium concentration in the extracellular fluid). Their results conformed very well to Figure 4-8. They found that the rate constant b of redistribution following a spike was between 10 and 33 $\sec^{-1}$, corresponding to a redistribution time constant in the range 30 to 100 msec. Furthermore, they computed the apparent size of the periaxonal space and found it to be approximately 200–300 Å. They tentatively attributed it to a barrier beyond the membrane of the axon.

In order to incorporate these observations into an equivalent circuit, one must have a separate branch representing the potassium-ion flow. Thus, one would be restricted to a mixed equivalent circuit, which would be valid for very slow signals, or for very fast signals if potassium permeation were restricted to channels with low mobilities for other ions. Such channels apparently occur in the squid giant axon, and the measurement of potassium reversal potential is accomplished by means of a fast voltage clamp. Figure 4-9 shows an equivalent circuit incorporating the variable reversal potential for potassium ions.

4.5. Summary

We have introduced four basic types of conductance for membrane equivalent circuits: steady-state, chord, slope, and instantaneous. We have also introduced two degrees of freedom in the specification of the signals in the system, namely, size and speed; and for each of the four extremes, we introduced a special equivalent circuit. So far, however, modelers have failed to deal with situations in which the speed of signals is comparable to the speed of ion-concentration profile shifts, and we have omitted them also. In the chapters that follow, we shall employ the various types of equivalent circuits as they have been employed by other modelers. The reader should bear in mind, however, the limiting assumptions underlying each model and its applications.

Suggested Further Readings

For further readings we refer the reader to Finkelstein and Mauro (1963); Frankenhaeuser (1960); Frankenhaeuser and Hodgkin (1956); Hodgkin and Huxley (1952); Mauro (1961); Noble (1972); and Schwartz (1971).

5

Models of Signal Generation in Neural Elements

In this chapter, we shall discuss together three topics that are often treated separately, namely, chemical synapses, sensory receptors, and spontaneous activity in neurons. The common feature of these three topics is the generation of signals in the nervous system. Since the nervous system generally is viewed as a processor of signals, it seems natural to present models concerning the generation of those signals early in our discussion. It is quite conceivable that certain signals in the nervous system arise through the actions of many neurons or other cells in concert. On the other hand, it is clear that many signals do in fact arise within individual cells, and it is these signals that are discussed in the present chapter. Discussion of the generation of signals through the concerted actions of many cells is reserved for later sections, covering neural networks. Strictly speaking, we might limit our discussion here to models of sensory receptors and models of spontaneous signal generation, considering synaptic action to involve the transmission of preexisting signals rather than the generation of new signals. Indeed, we shall consider direct electrical communication between neurons in just this way, since it generally involves the more or less direct transfer of preexisting signals. However, synapses apparently involve mediation by a chemical transmitter and a marked change in the nature of the signal as it is transferred to the receiving cell. Moreover, the interaction between the chemical mediator and the receiving cell membrane apparently is very similar to (perhaps identical with) the interactions between chemical stimuli and chemoreceptor membranes (e.g., the action at olfactory or gustatory receptors). In other words, the subsynaptic membrane apparently acts just as any other chemical receptor site. For these reasons, we have lumped the subsynaptic membrane with other receptor membranes as a location of signal generation.

This chapter begins with some general, theoretical considerations based on the mixed equivalent circuit model presented in Chapter 4. These are followed by models of subsynaptic membrane, then by models of various sensory structures and sensory receptor membrane, and, finally, by models of spontaneous signal generation in pacemaker neurons. For the most part, we shall begin our discussions at the level of the equivalent circuit rather than the underlying membrane mechanisms. Although many theories have been proposed concerning the structural changes in the membrane that might lead to signal generation, we shall not discuss them explicitly.

5.1. Some General Considerations

The equivalent circuit of Figure 5-1 includes a passive-flow path for each of the three common univalent ions permeating cell membranes plus a passive-flow path for calcium ions, which have been implicated recently in the actions of receptors (both synaptic and sensory), and active pumps for all three cation species. It represents a local patch of plasma membrane presumably capable of signal generation. To see how such signal generation might come about, consider the electrical potential diagram in Figure 5-2. When the membrane is at rest, and has been for some time, the resting potential will be determined by the pumped and passive flows of the ion species that are pumped. Any unpumped ions, on the other hand, will have zero net passive flow and reversal potentials just equal to the resting potential of the membrane.

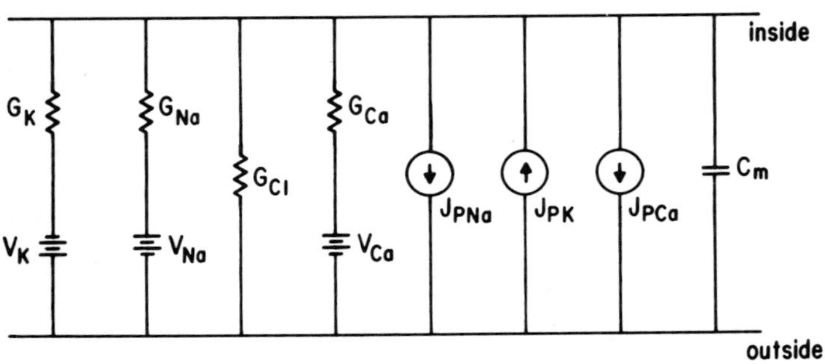

Figure 5-1. Mixed equivalent circuit for receptor membrane with four permeating ion species.

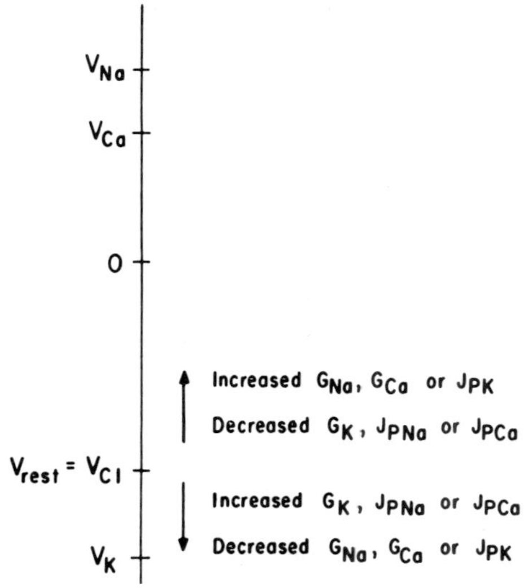

Figure 5-2. Electrical potential diagram for a cell membrane. The reference potential (0) is that of the extracellular fluid; the vertical scale represents the potential of the intracellular fluid, which resides at V_{rest} in the absence of signals. Upward arrow indicates changes that would produce depolarizations; downward arrow indicates changes that would produce hyperpolarizations.

The resting potential itself is given in terms of the equivalent-circuit parameters as follows:

$$(V_K - V_r)G_K + (V_{Na} - V_r)G_{Na} + (V_{Ca} - V_r)G_{Ca} + (V_{Cl} - V_r)G_{Cl} + qJ_{PK} - qJ_{PNa} - 2qJ_{PCa} - qJ_{PCl} = 0$$

$$V_r = \frac{qJ_{PK} - qJ_{PNa} - 2qJ_{PCa} - qJ_{PCl} + G_K V_K + G_{Na} V_{Na} + G_{Ca} V_{Ca} + G_{Cl} V_{Cl}}{G_K + G_{Na} + G_{Ca} + G_{Cl}} \tag{5-1}$$

where V_r is the resting membrane potential; G_K, G_{Na}, G_{Ca}, and G_{Cl} are the resting chord conductances of the four ions being considered; J_{PK}, J_{PNa}, J_{PCa}, and J_{PCl} are the pumped flow densities of the four ions (in the directions shown in Figure 5-1); V_K, V_{Na}, V_{Ca}, and V_{Cl} are the reversal potentials of the four ions (referred to the inside of the membrane); and q is the protonic charge. If all other parameters are fixed, an increase in the chord conductance of a given ion species will cause the membrane potential to shift toward the reversal potential for that species. Clearly, if the membrane is at rest, then a change in the chord conductance of an unpumped ion species will not alter the potential at all. A decrease in the chord conductance of a pumped ion species, on the other hand, will cause the resting membrane to move away from the reversal potential associated with that species. If all other parameters remain unaltered, a change in any one of the reversal potentials will cause a corresponding shift in the membrane potential. This could be brought about by the introduction or removal of ions of that species in the extracellular solution. It could also be brought about by a change in temperature, but that would affect all reversal potentials, not just one.

If the ion pumps were nonelectrogenic, involving no net transfer of charge across the membrane, then changes in pump rates would alter the resting potential rather slowly, requiring time for the accumulation of new concentration differences and acting through these on the ionic reversal potentials. On the other hand, if the pumps were electrogenic, providing some net transfer of charge as well as transfer of individual ion species, then changes in pump rates would have a much more rapid influence on the membrane potential, requiring only enough time for the accumulation of a small amount of charge.

Although electrogenic pumps have been implicated in the establishment of the resting potential in some neurons, their proposed role in certain photoreceptors has been thoroughly refuted, and there presently is no evidence that they are involved in signal generation. This by no means is meant to imply that such evidence might not be forthcoming in the future. Similarly, although changing reversal potentials might be excellent candidates for the sources of signals in certain osmoreceptors and temperature receptors, there is presently no evidence other than that derived from classical membrane theory. In fact, the only convincing evidence so far concerning receptor action is that related to chord-conductance changes. In those receptors where the action could be identified with reasonable certainty, it appeared to be a change in the chord conductance of one or more appropriate ions. The actions underlying spontaneous signal generation in neurons appear to be more difficult to discern, and at this point in time it may be best not to cloud the issue by favoring any of the candidates. In the case of receptors, on the other hand, one in which signal generation is mediated by any action other than a conductance modulation definitely would be considered exceptional in the light of present knowledge. The most compelling evidence in favor of chord conductance modulation as a source of signal generation in receptors is the appearance of a reversal potential for the signal itself. When the potential across the receptor membrane can be clamped by a feedback circuit, the reversal potential can be quite clear, with the current induced by the receptor's stimulus being of one polarity when the membrane is held at a greater potential, the opposite polarity when the membrane is held at a lesser potential, and precisely zero when the membrane is held at precisely the reversal potential itself. When a clamp cannot be achieved, the internal potential of the receptor can be altered by passage of a steady current into the cell, and a reversal potential may be seen, but not so sharply. In some cases, bridge circuits have been employed in attempts to detect changes in chord conductance; but the resulting observations are not as clear-cut as observations under voltage clamp.

5.2. The Eccles Model of a Chemical Synapse

After many years of occasionally heated debate, the neurophysiological community now seems to accept the evidence that in some cases communication between neurons is mediated by direct electrical coupling, while in other cases it is mediated by chemical intermediates. Eccles and his colleagues provided convincing early evidence concerning the nature of the chemical synapse when they observed reversal potentials for the synaptic currents in the postsynaptic or receiving cells, which happened to be motor neurons in the cat spinal cord.

Passing a steady current through one barrel of a double-barreled microelectrode inserted in a large motor neuron and with the other barrel observing postsynaptic potentials superimposed on the resulting bias, Eccles and his colleagues were able to find distinct reversal potentials for inhibitory and excitatory synaptic responses. Furthermore, by altering the ionic concentrations on either side of the receiving cell's membrane, Eccles and others have found that the synaptic reversal potentials depend systematically on ionic concentrations, implying that one step in chemical synaptic communication is the modulation of the permeability of the membrane of the postsynaptic cell to one or more ion species. Other, cumulative evidence indicates that the agent producing the modulation is a chemical substance released by the presynaptic, or "sending," cell.

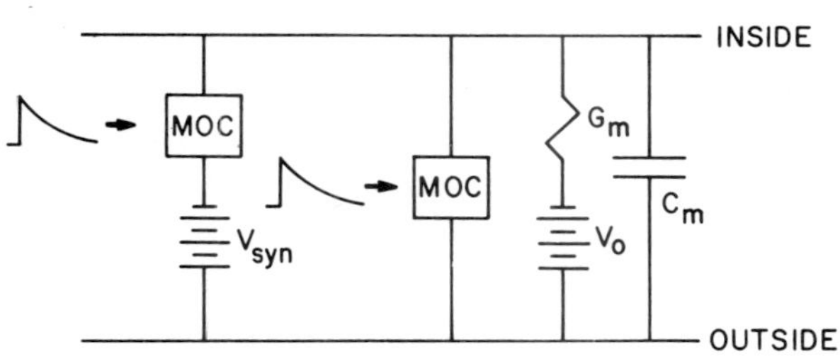

Figure 5-3. Eccles model of subsynaptic membrane. The abbreviation MOC represents *modulated ohmic conductance*, the time course of which is approximately exponential following a presynaptic spike. All the resting chord conductances of Figure 5-1 have been combined into the single conductance G_m, which usually is taken to be constant (i.e., slope and chord conductances are taken to be one and the same). The quantity V_0 is the resting potential of the subsynaptic membrane, and C_m is its capacitance.

Most circuit models of the postsynaptic membrane represent the modulated permeability as a modulated ohmic conductance in a mixed equivalent circuit, as shown in Figure 5-3. The reversal potentials associated with synaptic conductances vary from cell to cell, but inhibitory synaptic conductances are quite often associated with the reversal potential for potassium ions, while excitatory synaptic conductances are often associated with both potassium and sodium and exhibit reversal potentials close to zero. Because it is based on evidence first presented by Eccles and his colleagues, the model in Figure 5-3 occasionally is called the *Eccles model*, a practice we shall follow here.

The nature of the synaptic conductance itself apparently varies from one type of synapse to another. Using double-barreled electrodes, Takeuchi and Takeuchi (1960) found a linear relationship between all phases of synaptic (end-plate) current and the clamped postjunctional membrane potential of an excitatory synapse (end plate) between nerve and muscle in an amphibian, indicating that the equivalent conductance for that particular excitatory synapse is ohmic. In crayfish, Furshpan and Potter (1959) observed linear relationships between the voltage amplitudes of inhibitory synaptic responses and a steady-state ambient membrane potential established in a postsynaptic cell by passage of a steady current, indicating ohmic equivalent conductance for the synapse. In contrast with these observations, Coombs *et al.* (1955) found a distinctly nonlinear relationship between the voltage amplitude of the response of an inhibitory synapse and the ambient steady-state membrane potential of the postsynaptic cell in the cat, indicating a nonohmic synaptic conductance.

Measured at a soma connected to an extensive dendritic tree, as is the case in the motor neurons observed by Eccles, the shape of the synaptic response can exhibit a rather complex appearance. In order to explain the shape, Eccles (1957) at first proposed that the current induced in the postsynaptic cell has two distinct phases of exponential decline—a rapid phase and a slower phase. Employing the distributed passive membrane models described in the following chapter, Rall (1957–1962) was able to show that the complexities in the synaptic response could be accounted for by the electrical load imposed on the soma by the dendrites, and that a single exponential phase of synaptic current was all that was required. Rall simplified the Eccles model by replacing the modulated conductance with a current source, neglecting the nonlinearities that are inherent in any system with modulated conductances (whether or not those conductances are ohmic). The validity of Rall's approximation depends chiefly upon the magnitudes of the synaptic responses.

Considering the model of Figure 5-3 and assuming an ohmic synaptic conductance, we can write the following equation for the change (Δv) of membrane potential in response to synaptic conductance modulation:

$$\frac{d\Delta v}{dt} = \frac{1}{C_{\mathrm{m}}}(-G_{\mathrm{m}}\,\Delta v - G_{\mathrm{syn}}(t)[V_0 + \Delta v - V_{\mathrm{syn}}]) \tag{5-2}$$

If Δv is very small compared with $V_0 - V_{\mathrm{syn}}$, then this differential equation with a time-varying coefficient reduces to a nonhomogeneous, linear differential equation with constant coefficients

$$C_{\mathrm{m}}\frac{d\Delta v}{dt} + G_{\mathrm{m}}\,\Delta v = -[V_0 - V_{\mathrm{syn}}]G_{\mathrm{syn}}(t) \tag{5-3}$$

where the term on the right is equivalent to a time-varying current source, as shown in Figure 5-4. Employing this simplification, Rall (1962–1964) and others have gone on to study the theoretical effects of dendritic trees on interacting synaptic potentials.

Several modelers have included the Eccles model itself in more elaborate models for single neurons. MacGregor (1968), for example, included it directly in models of distributed dendritic trees and studied the nonlinear interactions inherent between modulated synaptic conductances; Lewis (1968) combined it directly with the Hodgkin–Huxley model to study the combined effects of synaptically modulated conductances and electrically excitable conductances. However, even without these elaborations, the Eccles model has interesting consequences.

For example, the calculations by Rall and subsequent observations by various neurophysiologists suggest that the time course of $G_{\mathrm{syn}}(t)$ following a spike in the presynaptic cell is a simple exponential

$$G_{\mathrm{syn}} = Ke^{-at} \tag{5-4}$$

If one assumes that the conductance modulation is cumulative from spike to spike, then he can calculate the theoretical time course of G_{syn} in

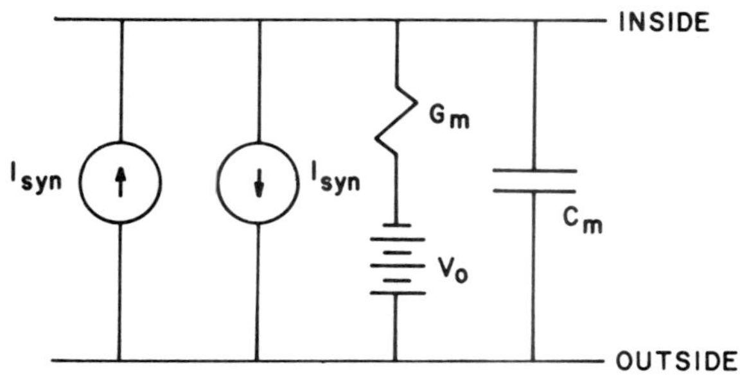

Figure 5-4. Linear approximation to Eccles' model. If the synaptic voltage responses are small relative to the differences between the membrane resting potential and the synaptic conductance reversal potentials, then this model should approximate the response dynamics quite well. The tacit assumptions of constant G_m and V_0 are also much better under these conditions.

response to a train of presynaptic spikes. Consider, for example, the case of a train of periodic spikes. If the spikes are separated by a constant interval T, then in steady state the conductance would fluctuate between $G_{\text{syn}_{\text{max}}}$ and $G_{\text{syn}_{\text{max}}} e^{-aT}$, where $G_{\text{syn}_{\text{max}}}$ is determined by the relationship

$$G_{\text{syn}_{\text{max}}} e^{-aT} + K = G_{\text{syn}_{\text{max}}} = K/(1 - e^{-aT}) \tag{5-5}$$

The transient growth of G_{syn} to this steady state level can be calculated quite easily by adding the response to each spike in a train, beginning with the first, which leads to the relationship

$$G_{\text{syn}}(t) = K \sum_{n=0}^{N} e^{-a(t-nT)} \tag{5-6}$$

where the first spike occurred at $t = 0$ and N is the number of spikes that have occurred prior to time t. This finite sum has the following generating function:

$$\sum_{n=0}^{N} x^n = \frac{x^{N+1} - 1}{x - 1} \tag{5-7}$$

$$\sum_{n=0}^{N} e^{-a(t-nT)} = \frac{e^{aT(N+1)} - 1}{e^{aT} - 1} e^{-at} \tag{5-8}$$

which leads to the following closed form of equation (5-6)

$$G_{\text{syn}}(t) = K \frac{e^{a\{t/T\}T} - 1}{e^{aT} - 1} e^{-at} \tag{5-9}$$

where $\{t/T\} = N$ is the largest integer less than t/T. This function is plotted in Figure 5-5 for a specific set of parameter values.

The resulting synaptic potential, Δv, could be calculated by substitution of equation (5-9) into equation (5-2) or (5-3), leading to an extremely complicated result. On the other hand, if as is often the case the time course of the synaptic potential is relatively slow in comparison with the membrane time constant ($C_{\text{m}}/G_{\text{m}}$), then equation (5-2) is approximated by

$$G_{\text{m}} \Delta v + G_{\text{syn}}(t)[V_0 + \Delta v - V_{\text{syn}}] = 0 \tag{5-10}$$

or

$$\Delta v = \frac{G_{\text{syn}}(t)[V_{\text{syn}} - V_0]}{G_{\text{syn}}(t) + G_{\text{m}}} \tag{5-11}$$

which is a saturating, nonlinear function of G_{syn}, as shown in Figure 5-6. Combining the curves in Figures 5-5 and 5-6, one obtains an estimate of the

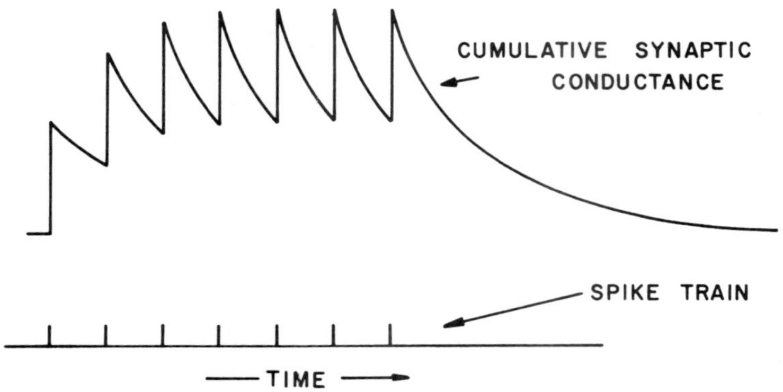

Figure 5-5. Linear accumulation of synaptic conductance in response to a train of evenly spaced spikes, according to the model of equation (5-4), in which synaptic conductance declines exponentially with time after a spike.

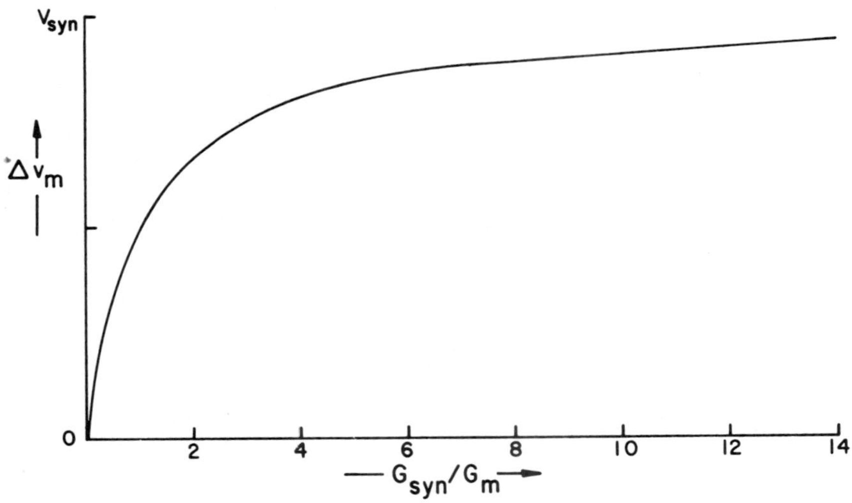

Figure 5-6. The nonlinearity imposed upon a subsynaptic membrane (or other receptor membrane) in which a modulated ohmic conductance mediates transduction [see equation (5-11)].

synaptic potential in response to a train of periodic spikes. An example is shown in Figure 5-7. In this case, the parameters were chosen to demonstrate an interesting possibility. If the cumulative synaptic conductance

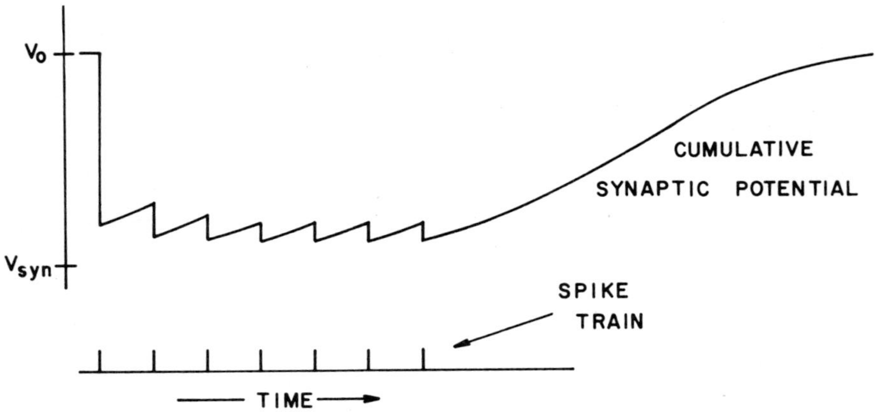

Figure 5-7. Approximation of a summed inhibitory postsynaptic response obtained by combining the summation in Figure 5-5 and the nonlinearity in Figure 5-6. The effects of the membrane capacitance have been neglected, otherwise the sharp peaks in the response would have been rounded.

increase in response to a train of spikes is sufficiently large to saturate the response (i.e., if G_{syn} is considerably larger than G_m in Figure 5-3), then the cumulative postsynaptic potential will exhibit a marked delay before beginning to fall after cessation of the presynaptic spike train. This is simply because G_{syn} takes a finite time to fall below the saturation level. Thus, almost directly from Eccles' model one has a possible explanation for one combination of the synaptic "degrees of freedom" described by Bullock (1959).

5.3. Bullock's Degrees of Freedom for a Chemical Synapse

Summarizing the observations of other physiologists and, particularly, his own observations with Hagiwara (1957) on the large cells of the lobster cardiac ganglion, Bullock proposed that synapses have at least three degrees of freedom, reflected in the postsynaptic response. A series of postsynaptic potentials may be excitatory or inhibitory; it may exhibit normal saturating summation, such as that depicted in Figure 5-7, or facilitated summation, in which the incremental responses become increasingly larger as a result of previous inputs, or antifacilitated summation, in which incremental responses become increasingly smaller [more so than expected from the saturating nonlinearity of equation (5-11)]; and it may exhibit excitatory or inhibitory aftereffects or neither or both. Bullock and Hagiwara observed most of the possible combinations of these features in the various synapses of the cardiac ganglion. The particular combination depicted in Figure 5-7 is inhibition, summing with normal saturation, with inhibitory aftereffects. It is found, among other places, in the inhibitory synapse of the crayfish stretch receptor.

Aftereffects with opposite polarity to the original synaptic responses generally are labeled "rebounds" and are quite common. Kater (1972) has shown that inhibitory synapses with excitatory aftereffects provide the required coordination among a group of motoneurons controling a sequence of movements in a freshwater snail. Undoubtedly, other concrete examples of the importance of aftereffects will be found as specific neural networks are studied in more detail. With a little reflection, one can see quite easily that rebound aftereffects could very well be the result of interaction between postsynaptic membrane and electrically excitable membrane (e.g., between the Eccles model and the Hodgkin–Huxley model) discussed in Chapter 7. During a subthreshold inhibitory postsynaptic potential, for example, the potassium conductance of the Hodgkin–Huxley model would decrease slightly while the sodium conductance would accommodate to the hyperpolarized level (e.g., the state variable "h" would increase slightly); as the synaptic potential returned toward the resting level following cessation of presynaptic spikes, the sodium conductance would increase while the potassium conductance remained low, and these two effects would combine to produce an excitatory afterpotential. With slight parameter adjustments in the Hodgkin–Huxley model, one can even obtain damped oscillatory aftereffects of excitatory or inhibitory synaptic responses (Lewis, 1968). Thus, two of Bullock's synaptic degrees of freedom seem to be inherent in the Eccles model (which alone provides excitation or inhibition, along with the possibility of aftereffects of the same polarity) and the Hodgkin–Huxley model (which provides rebound, a phenomenon akin to the "anode break" excitation discussed in Chapter 7).

Although forms of facilitatory and antifacilitatory summation can occur in the combined Eccles/Hodgkin–Huxley model, there is considerable evidence to suggest that these phenomena are presynaptic in origin. Hagiwara, for example, was able to apply a voltage clamp to the postsynaptic membrane, thus decoupling the synaptically induced current from electrically excitable portions of the membrane. He found antifacilitation in the current itself. If one assumes that the conductance change in the postsynaptic membrane is directly proportional to the concentration of the substance inducing that change (i.e., the chemical transmitter) in the synaptic cleft (see Figure 5-8), and that the transmitter is destroyed (or inactivated) by a first-order process, then the summation described by equation (5-9) can be taken to be that of the transmitter in the cleft, with a

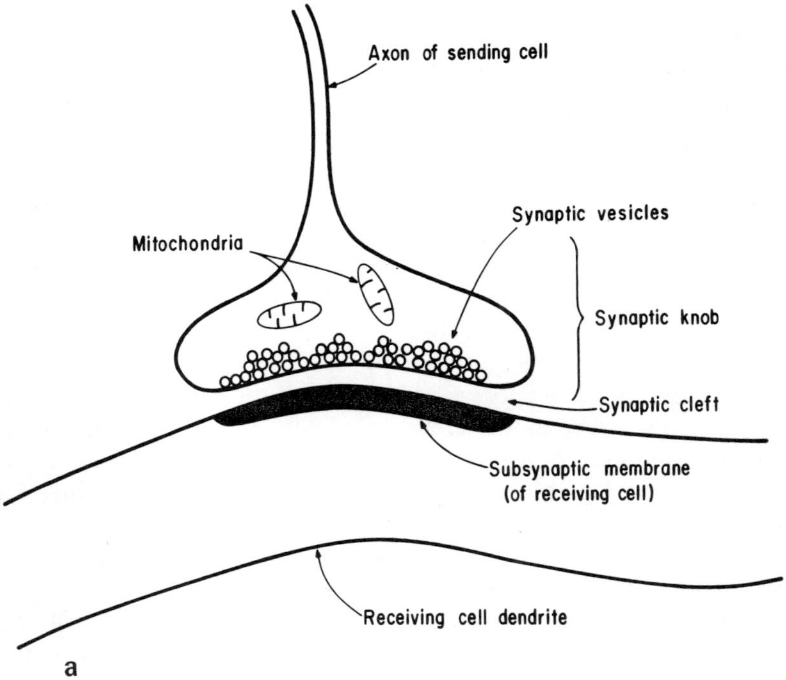

Figure 5-8. Synaptic structures. (a) Schematic diagram of an idealized synapse; (b) scanning electron micrograph of structures tentatively identified as synaptic knobs (magnified 30,000 times, reduced 45% for reproduction) in the abdominal ganglion of the sea hare (*Aplysia californica*). (From Lewis, Everhart, and Zeevi, 1969, *Science* **165**, 1140-1143.)

being the rate constant of inactivation. If one adds the further assumption that the transmitter supplied during each presynaptic spike is taken from a finite reservoir in the synaptic knob of the sending cell, with the amount supplied being in some way proportional to the amount available in the reservoir, then he has a perfectly reasonable basis for antifacilitatory summation.

As a specific example, consider the model depicted in Figure 5-9. Here the amount Q of transmitter supplied by a spike is directly proportional to the amount B available in the reservoir at the beginning of the spike, and the reservoir itself is supplied by a reversible, first-order process from a basic source. Now the summation process itself exhibits a saturation nonlinearity, which will compound with that inherent in the conductance modulation of the receiving-cell membrane. In the case of periodic presynaptic spikes with interspike interval T, one can obtain equations

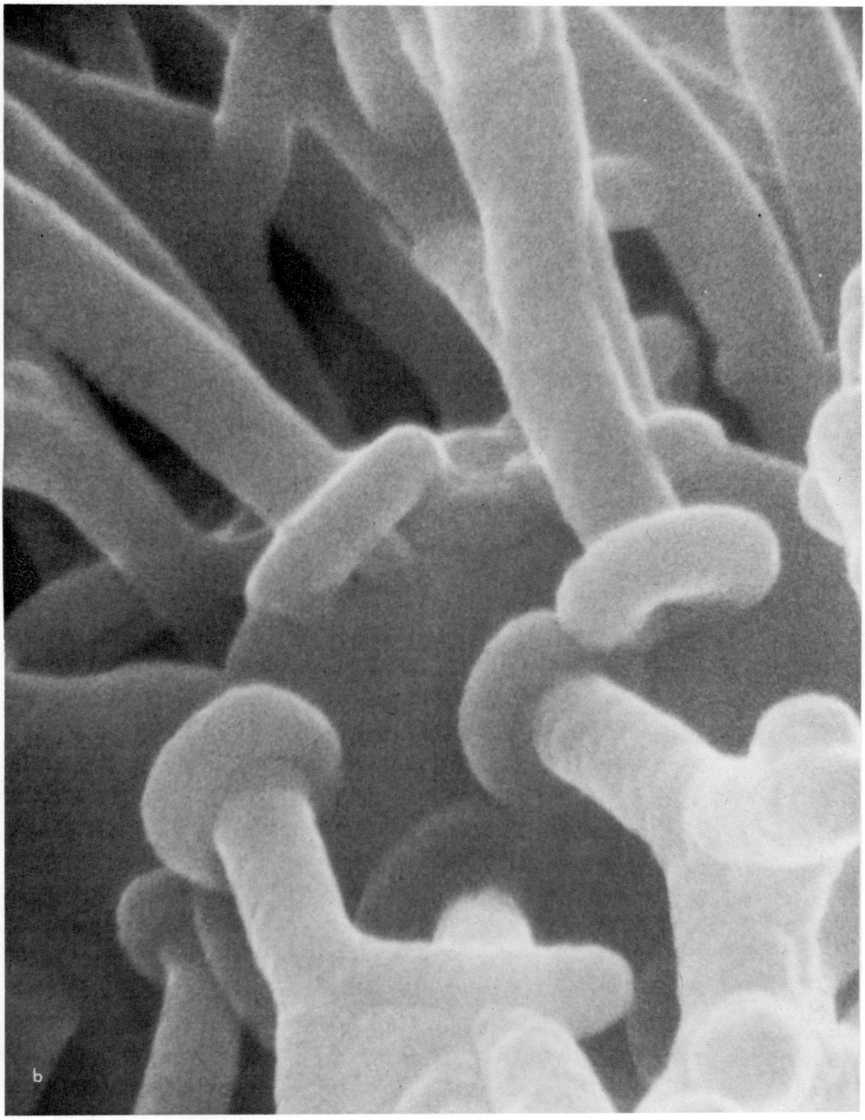

Figure 5-8b

analogous to equations (5-5) and (5-6). Using the notation in Figure 5-9, we have

$$B_N^+ = (1 - k'')B_N^- \tag{5-12}$$

$$B_N^- = B_0 - (B_0 - B_{N-1}^+)e^{-k'T} \tag{5-13}$$

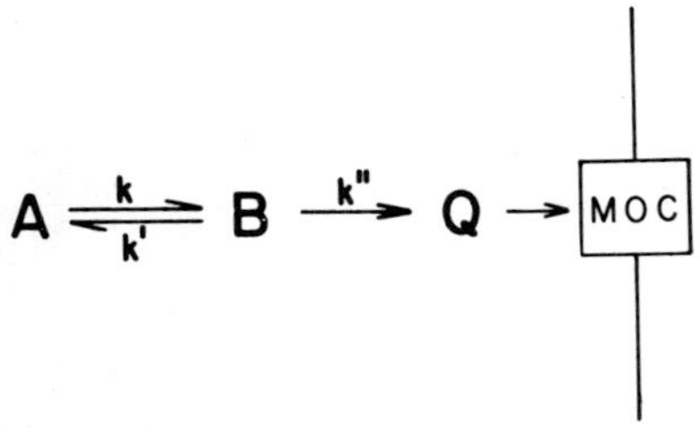

Figure 5-9. A simple model of transmitter depletion. The factor k'' is the proportion of transmitter (B) released during a spike; k and k' are rate constants associated with replenishment of transmitter in the presynaptic reservoir.

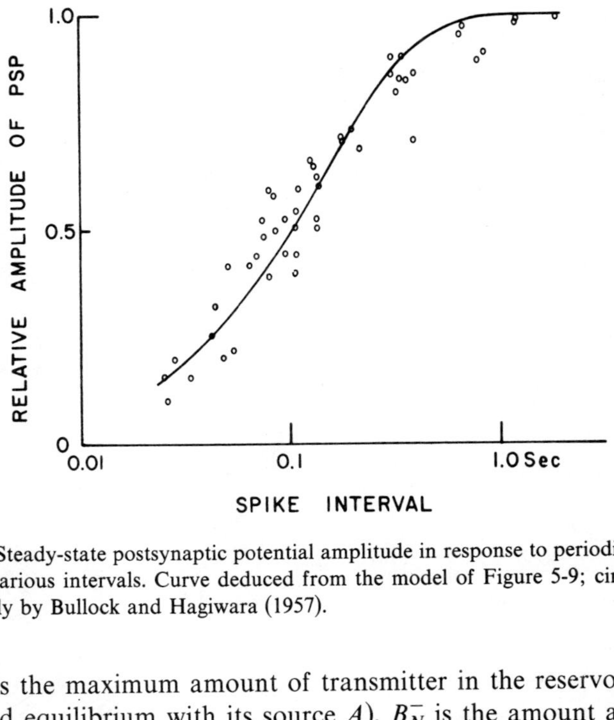

Figure 5-10. Steady-state postsynaptic potential amplitude in response to periodic presynaptic spikes with various intervals. Curve deduced from the model of Figure 5-9; circles obtained experimentally by Bullock and Hagiwara (1957).

where B_0 is the maximum amount of transmitter in the reservoir (when it has reached equilibrium with its source A), B_N^- is the amount available at the beginning of the Nth spike, and B_N^+ is the amount remaining just after the Nth spike in the train. These equations can be combined to yield the amount Q_N of transmitter released during the Nth spike:

$$Q_N = k''B_N^- = k''B_0\{(1 - e^{-k'T}) + [(1 - k'')e^{-k'T}]^{N-1}\} \quad (5\text{-}14)$$

and the steady-state magnitude of Q, approached by very long trains:

$$Q_{\text{steady state}} = \lim_{N\to\infty} Q_N = k''B_0(1 - e^{-k'T}) \quad (5\text{-}15)$$

For relatively slow postsynaptic potentials, such as those of the lobster cardiac ganglion, these two equations can be combined with equation (5-11) to provide a theoretical estimate of the antifacilitation inherent in the model. Figure 5-10 shows an example of the theoretical results along with the results obtained by Hagiwara and Bullock (1957). The similarity between these results tends to support the model, but it does not exclude all other possibilities.

Facilitatory summation is slightly more difficult to explain. One possiblity is the formation of a relatively persistent enzyme-substrate compound during the inactivation of transmitter, leading to a temporary depletion of inactivating enzyme. In this case, postsynaptic potentials should increase both in amplitude and duration as a result of previous activity.

5.4. The Quantum Model of Transmitter Release

For many years, the vesicles of the synaptic knob (see Figure 5-9) have been thought to be the reservoirs of available transmitter. A reasonable, but not necessary, corollary of this assumption is that release of transmitter into the cleft should occur in discrete steps, each corresponding to the simultaneous release of the entire contents of an individual vesicle. Some of the best evidence for such discrete events comes from muscles, where the end-plate potential (epp) is mediated by a structure (the myoneural junction) analogous to the chemical synapse. A full-sized end-plate potential occurs in response to a spike in the neuron driving the muscle, and in the absence of prejunctional spikes, sporadic, miniature end-plate potentials occur. Each miniature epp was thought to be the result of the release of a quantum of transmitter (del Castillo and Katz, 1954), possibly the contents of a single vesicle, into the cleft of the myoneural junction. Furthermore, as a result of the sporadic nature of the occurrence of miniature epp's, it was thought that quantum release was a stochastic process and that the probability of release somehow was increased greatly during a prejunctional spike.

In order to learn more about the nature of the release process, Boyd and Martin (1956), Katz and Miledi (1965), and others (see Martin, 1966; Katz, 1969) attempted to analyze its stochastic nature. The large electrical capacitance of the muscle cell membrane tends to prolong epp's and make them extremely difficulty to distinguish individually when many of them occur within a short period of time. This problem was partially solved by Katz and Miledi, who observed the underlying membrane currents extracellularly, before they had been integrated over the membrane capacitance. The remainder of the solution consisted of controlling the rate of release of quanta by cooling the preparation and by regulating the calcium concentration in the vicinity of the electrode. In these ways, Katz and Miledi made the individual miniature responses discernible and countable during a prejunctional spike. Thus, they were able to analyze the statistical nature of the responses directly, and they found that the responses followed the Poisson distribution. Assuming that each miniature response is in fact the result of a single vesicle releasing its contents, then the Poisson nature of the occurrence of miniature responses would imply that each vesicle effectively represents a Bernoulli trial during the prejunctional spike (either the contents of that vesicle are released or they are not) and that the probability of release is constant from vesicle to vesicle.

Thus, if we know the probability p of release of any given vesicle during a spike along with the total number N of vesicles present in the end plate, we can compute the entire distribution of release probabilities:

$$\begin{aligned} p &= \Pr[\text{given vesicle released during spike}] \\ N &= \text{number of vesicles in end plate} \\ pN &= \text{mean number of vesicles released per spike} \\ &\qquad \text{(taken over a large number of spikes)} \end{aligned} \tag{5-16}$$

$$\begin{aligned} \Pr[n] &= \Pr[\text{exactly } n \text{ vesicles released during a given spike}] \\ &= p^n(1-p)^{N-n}\binom{N}{n} \end{aligned}$$

When N is very large compared to n, as is the case in the motor end plate,

$$\binom{N}{n} \simeq N^n/n! \tag{5-17}$$

and when N is large compared to n and p is very small, which also is the case in the motor end plate,

$$(1-p)^{N-n} \simeq e^{-Np} \tag{5-18}$$

which leads to the Poisson approximation to the binomial distribution:

$$\Pr[n] \simeq (pN)^n e^{-Np}/n! \tag{5-19}$$

According to the observations of Boyde and Martin and of Katz and Miledi, equation (5-19) provides an excellent model of transmitter release during a spike. The probability p apparently increases with increasing calcium ion concentration in the extracellular fluid and also with increasing depolarization of the prejunctional membrane. Because of the structural similarities between motor end plates and chemical synapses, including the presence of vesicles in both, one might very well extend this model to include the synapses.

If the extension is made to chemical synapses in general, then the input–output relationships of synapses could be interpreted in terms of N, p, and the electrical properties of the postsynaptic membrane. For example, the input–output characteristics of the squid giant synapse (see Hagiwara and Tasaki, 1958) could be interpreted to imply that p increases monotonically with increasing depolarization over a fairly wide range, being very low at the resting potential and very high during large depolarizations. The saturation evident in the input–output relationship would be an expected result of conductance modulation [equation (5-11)] and does not necessarily reflect a limiting value of p. Clearly, however, p does have an upper limit of unity. When p approaches unity, one would expect rapid depletion of available transmitter (i.e., rapid reduction of N), and the synapse should exhibit marked antifacilitation. Furthermore, if the model of Figure 5-9 were correct, such depletion of available transmitter should bring the postsynaptic response to static depolarization of the presynaptic membrane very close to zero, making the input–output relationship essentially a horizontal line. On the other hand, a very plausible "inactivation" mechanism is the re-uptake of transmitter from the cleft, in which case inactivated transmitter might almost immediately be available for re-release, and a value of p close to unity would not necessarily flatten the static input–output relationship. Once again, with this model, facilitatory summation is a bit more difficult to explain. Perhaps p increases as a result of previous synaptic activity, or perhaps the effective value of N increases (i.e., more vesicles become available for immediate release). In either case, we have simply moved the questions to a different level; the underlying mechanisms remain unpostulated.

One very interesting aspect of the quantum model is the noise that is inherent in it. With it, for example, one can ask and attempt to answer such questions as the following: What is the probability that the synaptic response to a spike will exceed the threshold of the postsynaptic cell? What is the probability that the threshold will be exceeded during a certain interval of time in which there is no presynaptic spike? Such questions would be rather simple to attack if threshold could be defined in terms of n, the number of vesicles released during a certain increment of time. On the other hand, the accommodative mechanisms inherent in excitable membrane (see Chapter 7) might very well make such a definition unrealistic.

Finally, S. J. Redman and his co-workers (personal communication) have recently applied deconvolution routines to responses of Ia synapses and have deduced tentatively that transmission is not quantal. Here, at least, the transmission mechanisms apparently are significantly different from those in the myoneural junctions studied by Katz and his colleagues. It should be very interesting to observe future developments in this field.

5.5. Discrete Inputs to Other Receptors

Clearly, synaptic receptors are not the only ones subjected to discrete (or quantal) stimuli. Many physical entities (e.g., electromagnetic radiation, chemical substances, electrical currents) comprise discrete components (photons, molecules, ions), and when these entities occur in very small quantities, their discrete natures often become quite apparent. Thus, for example, at low stimulus intensities, chemoreceptors, photoreceptors, and electroreceptors all could be subject to discrete inputs. One question that arises is: How many such discrete units are required to excite a given receptor? Once again, the binomial distribution and its Poisson approximation can be used to help answer the question. If the receptor cell itself can be penetrated with a microelectrode, then individual responses might be observed and their occurrences compared with the expected occurrences of the discrete inputs. This has been done in photoreceptors, where individual receptor responses thus have been correlated with the occurrence (capture) of single photons. Often, however, the experiments are not so direct, and in such cases unambiguous interpretation depends upon a simple outcome.

Consider, for example, the experimental arrangement depicted in Figure 5-11 which is analogous to that employed in several studies of photoreceptors and chemoreceptors. An isotropic source of particles on the

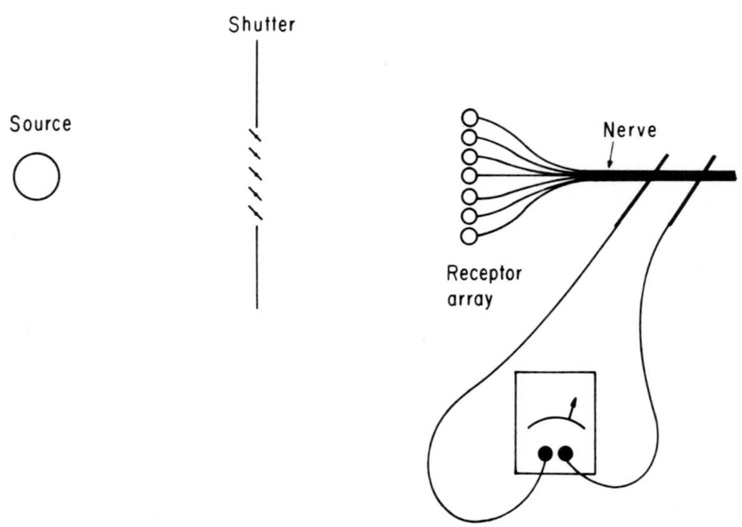

Figure 5-11. Experimental arrangement by which receptors are often tested for quantal sensitivity. The source provides a quantal stimulus (e.g., molecules of chemical stimulus, photons), and the shutter provides a gating mechanism.

left is isolated by a gate from an array of receptors. The bundle of receptor axons is monitored for spikes triggered by the stimuli. When the gate is opened for a brief interval of time, a certain number N of particles is expected to pass through and be captured by the receptor array. The quantity N will increase with the intensity of the source. Assuming that each of the N particles is emitted in a random direction, independent of the direction of any of the others (i.e., the particles are not emitted in chunks or bunches), then each particle can be considered a Bernoulli trial for any given receptor. In other words, the receptor either will catch that particle or it will not, and the probability of capture by a given receptor is the same for all N particles. Let that probability be p; and let its complement $(1 - p)$ be q. Given a brief pulse of N particles, the probability that all of them miss the receptor is simply

$$\Pr[0 \text{ hits}] = q^N \tag{5-20}$$

and the probility of at least one hit is its complement

$$\Pr[1 \text{ or more hits}] = 1 - q^N \tag{5-21}$$

The probability of exactly one hit is

$$\Pr[1 \text{ hit}] = Npq^{N-1} \tag{5-22}$$

and the probability of at least two hits is

$$\begin{aligned} \Pr[2 \text{ or more hits}] &= 1 - \Pr[0 \text{ hits}] - \Pr[1 \text{ hit}] \\ &= 1 - q^N - Npq^{N-1} \end{aligned} \tag{5-23}$$

Similarly, the probability of at least m hits is derived from the binomial distribution [equation (5-16)]:

$$\begin{aligned} \Pr[m \text{ or more hits}] = 1 - q^N - Npq^{N-1} - \frac{N(N-1)}{2!}p^2q^{N-2} - \cdots \\ - \binom{N}{m-1}p^{m-1}q^{N-m+1} \end{aligned} \tag{5-24}$$

If, during our very brief stimulus pulse, m particles are required to strike a given receptor in order to elicit spikes from it, then the probability of one or more spikes appearing in a given receptor axon in response to a given stimulus pulse is described by equation (5-24). Thus, one can attempt to determine m by observing the proportion of repeated stimulus pulses that produce at least one spike in the receptor, and repeating the observation for several values of N.

Alternatively, if one is observing the axons from a large number of essentially identical receptors, and if N is small compared to the total number of receptors, then the observer simply can determine the expected number of spikes in the bundle of axons as a function of N. In either case, however, if a value of m greater than unity is indicated, then that value will be difficult to determine. Fortunately, in some of the interesting photoreceptors and chemoreceptors where this method has been employed (see Reichardt, 1970; Schneider, 1970), the observations clearly indicated that m was unity. In other words, a single molecule or a single photon was sufficient to elicit a spike from the individual receptor. This is very fortunate with respect to interpretation. If two or more particles had been required, the question would remain open as to whether those particles had to strike the same receptor site or could strike any of the sites on a given receptor cell. Because only one particle was required, the threshold at every level is clear: One particle is required to elicit a response from a given receptor site; one receptor-site response is sufficient to elicit a spike. Figure 5-12 shows the experimental results that should occur if one particle is sufficient and compares them with the results expected if more than one particle is required. The most obvious difference is the lack of inflection in the single-particle-response curve.

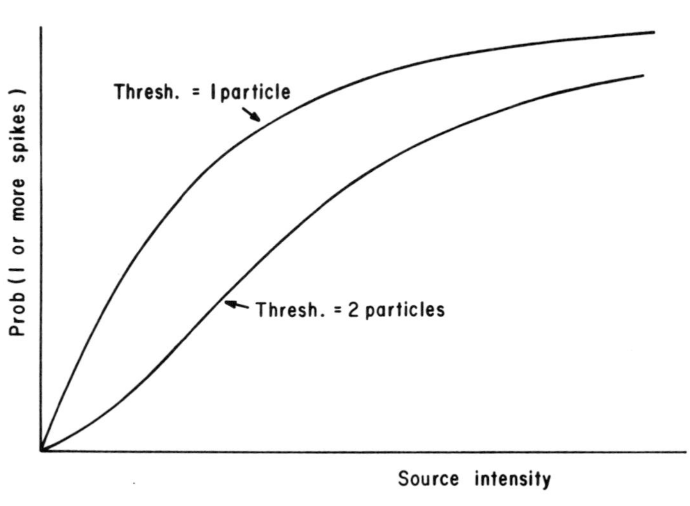

Figure 5-12. Theoretical responses of receptors receiving particulate (quantal) stimuli. If each stimulus particle is emitted in a manner independent of all other particles, so that it represents one of many identical Bernoulli trials, then the probability of one or more spikes occurring during a given stimulus pulse should follow the dependence on stimulus source intensity shown here.

5.6. Reliable Detection of Weak Signals in the Presence of Noise

Relatively early in the game of neural modeling, Cowan and Winograd (1963) and others became concerned with the problem of reliable signal processing by inherently unreliable elements. More recently, however, neurobiologists have become increasingly inclined to consider neurons as fairly reliable elements. In fact, as we shall see in the discussion of pacemakers, neurons apparently can exhibit extraordinary stability, which is an important facet of reliability. Another interesting question has surfaced, however. How can these apparently reliable elements function in the face of noisy (and therefore unreliable) signals? This question was attacked head-on by Marr (1969) in his model of the cerebellum, and it has surfaced numerous times in studies of sensory systems.

Since we have just discussed the inherent stochastic nature of particulate stimuli, this seems to be an appropriate place to introduce the problem of detection, although it also could be discussed very properly under neural networks.

Consider a single receptor being bombarded by a random influx of stimulus particles from an essentially constant background of stimulus sources in the environment. Suddenly, another source appears in the environment; how can the central nervous system (CNS) detect it reliably on the basis of spikes coming from the receptor? As a result of the background bombardment, the receptor will produce x spikes in a certain interval of time Δt, where x is a random variable. Since spikes are discrete events, we can treat x as a discrete variable, taking on only integral values; and we can determine the probability for each of those values:

$$\Pr[x = i] = p_i \tag{5-25}$$

In response to the background bombardment *plus* the new source, the receptor will produce y spikes in Δt, where y also is a discrete random variable:

$$\Pr[y = i] = p_i' \tag{5-26}$$

If the stimulus particles are independent, then one should expect p_i and p_i' each to exhibit a single maximum with respect to i and to decrease monotonically as i is increased beyond its value at the maximum. Furthermore, one should expect that p_i' is maximum at a value of i that is equal to or greater than that corresponding to the maximum of p_i. In the following discussion, we shall operate under these assumptions. The arguments could be expanded quite easily, however, to include other situations (e.g., multimodal distributions of x, and stimuli that subtract from the background rather than adding to it).

If, in the face of unimodal background, the CNS must decide during each Δt whether or not a new additive source was present, it can base its decision on the number of spikes emanating from the receptor during that interval. To do so, it must establish a threshold θ. If the number of spikes is less than θ, the decision is negative; if the number of spikes is equal to or greater than θ, the decision is positive. Clearly, the optimum value of θ will be one which minimizes the probability of an incorrect decision. However, there are two types of incorrect decisions that can be made: The CNS could decide that the new source was present, when indeed it was not; or the CNS could decide that it was not present when in fact it was. These two types of errors may not be equally important (e.g., in the extreme, one might be fatal, the other merely an inconvenience). Therefore, the optimum value of θ also must be based on the relative importance of the two types of errors. Formally, we can state that θ should minimize appropriately the probability of x exceeding θ

$$\Pr[x \geqslant \theta] = \sum_{i=\theta}^{\infty} p_i \tag{5-27}$$

while appropriately maximizing the probability of y exceeding θ

$$\Pr[y \geqslant \theta] = \sum_{i=\theta}^{\infty} p_i' \tag{5-28}$$

If the source to be detected is weak relative to the background, then p_i' will be very nearly equal to p_i, and the minimized probability of error will be quite large. If the new source remains present for a long time, and the CNS is not in a hurry to detect its presence, then it can reduce the probability of error by counting the spikes from the receptor over a large number of intervals and establishing a decision threshold for the total count. Let x_j be the number of spikes that would result from the background bombardment alone in the jth interval of the count, and y_j be the number that would result from the background plus the new source in the jth interval. In the absence of the new source, the total count would be the sum of n random variables (n being the total number of intervals in the count period):

$$S_x = x_1 + \cdots + x_n \tag{5-29}$$

which itself, of course, would be a random variable. Similarly, in the presence of the new source, the total count would be

$$S_y = y_1 + \cdots + y_n \tag{5-30}$$

As n increases, the distributions of the two random variables, S_x and S_y, will become more and more disparate, and the minimized probability of decision error can be made smaller and smaller. To see that this is so, one

TABLE 5-1. Poisson-Distributed Spike Probabilities in the Presence of Background Bombardment, with and without the Source to Be Detected[a]

Number of spikes in a single receptor in observation period Δt	Probability in absence of new source	Probability in presence of new source
0	0.904837418	0.895834135
1	0.090483742	0.098541755
2	0.004524187	0.005419797
3	0.000150806	0.000198725
4	0.000003770	0.000005465
5	0.000000075	0.000000120
6	0.000000001	0.000000002

[a]The source intensity is approximately one-tenth the intensity of the background.

can compare the means and the standard deviations of those distributions as n increases. The means will increase directly as n, while the standard deviation will increase at a lesser rate (a rate lying somewhere between $\sqrt{n}$ and n, depending upon the correlations between the y_j and between the x_j). The square-root relationship holds if the random variables of each sum (S_x and S_y) are independent (e.g., x_j is independent of x_{j+1}), and the linear relationship holds only if the random variables are completely correlated (e.g., x_j is perfectly correlated with x_{j+1}), in which case they are no longer random.

If the CNS is in a hurry to make its decision, then instead of summing counts over many intervals of time, it could sum over many separate receptors of the same type. Equations (5-29) and (5-30) would still apply, but in this case n would be the number of receptors over which the count was taken. Whereas a stimulus is likely to leave residual effects in the cell receiving it, which could survive from one interval of time to the next and thus lead to some degree of correlation between successive values of x or y for a single receptor, such effects may be absent between separate receptors, in which case the summed random variables in equation (5-30) and in equation (5-29) are very likely to be virtually uncorrelated. On the other hand, the connections (synaptic and direct electrical connections) that apparently exist among adjacent or nearly adjacent receptors definitely would provide some degree of correlation.

As an example of the effectiveness of summation, consider the following situation. The background bombardment arises from a Poisson source, which generates a Poisson distribution for x:

$$p_i = \frac{\lambda^i}{i!} e^{-\lambda} \tag{5-31}$$

TABLE 5-2. The Spike Probabilities of Table 5-1, Taken over 10,000 Samples in the Presence of Background Bombardment, with and without the Source to Be Detected[a]

Number of spikes in 1000 receptors in an observation period of $10\Delta t$	Probability in absence of new source	Probability in presence of new source
Less than 875	Less than 0.0001	—
Between 875 and 900	0.0008	—
Between 900 and 925	0.0081	—
Between 925 and 950	0.0482	—
Between 950 and 975	0.1577	0.0001
Between 975 and 1000	0.2852	0.0012
Between 1000 and 1025	0.2852	0.0106
Between 1025 and 1050	0.1577	0.0536
Between 1050 and 1075	0.0482	0.1611
Between 1075 and 1100	0.0081	0.2734
Between 1100 and 1125	0.0008	0.2734
Between 1125 and 1150	Less than 0.0001	0.1611
Between 1150 and 1175	—	0.0536
Between 1175 and 1200	—	0.0106
Between 1200 and 1225	—	0.0012
Greater than 1225	—	0.0001

[a]The source intensity is approximately one-tenth the intensity of the background. The values in Table 5-2 were obtained by means of the normal-distribution approximation, which is quite applicable in this case.

where $\lambda = 0.1$. With the new source present, the distribution for y also is Poisson:

$$p'_i = \frac{\lambda'^i}{i!} e^{-\lambda'} \tag{5-32}$$

where $\lambda' = 0.11$. Table 5-1 displays the corresponding probability values. It is clear from Table 5-1 that regardless of where the CNS sets its threshold of detection, an extremely high proportion of errors will be made. If it attempts to avoid false detection of the new source, then the threshold would have to be set somewhat beyond six spikes, in which case the presence of the new source would be detected only a minuscule proportion of the times it actually occurred. However, if the spikes were summed over ten units of time and over 1000 identical receptors, providing a total of 10,000 samples (which we shall assume to be essentially independent), then, as Table 5-2 shows, detection becomes considerably less problematical.

In order to use the information in Table 5-2 to develop an optimum threshold of detection, some sort of cost function must be determined and a variable to be maximized or minimized must be established. For example, if failure to notice the new source when it was present were fatal 50% of the time (e.g., the source is a predator) and the actions taken when the source is detected (e.g., the escape response) were fatal 1% of the time, regardless of whether or not the source actually was present, then the optimum threshold simply could be that which maximized the probability of survival of the individual. Clearly, however, in order to carry out the calculations, we need one more piece of information, namely, the probability that the source actually is present during the interval in question (10 Δt's). Thus, it is clear that the optimum design of a nervous system is critically dependent on environmental parameters (as one should expect). If the source (e.g., the predator) is very unlikely to be present, then a high threshold (e.g., greater than 1100 spikes) would very likely maximize survivorship, since it would minimize the number of potentially fatal false alarms. On the other hand, suppose that the probability is 0.5 that the source will be present throughout a 10 Δt observation period and 0.5 that it will be absent throughout the period. In that case, the optimum threshold would lie between 1000 and 1025 spikes (the resolution of Table 5-2 is not adequate for more precise specification). If the threshold were reduced to values below 1000 spikes, false alarms would be the dominant mortality mechanism, and the mortality rate would increase toward 1%. On the other hand, if the threshold were increased to values above 1025, failure to notice the source would be the dominant mortality mechanism, and the mortality rate would increase toward 50%. In the range 1000 to 1025 spikes, the mortality will be minimum, lying somewhat below one-half of 1%.

Not all stimuli exhibit an obvious discrete nature and thus a minimum (quantal) amplitude. In the case of the vertebrate inner-ear and lateral-line receptors, for example, the effective stimulus (after mechanical translation by various accessory structures) is apparently a shearing motion, horizontal to the cell surface and directed against the array of fingerlike projections from the surface (the stereocilia, which probably are modified microvilli, and the true cilium, or kinocilium; see Figure 5-13). In some of the inner-ear organs, the threshold for detection of shearing movement by the CNS has been estimated to be a small fraction of 1 A, but no quantal nature has been indentified in the stimulus. Apparently, as far as the individual receptor is concerned, the threshold stimulus is vanishingly weak, utterly incapable of producing a response sufficiently large to generate a spike in any of the neuronal fibers with which the receptor communicates synaptically (spikes apparently do not occur in these receptors themselves, even in response to large stimuli). However, the weak stimuli are capable, in principle, of modulating ongoing spike trains in the sensory fibers simply by altering (very slightly) the intervals between successive spikes. Thus, one would expect the fibers responsible for transmitting these very weak signals to the CNS to exhibit spontaneous spike activity, even in the total absence of stimuli.

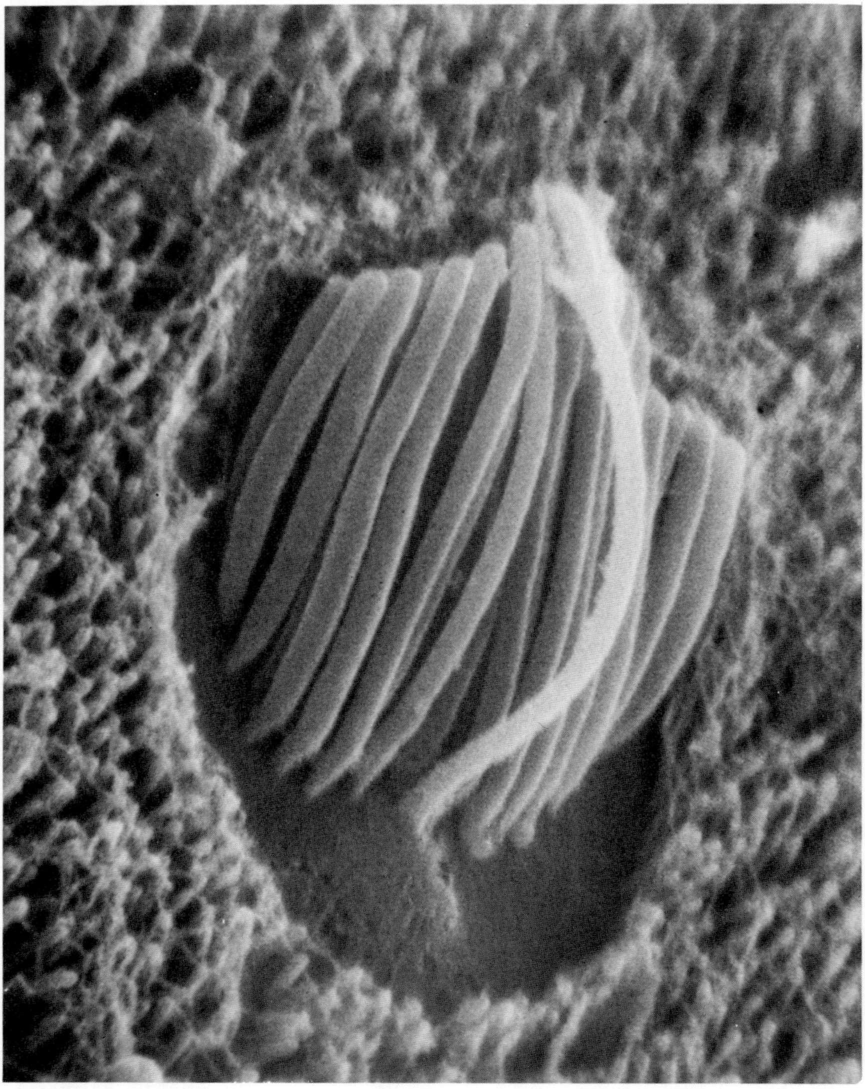

Figure 5-13. Surface of a receptor cell (magnified 24,000 times, reduced 40% for reproduction) from the inner ear of a salamander (*Necturus maculosus*). This particular cell is from the sacculus (whose function is not known in this animal), but it illustrates features common to many auditory and vestibular receptors. A single true cilium projects from the side of the cell nearest the viewer and is connected at its distal end to the longest of numerous elongated microvilli ("stereocilia"), each of which has a markedly constricted base. (From Lewis and Nemanic, 1972, *Zeitschrift für Zellforschung und Mikroskopische Anatomie* **123**, 441-457.)

In general, such spontaneous spike activity can be expected to exhibit random fluctuations (noise) against which the CNS must contend in its attempts to detect weak stimuli. Once again, however, the principles represented in Tables 5-1 and 5-2 apply. In the case of the inner-ear or lateral-line receptor, the threshold displacement may by extremely small even in comparison to the thermal motion of any single molecule in the receptor's plasma membrane; yet when that thermal motion is summed over all the molecules in the membrane, its distribution will become extremely narrow, and the threshold displacement will no longer seem so minuscule. Furthermore, especially in the inner ear, there are often very large numbers of apparently identical receptors over which ongoing spike trains could be summed in order to separate the effects of the random thermal noise and the stimulus displacement. Clearly, such summation over many receptors increases in effectiveness as the noise apparent in each individual receptor is reduced. Thus, one might expect that the spontaneous activity in a receptor of very weak signals would exhibit a rather low noise component. This generally is the case, and it is accompolished simply with spike trains that are nearly periodic, exhibiting relatively constant intervals between spikes. For example, the distribution of spikes given in the middle column of Table 5-1 exhibits a mean frequency of 0.1 spikes per Δt and a root-mean-square (rms) deviation of 0.316 spikes per Δt from that mean, which represents a very large noise component. If a sensory fiber is inherently capable of producing a more periodic spike train, with a mean frequency of 0.1 spike per Δt and an rms deviation of 20%, then as far as separation of stimulus response from noise is concerned, that fiber is equivalent to approximately 25 of those represented in Table 5-1; it would require summation over 25 of the more noisy fibers to accomplish the separation inherent in the more periodic fiber. In fact, for every halving of the rms deviation (the noise) in the periodic fiber, the equivalent number of noisy fibers represented is increased fourfold. Thus, a 10% rms deviation in the periodic fiber of our example would increase its value to 100 of the noisy fibers.

5.7. The Fuortes–Hodgkin Model

Perhaps the most thoroughly modeled receptor is the photoreceptor of the lateral eye of the horseshoe crab (*Limulus*). Here, the input–output relationships of a very complicated system are reasonably well known, and models are used in attempts to identify the physical mechanisms at work within the system.

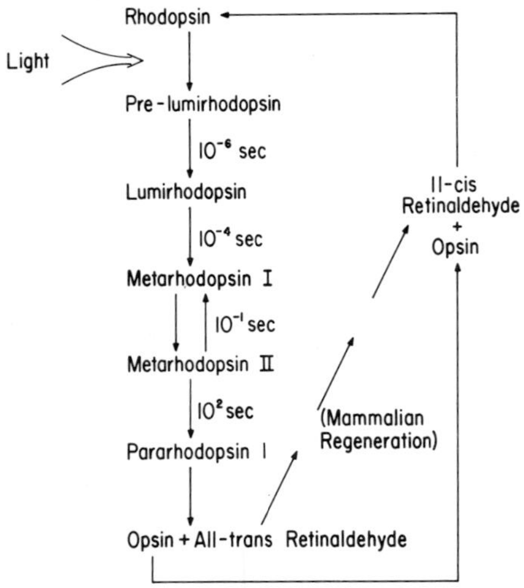

Figure 5-14. Rhodopsin cycle. One forward pathway and three reverse pathways (not shown) are known to require light. The remaining reactions take place in the dark. All the reaction half-times shown are from bovine photochemistry. Hubbard and Wald (1960) showed that a similar system is operating in the *Limulus* lateral-eye photoreceptors. The figure is taken from Charlton (1973), and is based on Brindley (1970) and Hubbard and Wald (1960).

The photopigment in the receptors of the lateral eye has been identified as rhodopsin, and the photochemical cycle apparently is similar to that of other rhodopsin photoreceptors, including a dissociation step (bleaching), as shown in Figure 5-14. Following a photon capture, the chromophore apparently undergoes isomerization, after which the photopigment molecule apparently undergoes a series of conformational changes. The early changes are extremely rapid, with half-reaction times of microseconds; the later changes are quite slow, with half-reaction times of minutes.

At the output end of the phototransduction process is the photoreceptor membrane, as depicted in Figure 5-15. Here, in response to a very brief flash of light, one observes a deeply inflected transient depolarization. The peak of this response occurs 100 to 200 msec after the flash, and the general shape of the response indicates that the underlying process is of high order,

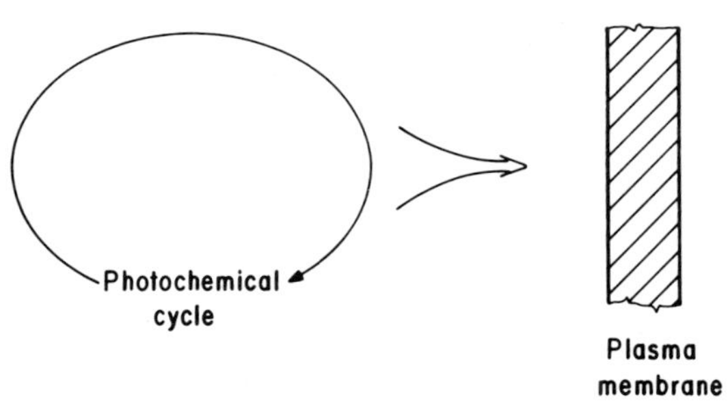

Figure 5-15. A partial model of a photoreceptor. The photochemical cycle somehow is coupled to the receptor membrane, where responses are observed.

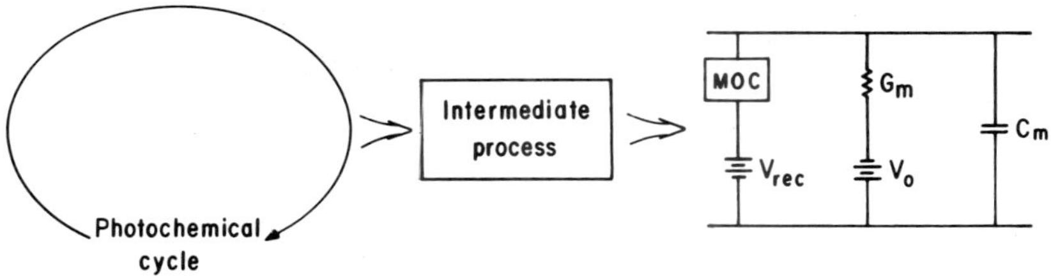

Figure 5-16. A more complete model of a photoreceptor. Apparently, between the photochemical cycle and the receptor membrane lies an intermediate process with interesting dynamic properties and with an output that leads to modulation of a conductance across the membrane.

consisting of several coupled subprocesses each with a time constant comparable to the fall-time of the response itself. The time constant of the receptor membrane is more than an order of magnitude too small to account for the dynamics of the response, and the photochemical cycle of Figure 5-14 has only one reaction with a time constant even close to the correct range. Evidently, therefore, some sort of intermediate process must exist between the photochemical cycle and the membrane.

Fuortes (1959) found good evidence to indicate that the transient response of the photoreceptor is mediated by a chord-conductance change in the membrane, and that the modulated conductance has a reversal potential that is slightly positive. This observation is incorporated in the model of Figure 5-16. Here we have attributed the capture of photons to the photochemical cycle at the input end of the system, we have attributed the system dynamics to the intermediate process, and we have attributed the final transduction step to a modulation of membrane conductance. Since it embodies the dynamics of the system, it is the intermediate process that has interested most modelers.

As is true of most receptors, the photoreceptor of the lateral eye exhibits a form of adaptation, i.e., the sensitivity of the receptor is adjusted according to the background level of light, so that the eye in the dark is much more sensitive than it is in the light. Various descriptive models have been proposed for adaptation, perhaps the most common one being a logarithmic relationship:

$$R \propto \log I \tag{5-33}$$

where R is the amplitude of the response and I is the intensity of the stimulus. According to this model, the incremental response ΔR to an incremental stimulus ΔI superimposed on a steady background stimulus I_0 would be

$$\Delta R \propto \Delta I/I_0 \tag{5-34}$$

However, no dynamics are implied in this description of adaptation. In other words, the receptor very well might be operating on a static (no-memory) nonlinear curve, such as that shown in Figure 5-6. Such a nonlinearity is certainly inherent in the conductance modulation assumed to occur in the final stages of transduction. If the response of the photoreceptor is slow relative to the membrane time constant, then as a first approximation one can describe the no-memory nonlinearity inherent in the receptor membrane as follows:

$$\Delta V = \left(\frac{G_{r0} + \Delta G_r}{G_{r0} + \Delta G_r + G_m} - \frac{G_{r0}}{G_{r0} + G_m}\right)(V_{rec} - V_0) = \frac{G_m \Delta G_r (V_{rec} - V_0)}{(G_{r0} + \Delta G_r + G_m)(G_{r0} + G_m)} \tag{5-35}$$

where ΔV is the incremental response to an incremental change ΔG_r of the modulated conductance superimposed on a steady background G_{r0}; V_{rec} is the reversal potential associated with the modulated conductance; V_0 is the resting potential of the receptor when it is unstimulated; and G_m is the chord conductance of the unstimulated receptor. If G_{r0} were directly proportional to I_0 and ΔG_r were directly proportional to ΔI, then we would have

$$\Delta R \propto \frac{\Delta I}{(I_0 + A)(I_0 + \Delta I + A)} \tag{5-36a}$$

or

$$R \propto I/A(I + A) \tag{5-36b}$$

which is similar to the logarithmic relationship, but not identical to it.

For a short while recently, it was believed that photoreceptors did in fact operate along such static nonlinearities. Shifts in the operating curves, such as those shown in Figure 5-17, which were observed in the higher-order neurons of the retina and which provide increased contrast range about any ambient operating point, were thought to be mediated by neurons other than the receptors themselves. However, Fuortes and Hodgkin (1964) had demonstrated that the process of adaptation in the photoreceptor of the lateral eye did in fact exhibit dynamics, that it could not be the reflection of a static nonlinearity. Charlton (1973) verified this when he obtained the shifting curves of Figure 5-17 for the photoreceptor itself. Thus, adaptation in the lateral-eye photoreceptor is considerably more complicated than a no-memory nonlinearity and cannot be accounted for by the conductance modulation of the final step in transduction.

Fuortes and Hodgkin, in fact, placed adaptation in the intermediate process. They came to this conclusion after observing that there is an

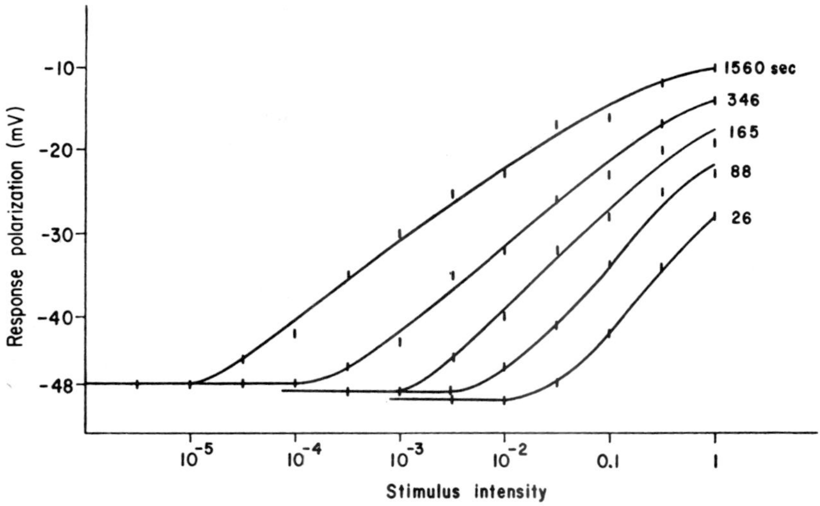

Figure 5-17. Dynamic adaptation in a photoreceptor of the lateral eye of *Limulus* [from Charlton (1973)]. Instantaneously, the receptor operates along a single nonlinear curve, similar to that deduced from the models of equations (5-33) and (5-36). However, as the background illumination changes, the operating curve gradually shifts (to the right for increasing background intensity, to the left for decreasing intensity). In this particular case, the receptor was adapted to a fairly strong background intensity, then allowed gradually to readapt to darkness. The operating curve gradually shifted to the left as dark adaptation proceeded (time since the termination of the background light is given on the right).

apparent concomitancy between the sensitivity of the photoreceptor and the dynamics of its response. Thus, whatever parameters are adjusted during the process of adaptation not only affect sensitivity, but also response dynamics, and the response dynamics are embodied in the intermediate process. Therefore, one might very reasonably add another component to the model of Figure 5-16 to represent the coupling of input light intensity to the parameter change in the intermediate process. This component too embodies dynamics, the dynamics of the process of adaptation. Although the dynamics of adaptation and the source of its coupling to the light input have been of some interest to modelers, most of their attention has been focused on the intermediate process itself. The reason for this apparently is the fact that the intermediate process is quite linear during the initial phases of its response, and the nonlinearity inherent in adaptation is sufficiently delayed to allow this initial linear response to be observed. Furthermore, if the input signals are kept small, the response remains very nearly linear even in steady state (Pinter, 1966; Dodge *et al.*, 1968).

Fuortes and Hodgkin demonstrated that the impulse response of the intermediate process had a shape reasonably well described by the equation

$$V_{\text{resp}} = At^{n-1}e^{-\alpha t} \tag{5-37}$$

which is the impulse response of a system with n identical, negative-real natural frequencies, in other words, a linear system of order n and described by the transfer function

$$V_{\text{resp}}/I = A'/(s + \alpha)^n \tag{5-38}$$

where s is the complex variable of the Laplace transform. Furthermore, they noted that as the sensitivity of the receptor decreased in strong ambient background illumination, the change in receptor dynamics was reflected in the time between the flash (impulse) of light and the peak of the response. In fact, as the sensitivity decreased by a factor of approximately 1000, the time between the flash and the response peak t_{p} decreased by a factor of approximately 2.5. This concomitancy between sensitivity and dynamics could be accounted for by their model [equation (5-38)] if the parameter α were dependent on ambient light intensity. According to equation (5-37), the peak response occurs at

$$t_{\text{p}} = (n - 1)/\alpha \tag{5-39}$$

units of time after the flash, and the amplitude of the peak is

$$v_{\text{p}} = A(n - 1)^{n-1}e^{-(n-1)}/\alpha^{n-1} \tag{5-40}$$

Thus, t_{p} is inversely proportional to α and v_{p} is inversely proportional to α^{n-1}. If the model of equation (5-38) is correct, then the order of the intermediate process should be given by

$$n = 1 + \frac{\log[v_{\text{p}}(1)/v_{\text{p}}(2)]}{\log[t_{\text{p}}(1)/t_{\text{p}}(2)]}$$

or

$$n \simeq 1 + \frac{\log 1000}{\log 2.5} = 8.538 \tag{5-41}$$

In other words, the order of the system is approximately 8 or 9. This seemed to be confirmed by the fact that the model provided the best fit to the photoreceptor response waveform when n was between 8 and 12.

In attempting to find topographical realizations of their model, Fuortes and Hodgkin devised the scheme depicted in Figure 5-18a. Here we have n identical, unilateral linear processes, coupled to form a ladder network. The transfer ratio of this network model is

$$V_{\text{resp}}/I = \alpha'^{n-1}/(s + \alpha' + \alpha'')^n \tag{5-42}$$

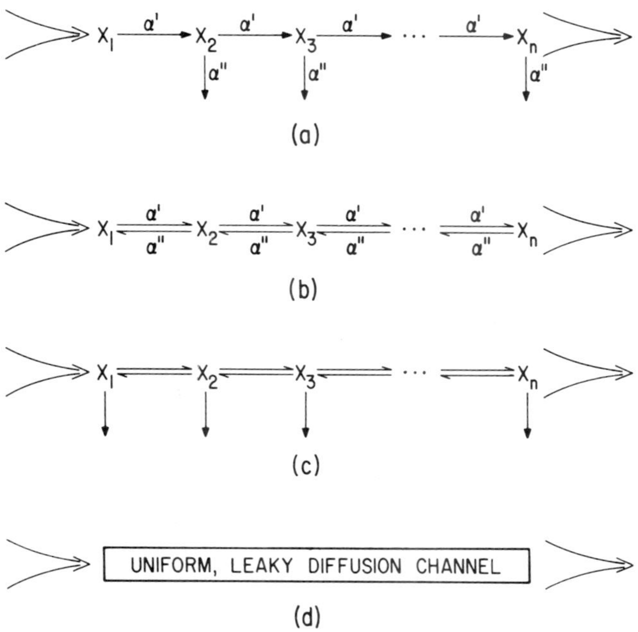

Figure 5-18. Four topographies of the basic Fuortes–Hodgkin model. (a) The original Fuortes–Hodgkin topography: n identical, unidirectional processes between the light input and the modulation of the receptor membrane conductance; n in this topography is restricted to the range between 8 and 12. (b) The topography of Borsellino *et al.*: n (8 to 12) bidirectional processes between input and conductance modulation; the forward rate constant is very small compared to the reverse rate constant. (c) The topography of Lewis and Charlton: n (8 to infinity) reversible processes between input and output, with light-dependent shunting process at each stage. (d) The Lewis and Charlton topography with n equal to infinity, a leaky diffusion process.

which, if α'' is taken to be dependent on ambient light intensity and α' is taken to be a constant, is equivalent to the model of equation (5-38). Borsellino, Fuortes, and Smith (1965) proposed the alternative configuration of Figure 5-18b, whose transfer ratio reduces to that of equation (5-42) if α'' is very large compared to α'. In this case, we have a sequence of n states coupled bilaterally, with the reverse path in each case being dependent on ambient light intensity and being much stronger than the forward path. True photocatalysis in this topography would require that both α' and α'' be light dependent if either were.

Lewis and Charlton (1972) proposed another topographical realization, shown in Figure 5-18c. In this case, we have a sequence of n states coupled bilaterally to form a ladder network, but each state also has a leakage path associated with it. This topography has three distinct advantages over the other two: First, the models of Borsellino *et al.* and of Fuortes and Hodgkin both exhibit a linear relationship between $\log[v_p]$ and $\log[t_p]$, whereas the observed data always exhibit marked curvature. Very similiar curvature is exhibited by the Lewis–Charlton topography. Second, the observed v_p vs. t_p relationships in the receptor strictly limit the order of the system in the first two models (i.e., n is not much greater than 8 or 9), whereas those same data do not impose upper limits on the order of Lewis–Charlton topography. Finally, the impulse response of the first two models are very nearly maximally symmetrical for a given system order, whereas the observed impulse response of the photoreceptor is not quite so symmetric, exhibiting a slight but consistent skewing toward longer times. Essentially the same degree of skewing occurs with the Lewis–Charlton topography.

Although the Lewis–Charlton topography provides some subtle improvements over the earlier models, it definitely does not provide the last word on the subject. In fact, the most important conclusion that one can draw from it is that a transfer ratio and an impulse response with the following forms seem to represent the intermediate process very well:

$$V_{\text{resp}}/I = A/\prod_{i=1}^{n} (s + \alpha_i) \tag{5-43}$$

$$V_{\text{resp}} = A' \sum_{i=1}^{n} \left[e^{-\alpha_i t} / \prod_{\substack{i=1 \\ i \neq j}}^{n} (\alpha_i - \alpha_j) \right] \tag{5-44}$$

The realizations of these processes are constrained in that n must be greater than 8 and that, owing to the symmetry of the observed waveform, the dominant natural frequencies (those determining dynamics) must be reasonably well clustered (but not identically situated) on the negative-real axis of the complex frequency plane. Furthermore, since the symmetry is not reduced by adaptation (if anything it is slightly increased), the clustering of the natural frequencies (the α_i's) must be maintained as they are shifted by the presence of ambient light. The Lewis–Charlton topography is only one of an infinity of realizations that meet these constraints.

On the other hand, most of these realizations have a flaw. One ordinarily does not expect to find a large number of sequential state changes exhibiting rate constants that are even relatively close to one another. In other words, the clustering of natural frequencies provides a major conceptual difficulty. One natural process that does exhibit an apparent clustering of natural frequencies, however, is diffusion over a nearly uniform distributed channel. The order of a diffusion process is infinite, but this provides no difficulty since there is now no upper bound on n. Thus, we should introduce one final model, which is depicted in

Figure 5-18d. Here the intermediate process is depicted as diffusion over a leaky distributed channel. The leak itself is distributed, increases with ambient light, and might represent diffusion out of the main flow path or some other change of the state of the flowing substance, rendering it ineffective (e.g., a chemical reaction whose rate depends on ambient light intensity). One likely candidate for the diffusing substance now seems to be calcium (Brown and Blinks, 1974).

All the models described in the previous paragraphs account very well for the linear dynamical aspects and for the static aspects of adaptation in the intermediate process of the photoreceptor of the lateral eye. When the light stimulus is especially weak, the response of the photoreceptor becomes stochastic (Fuortes and Yeandle, 1964; Adolph, 1964; Dodge *et al.*, 1968). Dodge *et al.* (1968) showed that the nature of these stochastic responses was consistent with the nature of the more deterministic responses to stronger stimuli, under the assumption that the latter comprised large ensembles of the former and that the former exhibited a Poisson distribution. Similarly, in the model of Fuortes and Hodgkin and in all of the topographical variations of it presented in the previous paragraphs, one can imagine that each of the *n* states is occupied by a population of particles of some sort. The rate constant associated with each path in a given topographical realization then represents the mean rate of a first-order Poisson process; and when the number of particles occupying a given state becomes small, the stochastic nature of this Poisson process would become apparent. Thus, the model of Fuortes and Hodgkin is completely consistent with the model of Dodge *et al.*

However, none of the models takes into account the dynamics of the adaptation process (i.e., the temporal aspects of the shifting of the natural frequencies). This was attempted by Marimont (1965), using a computer simulation of the Fuortes–Hodgkin model, with parameter modulation

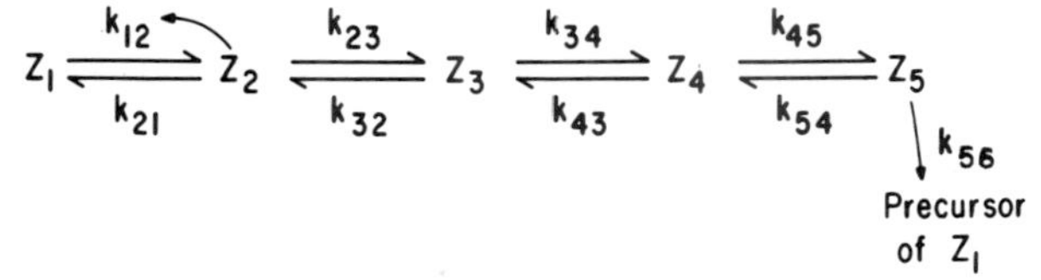

Figure 5-19. Second five-stage model of Baylor and Hodgkin, designed to account for the dynamics of adaptation. The quantity k_{12} represents the light-dependent shunt from the last stage of a Fuortes–Hodgkin-like model. The arrow from Z_2 represents autocalalysis, which in effect produces light dependence, since the formation of Z_1 and, from it, Z_2 depends on light input to the system. The forward rate constants are taken to be large in comparison to the reverse rate constants, and increase from left to right as follows: $k_{12} = 0.01$ to 0.1; $k_{23} = 0.1$; $k_{34} = 1$; $k_{45} = 10$; $k_{56} = 100$.

made a function of the various state variables. She concluded that the parameters probably were modulated by one of the final two or three state variables of the system. On the other hand, she was attempting to match nonlinear transients, some of which very well may have resulted from membrane nonlinearities and voltage-dependent conductances.

Recently, Baylor and Hodgkin (1974) have observed responses in turtle photoreceptors that indicate underlying phenomena that are quite similar to those in *Limulus*. To account for these, they proposed a model similar to that of Fuortes and Hodgkin, but with only five stages and with a single rate constant (that leading out of the fifth stage) being dependent on ambient light intensity. Coupled to the fifth stage of their model is another five-stage process, designed to account for the dynamics of the adaptation process and the nonlinear transient responses of the system, with the instantaneous nonlinearities accounted for. The first stage of the latter process is taken to be autocatalytic, thus providing the rate constant that depends upon ambient light (see Figure 5-19).

5.8. Spontaneous Activity in Neurons

Receptors and sensory fibers are not the only neural elements in which spontaneous activity may be present and useful. It apparently occurs in abundance in the CNS. Its utility, however, as with most aspects of the nervous system, is most apparent when it occurs near the periphery. Thus, for example, the function of a spontaneously active group of nine nerve cells in the wall of the crustacean heart seems quite clear: It is the pacemaker for the heart. The output from this group of nerve cells consists of periodic bursts of spikes that are transmitted down the axons of five of the cells and drive the muscles of the heart into contraction. Although the functional implication is not quite as clear on the sensory side, it follows from our previous discussion that a single sensory channel that exhibits reasonably periodic, spontaneous activity upon which incoming sensory signals are superimposed is equivalent to many noisy channels as far as the enhancement of signals with respect to noise is concerned.

The mechanisms underlying spontaneous signal generation in individual neurons are not well established. From this point of view, perhaps the most thoroughly studied examples are the various spontaneous neurons of the abdominal ganglion of *Aplysia*. Some of these cells produce more-or-less periodic, spontaneous single spikes. These could be explained in terms of the active mechanisms described in Chapter 7. For example, a reasonable value of the sodium-ion chord conductance at what otherwise would be the resting state of the cell could produce a gradual depolarization, leading to a spike. The spike recovery processes then could reset the membrane potential, gradually allowing the depolarizing current to become dominant again and lead to another spike.

Other cells in the abdominal ganglion exhibit spontaneous bursts of spikes, separated by long periods of quiescence. In these cells, the spike bursts are triggered by the depolarizing phases of slow oscillations generated by the cells themselves. Among the mechanisms postulated to account for these slow waves are negative conductances (Wilson and Wachtel, 1974), time-varying conductances, and time-varying electrogenic pumps (see Plant and Kim, 1975). One plausible hypothesis proposed very recently involves a gradual accumulation of calcium (the principal inflowing ion species during spikes in *Aplysia* cell bodies) during a burst. This in turn is assumed to bring about a temporary increase in potassium-ion efflux, turning off the burst (Eckert, personal communication). Of course the essence of spontaneous activity in these cells is the time-varying phenomena themselves, and it is their time variations that still require explanation. In one of the most thoroughly studied cells, R-15, the slow wave pattern exhibits diurnal and seasonal modulation, and these too invite explanation. At this point, we clearly have moved into the realm of biological clocks, the theory of which is still in a preliminary state. The key questions include: What are the sources of these long-term oscillations? How are they entrained by each other and by external stimuli? Similar questions arise concerning the more short-term activity of spontaneously active neurons. For example, the work by Perkel *et al.* (1964) on the theory of entrainment of spontaneous neurons is closely analogous to the work of Winfree (1967) on entrainment of biological clocks.

One of the remarkable aspects of spontaneous neural activity is the high degree of stability that it often exhibits. Maynard (1955) noted that the firing sequence of the nine cells in the early phase of the burst in the cardiac ganglion of the lobster was quite consistent over very long times (i.e., over 3500 observed sequential bursts). Mayeri (1969), carefully controlling the ambient temperature of the ganglion, observed similar consistency even in the later phases of the burst. Indeed, temperature appears to be a key parameter in neural oscillators. Perhaps the most stable neural oscillator observed so far is the pacemaker for the electrical discharge of a gymnotid fish. Even in this cell, however, the frequency is quite sensitive to temperature, exhibiting a Q_{10} of approximately 1.7 (see Bullock, 1970). In a *Sternarchus*, the observed firing frequency of the electrical organs (which reflects on a one-for-one basis the firing frequency of the pacemaker cell) was approximately 850 Hz, and its standard deviation was 0.012%. As far as the separation of signal and noise is concerned, a sensory channel whose signals were superimposed upon a periodic background of this stability would be equivalent to approximately 10^8 Poisson channels of the type represented in Table 5-1.

Suggested Further Readings

For further readings we refer the reader to Beament (1962); Beidler and Reichardt (1971); Eccles (1964); Pfaffmann (1969); Stevens (1968); *Sensory Receptors* (1965); Møller (1973); Bennet (1974); and Hubbard, Llinas, and Quastel (1969).

6

Models of Distributed Passive Membrane

In Chapters 3, 4, and 5, we have presented models of local membrane phenomena. In other words, we have dealt with gradients and ion flows across the membrane, but we have ignored gradients and flows directed tangential to its surfaces, and in doing so we have ignored mechanisms by which receptor responses and synaptic responses can spread from one region of the neuron to another. As we mentioned in Chapters 1 and 2, local regions of a neuron are often apparently adapted to specific subtasks with respect to signal processing. Quite often, for example, sites of sensitivity to external stimuli (receptor sites) are separated spatially (and are often quite remote) from spike-triggering sites. The influencing of spike generation by receptor-site responses is generally accepted as being a key link in the neuron's signal processing; and, of course, the spread of the receptor response from the receptor site to the triggering site is *the* crucial phenomenon in this link.

In this chapter we present the basic theory of *passive* spread of responses, where all the energy required for the spreading process is provided by the response itself. In Chapter 7, we develop the theory of *active* spread of responses, where energy is added to the response as it progresses over the membrane. Apparently, both types of spread occur together or interact to some extent in many neurons, in which case the models presented in Chapter 6 are incomplete. In certain regions of many neurons, on the other hand, passive spread apparently is overwhelmingly predominant, and the models presented here are apparently quite applicable.

6.1. The Basic Model

The basic situation that concerns us here is depicted graphically in Figure 6-1. We can picture two patches of membrane, one at a receptor site and the other remote from it, and ask how a response at the former can spread to the latter. There seem to be three regions of space through which the spread could occur—the intracellular fluid, the extracellular fluid, and the membrane itself. It is quite possible that the fundamental phenomena involved in signal processing in the neuron have gone unnoticed so far, and that the observed ionic and electrical events actually are epiphenomena. For example, the quintessence of the local receptor response might be configurational changes in macromolecules of the membrane, and these might spread passively to neighboring macromolecules and thus be transmitted to the remote site. On the other hand, all evidence to date supports the notion that the observed ionic and electrical events are not epiphenomena, but are fundamental in themselves. If we accept this common tenet, then we are left with two possibilities for the driving of passive spread: It can be driven by tangential ionic concentration gradients, and it can be driven by tangential electrical potential gradients, both of which are established by the flow of ions and the resulting change in electrical potential across the local membrane at the receptor site. Given these two possibilities, it is quite easy to relegate one very logically to a minor role. According to the doctrine of space-charge neutrality and the arguments presented in Chapter 3 in support of it, the relative change in ionic concentrations on either side of the membrane in response to receptor activity will be minuscule in comparison to the relative changes in electrical potential across the membrane. Thus, with respect to sites remote from the receptor site, where electrical potentials and ionic concentrations have not been affected directly by the receptor activity, marked proportional differences in electrical potential but only minute proportional differences in ionic concentrations will exist.

Thus, since the Einstein relation dictates that diffusivity and conductivity must be comparable, drift down the electrical potential gradient should be far greater than diffusion down the concentration gradient and definitely should dominate.

The next question, relative to Figure 6-1, is: Where do the paths lie along which tangential drift occurs? Two empirical observations must influence our answer. First of all, the volumes of the extracellular space and the intracellular space are almost always very much greater than the volume of the membrane; so even if the electrical conductivities and the electrical potential gradients were comparably large in all three spaces, the total tangential drift of ions in the membrane would be minute compared to that in the other two spaces. Accentuating this is the observation that, at least

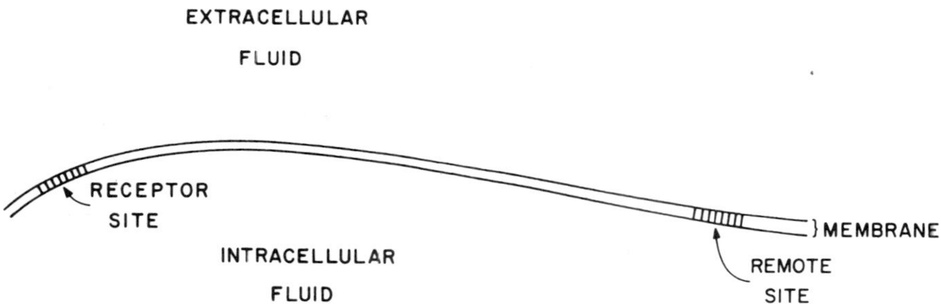

Figure 6-1. Idealization of the physical setting for conduction of signals in a neuron. Electrical responses generated by transduction mechanisms at a receptor site are conducted to a remote region (hatched) of the neuronal membrane.

with respect to potential gradients across it, the membrane has a very low conductivity in comparison to those of the intracellular and extracellular fluids. Furthermore, if as some evidence indicates, the flux of ions through the membrane is limited to transverse channels of some sort, then one would expect an anisotropy with respect to membrane conductivity that would favor tranverse flow of ions and preclude tangential flow. Taking these things into account, one is compelled to favor the common tenet that tangential drift occurs almost exclusively in the ionic solutions on either side of the membrane.

Thus, we are led directly to the problem of devising a model for ionic drift in those two solutions. The model that seems most reasonable and most generally accepted is that of simple ohmic conduction. Some modelers have assumed that ionic flows in these solutions should have significant inertia, perhaps from induced magnetic fields, and have included inductive effects in their representations. However, there seems to be little or no justification for this in the body of physiological data; so most modelers operate under the assumption that inertia or inductive effects are insignificant in comparison to entropic, or resistive, effects, and we shall do so here.

Having decided to represent the two ionic solutions as simple ohmic conductors, we are left with the problem of selecting representations for the patches of membrane. Although the various equivalent circuits presented in Chapters 4 and 5 may be useful when one wishes to analyze candidate modes of local membrane action, they are usually considered unnecessarily complicated for the analysis of passive spread of receptor response in neurons. In such cases, the models of the receptor sites themselves may retain their complexity, but the models of the remaining membrane sites can logically be simplified considerably by the removal of any batteries (and considering only *changes* in transmembrane potential, ignoring the

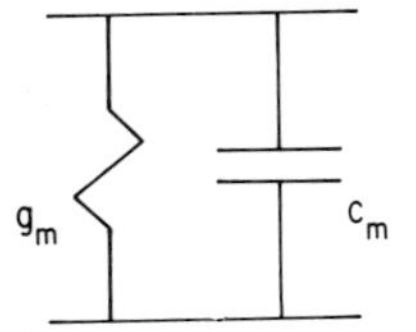

Figure 6-2. Simple slope-conductance membrane model usually invoked in the theory of passive conduction of signals in neurons. The quantity g_m is the slope conductance of the membrane around the operating point (usually the resting potential).

resting potential) and the lumping of all conductances into a single element. Usually, any nonlinearities or hysteretic effects that might be present in the membrane are also ignored, and the local nonreceptor membrane is represented by linear (slope) conductance in parallel with constant capacitance, as depicted in Figure 6-2.

By extending this model to a distributed version, one can draw a number of interesting but tentative conclusions concerning the passive spread of responses in neurons. These conclusions will remain tentative until one can demonstrate that the basic model itself is valid (i.e., that the instantaneous and steady-state slope conductances of the membrane in question are one and the same, and constant). Distributed models incorporating nonlinear or hysteretic conductances have been simulated and even analyzed in some cases. The results are interesting but, unfortunately, have not been applied significantly to neural theory. Furthermore, the results depend rather strongly on the nature of the nonlinearity or hysteresis employed, so the applicability of any one of these models to any particular passive neuronal structure would also be quite tentative, perhaps even more so than the linear model. Nonetheless, it might be interesting and, probably, insightful to list the most genral properties of passive distributed systems based on models such as that shown in Figure 4-2. However, for reasonably small, passively conducted signals, the scant physiological evidence that does exist (mostly obtained from cell bodies and axons sinçe dendrites remain inaccessible to the required measurements) indicates that the model of Figure 6-2 is a reasonably good one, and most modelers are apparently willing to accept it as the basis for analysis of passive conduction.

6.2. Dipole Annihilation and Redistribution

Developed from the simplified membrane model of Figure 6-2, the basic distributed model consists of an electrical capacitance, distributed over the membrane and over which charge is stored, a conductance between the intracellular solution and the extracellular solution, distributed over the membrane and through which charge flows across the membrane, and resistive flow paths, distributed over the intracellular and extracellular solutions and connecting every point on either side of the membrane to every other point on that same side. If it were not for the distributed capacitance, excess charge anywhere in the system would tend to be distributed uniformly over the surfaces of the conductors in exceedingly short time (the dielectric relaxation time for matter), with redistribution velocities approximating the speed of light. With the capacitance, however, an excess charge on one side of the membrane can form a more-or-less stable association with an equal but opposite charge on the other side. If it is pulled away momentarily from that association, extremely large forces develop that tend to pull it back. Thus, charge is stored not in the form of independent, individual excess ions, but in the form of ion pairs, which are separated by the membrane but are in strong association with one another and tend to move together from place to place. It is these dipole pairs rather than individual ions that drift tangentially along the membrane (note that they may jump essentially instantaneously between any two places connected by both intracellular and extracellular resistive pathways), and it is these dipole pairs that are formed or annihilated by charge flowing across the membrane.

Basically, we have two processes with which to contend in our model of passive conduction. We have the annihilation of excess dipole pairs (those in excess of the number required to establish the resting potential, which is ignored in the model), which is mediated by flow through the distributed membrane conductance, and we have redistribution of dipole pairs, which is mediated by flow through the distributed intracellular and extracellular resistances. The relative importance of these two processes is determined by the geometry of the system.

In situations where the rate of dipole redistribution over a given region is very high compared to the rate of dipole annihilation, the redistribution process can be ignored and the entire region can be approximated quite accurately as having the same transmembrane potential throughout. In other words, the membrane in such a region acts as a single patch. The rate of dipole redistribution between two sites is inversely proportional to the sum of the intracellular and extracellular electrical resistances connecting those sites; the rate of dipole annihilation at a given site is directly proportional to the membrane conductance at that site. Thus, we have a reasonably good test for a region of simple geometry suspected of having a high rate of redistribution. We can simply consider the membrane in the region as being divided into several patches and then estimate the sum of the intracellular and extracellular electrical resistances between the two most remote patches. If the reciprocal of that sum is very large compared to the conductances of the two patches, then the approximation holds.

In the case of a spherical cell surrounded by a large volume of extracellular fluid, for example, one might consider two small circular patches at opposite poles, as depicted in Figure 6-3. Imagine a cylindrical

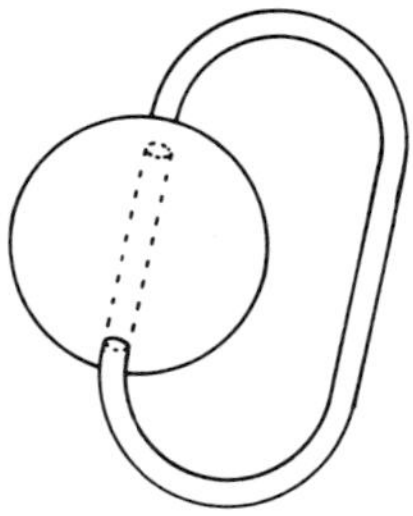

Figure 6-3. Idealized spherical cell with cylindrical current paths connecting patches of membrane on opposite sides of the cell. The intracellular path is indicated with dashed lines, the extracellular path with solid lines. These are just two of many current paths in parallel between the two patches of membrane.

tube connecting these two patches intracellularly and a curved cylindrical tube connecting them extracellularly. The total resistance measured axially around the combined tubes is clearly considerably greater than the actual resistance between the two patches. Therefore, its reciprocal will be considerably less than that of the actual resistance, and we can use it to perform our test in a very conservative manner. The total resistance around the combined tubes is simply

$$r = \rho l / a \tag{6-1}$$

where ρ is the resistivity of the ionic solutions (assumed for simplicity to be the same inside and outside the cell); a is the cross-sectional area of the tube, and also the area of each membrane patch; and l is the total length of the tube, which is perhaps three diameters of the sphere. The conductance of each membrane patch is simply

$$g = G_m a \tag{6-2}$$

where G_m is the specific membrane conductance (mho/cm^2). The appropriate test ratio is

$$\frac{1/r}{g} = \frac{1}{G_m \rho l} \tag{6-3}$$

which for typical cells will be quite large. For example, if $G_m = 10^{-3}$ mho/cm^2, $\rho = 250$ ohm-cm, and $l = 4 \times 10^{-2}$ cm, all of which are conservatively high, then the test ratio will equal 100, which is sufficiently large to make the approximation quite good, with redistribution being at least 100 times as fast as annihilation. Rall (1969) carried out a similar, but more precise, test for a cylindrical part of a cell 20 μm in diameter (e.g., the approximate geometry of a moderately large axon or dendrite) and found the value of the test ratio to be 10^4 for circumferential redistribution. In other words, redistribution over a spherical cell and redistribution around the circumference of a cylindrical cell are both very much faster than annihilation and, in comparison with that process, can be considered instantaneous.

In effect, we have said that there are certain geometries for which the resistances connecting the various patches of membrane are so small that they can (and should) be ignored, and the entire structure can be treated as a single patch of membrane separating an isopotential intracellular solution from an isopotential extracellular solution. In such cases, the spread of response can be treated as being instantaneous (compared with the time courses of other neuronal phenomena, such as dipole annihilation), and the entire membrane acts as an instantaneous load on the receptor site and the ion flow stimulated there.

6.3. Continuous Model for Response Spread over Neuronal Fibers

On the other hand, there are certain geometries for which response spread is not instantaneous, specifically, the long, more-or-less cylindrical neuronal structures such as axons and dendrites. Although Rall's calculations demonstrate that the spread of dipoles around the circumference of such a structure should be essentially instantaneous, the spread in the axial direction will generally not be so, since the paths involved generally are very long. The most direct way to analyze the passive spread in such a system is through application of a simple pair of partial differential equations known as the "cable equations." These have been applied widely in neural modeling for at least 100 years (Weber, 1873; Weinberg, 1941) and are reviewed in detail by Taylor (1963). They originally were applied to oceanic cables, where the enormous lengths involved made the cumulative axial resistance and the cumulative insulator leakage quite significant. Because they deal with axial flow of currents but ignore circumferential variations, they are ideal for describing the passive spread of responses in neuronal fibers.

They can be derived as follows: Consider a very short section (of length ∂x) of a neuronal fiber, as depicted in Figure 6-4. A receptor response is spreading through this section from left to right and, as it passes, excess dipoles flow into the left end of the section and out the right end. However, some of the excess dipoles entering the section are stored across its membrane capacitance, and some of them are annihilated as they attempt to pass through. Thus, the net axial flow i of excess dipoles is diminished very slightly as it passes through the section. The rate I_g of excess dipole annihilation is given directly by the membrane model of Figure 6-2:

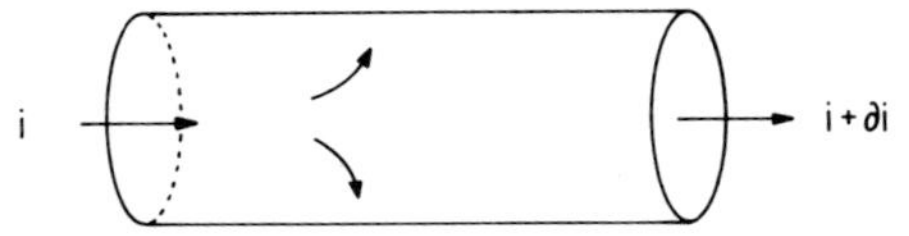

Figure 6-4. Idealized segment of a neuronal fiber used to derive the current equation [equation (6-7)].

$$I_g = v_m g_m \partial x \tag{6-4}$$

where $g_m \partial x$ is the total conductance of the membrane around the section (g_m being the membrane conductance per unit length of fiber), and v_m is the excess electrical potential across the membrane of the section (the amount by which the transmembrane potential differs from the resting potential). The rate I_c at which dipoles are being stored in the section is also given by the model

$$I_c = \frac{\partial v_m}{\partial t} c_m \partial x \tag{6-5}$$

where $c_m \partial x$ is the total capacitance of the membrane around the section. Therefore, the total amount $-\partial i$ by which the flow of dipoles is diminished as it passes over the length ∂x is given by

$$-\partial i = \left(g_m v_m + c_m \frac{\partial v_m}{\partial t}\right) \partial x \tag{6-6}$$

and the rate at which the flow diminishes with distance along the fiber is

$$\frac{\partial i}{\partial x} = -\left(g_m v_m + c_m \frac{\partial v_m}{\partial t}\right) \tag{6-7}$$

The axial flow of dipoles is accompanied by axial electrical potential gradients both inside and outside the membrane, which are given by

$$\begin{aligned} \partial v_i &= -(r_i \partial x) i \\ \partial v_o &= (r_o \partial x) i \end{aligned} \tag{6-8}$$

where $r_i \partial x$ is the axial resistance of the intracellular solution over length ∂x (r_i being the intracellular resistance per unit length of the fiber), and $r_o \partial x$ is the total resistance of the extracellular solution over length ∂x of the fiber (r_o being the extracellular resistance per unit length of the fiber). From Kirchhoff's voltage law (see Figure 6-5) we have

$$\partial v_m = \partial v_i - \partial v_o = -[(r_i + r_o) \partial x] i \tag{6-9}$$

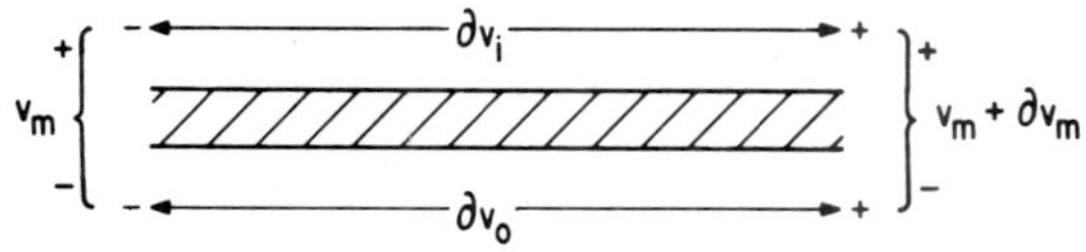

Figure 6-5. Idealized patch of membrane used to derive the voltage equation [equation (6-10)].

and the rate at which the excess membrane potential diminishes with distance along the fiber is

$$\frac{\partial v_m}{\partial x} = -(r_i + r_o)i = -ri \tag{6-10}$$

Taken together, equations (6-7) and (6-10) form the basis of cable theory. If r_o, r_i, g_m, and c_m are constant (i.e., do not vary either with time or with position along the fiber), then these two equations can be solved rather easily for a wide variety of specific situations. On the other hand, if the parameters are not constant, then analysis can be carried out by means of repeated finite-difference approximations on a digital computer. In other words, the fiber can be represented as a finite sequence of segments, each of which is sufficiently short that redistribution of dipoles over it (both axially and circumferentially) is essentially instantaneous. Thus, each segment can be represented as a single patch of membrane and modeled by one of the equivalent circuits in Chapters 3 or 4, or by the equivalent circuit of Figure 6-2. In this way, the distributed fiber is represented by a ladder-network model, such as that of Figure 6-6. The ladder approximation improves as the number of segments is increased, but any finite error criterion can be met with a finite number of elements. Rall (1960) showed that branching fiber trees (e.g., dendritic trees) meeting certain rather restrictive (but not unreasonable) geometric constraints behave precisely the same as a single, uniform fiber and can be analyzed as such directly by means of the cable equations [equations (6-7) and (6-10)]. In general, however, analysis of branching trees is easiest if each branch is represented by a ladder approximation.

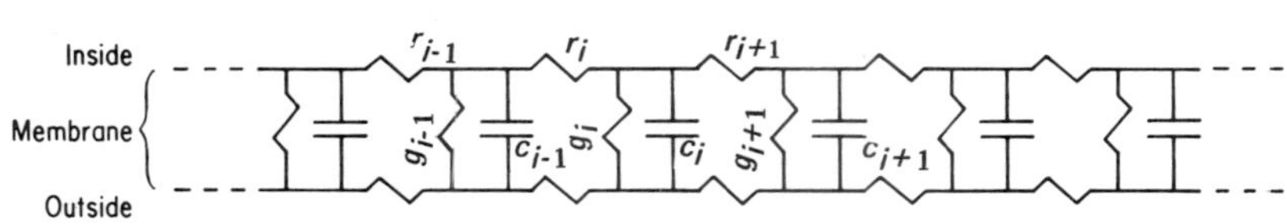

Figure 6-6. Lumped or ladder-network model of a passively conducting fiber.

6.4. Continuous Analysis of the Uniform, Passively Conducting Fiber with Time-Invariant Parameters

In the case of a single fiber with constant parameters, equations (6-7) and (6-10) can be solved analytically either by separation of variables or by application of Laplace transforms.

The forms of the solutions are quite different in the two cases, particularly with respect to convergence, those obtained by the separation-of-variables technique converging most rapidly for short pathways of signal conduction, and those obtained by Laplace transform converging most rapidly for long pathways. Even more important is the fact that the solutions generated by separation of variables are easy to match to the spatial distributions of input stimuli, but not to their temporal distributions, while the solutions generated by the Laplace-transform method are easy to match to the temporal aspects of a stimulus, but not to its spatial aspects. Therefore, if the stimulus of interest is a complex temporal waveform delivered at a single point in space, the Laplace-transform procedure is generally the method of choice. On the other hand, if the stimulus of interest is a complex spatial waveform delivered at a single point in time, the separation-of-variables method is preferable.

The separation-of-variables method, as applied to this problem, is discussed quite thoroughly by Rall (1969). Basically, differentiation of both sides of (6-10) with respect to x and substitution of equation (6-7) for $\partial i/\partial x$ leads to a second-order partial differential equation that can be normalized to become

$$\frac{\partial^2 v_m}{\partial X^2} = v_m + \frac{\partial v_m}{\partial T} \tag{6-11}$$

where $X = \sqrt{rg_m}\,x$, $T = (g_m/c_m)t$, and v_m is the voltage signal of interest. The assumption is made that the solutions of this equation can be expressed as products of functions:

$$v_m = F(X)G(T) \tag{6-12}$$

When equation (6-12) is subsituted into equation (6-11), one obtains

$$F''(X)G(T) = F(X)G(T) + G'(T)F(X) \tag{6-13}$$

in which the separation of variables is quite simple:

$$\frac{F''(X)}{F(X)} = 1 + \frac{G'(T)}{G(T)} = -\alpha^2 \tag{6-14}$$

where α^2 is dependent neither on X nor on T. Therefore, we now have two ordinary differential equations:

$$F''(X) = -\alpha^2 F(X) \tag{6-15}$$

$$G'(T) = (-\alpha^2 - 1)G(T) \tag{6-16}$$

whose general solutions are

$$F(X) = A \sin \alpha X + B \cos \alpha X \tag{6-17}$$

and

$$G(T) = Ce^{-(\alpha^2+1)T} \tag{6-18}$$

respectively.

The specific solutions depend on the boundary values appropriate to the situation being considered. For example, if a dendrite is connected to a voltage-clamped soma, which provides an electrical short circuit at one end, as shown in Figure 6-7, then one might reasonably apply the following boundary values:

$$v_{\mathrm{m}}(X = 0, T) = 0; \qquad i(X = L, T) = 0 \tag{6-19}$$

the second condition is equivalent to

$$\left.\frac{\partial v_{\mathrm{m}}}{\partial X}\right|_{X=L} = 0 \tag{6-20}$$

The first condition is met only if $B = 0$, which leads to

$$v_{\mathrm{m}}(X, T) = A(\sin \alpha X)e^{-(1+\alpha^2)T} \tag{6-21}$$

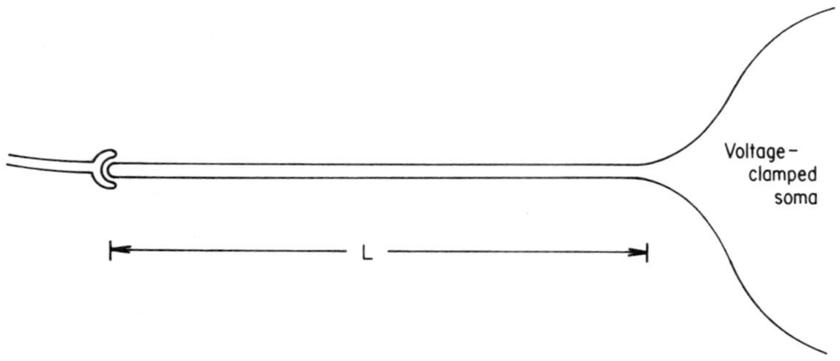

Figure 6-7. Idealized geometry for simple calculations with the cable equations. The synaptic end of the dendrite represents an open circuit, while the clamped soma represents a short circuit on the dendritic cable.

from which

$$\frac{\partial v_m(X,T)}{\partial X} = A\alpha(\cos \alpha X)e^{-(1+\alpha^2)T} \tag{6-22}$$

and

$$\frac{\partial v_m(L,T)}{\partial X} = A\alpha(\cos \alpha L)e^{-(1+\alpha^2)T} = 0 \tag{6-23}$$

or

$$\alpha = \frac{(2n+1)\pi}{2L} \tag{6-24}$$

where n is any integer. Thus, the solution becomes a series of sinusoids in X, each with its own damping factor:

$$v_m(X,T) = \sum_{n=0}^{\infty} A_n(\sin \alpha_n X)e^{-(1+\alpha_n^2)T} \tag{6-25}$$

The remaining boundary-value problem consists of matching $v(X,T)$ with a stimulus waveform (or any other arbitrary spatial waveform) at an arbitrary time (usually defined to be $T = 0$ for convenience). The weighted Fourier coefficients of the spatial waveform become the A_n in equation (6-25). Note that the natural frequencies of this system are given by

$$k_n = -\left[1 + \frac{(2n+1)^2\pi^2}{4L^2}\right]\frac{g_m}{c_m} \qquad 0 \leqslant n \leqslant \infty \tag{6-26}$$

The lowest possible value of k_n(corresponding to the longest time constant) is $-g_m/c_m$, the reciprocal of the "membrane time constant." This time constant is independent of the size of a patch of membrane and typically ranges between 2 and 10 msec. This will always be the longest time constant associated with any passive membrane system, regardless of the complexity of its geometry. Other boundary values can be applied to equations (6-17) and (6-18), leading to a fairly wide variety of solutions similiar to equation (6-26). The methods are presented quite clearly by Rall (1969).

The Laplace-transform method of solving equations (6-4) and (6-5) has the advantage of being simpler to apply in cases of stimulus waveforms with conventional temporal forms, but it has the marked disadvantage of generating solutions that converge slowly if the geometric extent of the system is not large. The transformation will take us from the time domain into the complex frequency domain, replacing $v_m(x,t)$ with $v_m(x,s)$, where s is complex frequency (the independent variable of the Laplace transform).

After transformation, equations (6-7) and (6-10) can be written as ordinary differential equations:

$$\frac{dv_m(x,s)}{dx} = -ri(x,s) \tag{6-27}$$

$$\frac{di(x,s)}{dx} = -g_m v_m(x,s) - sc_m v_m(x,s) \tag{6-28}$$

These can be combined to yield the general equations

$$\frac{d^2 v_m(x,s)}{dx^2} = (rg_m + rc_m s)v_m(x,s) \tag{6-29}$$

$$\frac{d^2 i(x,s)}{dx^2} = (rg_m + rc_m s)i(x,s). \tag{6-30}$$

the general solutions to which can be written

$$\begin{aligned} v_m(x,s) &= v(0,s)\cosh[(rg_m + rc_m s)^{1/2}x] \\ &\quad - [r/(g_m + c_m s)]^{1/2} i(0,s)\sinh[(rg_m + rc_m s)^{1/2}x] \end{aligned} \tag{6-31}$$

$$\begin{aligned} i(x,s) &= i(0,s)\cosh[(rg_m + rc_m s)^{1/2}x] \\ &\quad - [(g_m + c_m s)/r]^{1/2} v(0,s)\sinh[(rg_m + rc_m s)^{1/2}x] \end{aligned} \tag{6-32}$$

To experience the difficulty in applying arbitrary spatial constraints to these solutions, consider the boundary values derived from Figure 6-7 and applied to equations (6-17) and (6-18). When applied to equations (6-31) and (6-32), these boundary values lead to the totally useless solutions

$$v_m(x,s) = 0; \qquad i(x,s) = 0$$

On the other hand, limited spatial constraints *can* be applied to equations (6-31) and (6-32). Normally, one defines $x = 0$ to be the point at which a temporal input waveform is known. Suppose, for example, that a synapse delivered a known current across the membrane at the distal end of the dendrite in Figure 6-7. Assuming that the soma was nearly a perfect short circuit at $x = l$, one could write

$$v_m(l,s) = 0; \qquad i(0,s) = I_{syn}(s) \tag{6-33}$$

Combining the left-hand constraint with equation (6-31), we have

$$v_m(0,s) = [r/(g_m + sc_m)]^{1/2} i(0,s)\tanh[(rg_m + rc_m s)^{1/2} l] \tag{6-34}$$

which can be combined with equation (6-31) to give $v(x, s)$ in response to $i(0, s)$, or with (6-32) to give $i(x, s)$ in response to $i(0, s)$.

Suppose one is interested in the current flowing into the soma in response to the synaptic current at the distal end of the dendrite. Using equation (6-32), one would have

$$i(l, s) = I_{\text{syn}}(s)\,\text{sech}[(rg_{\text{m}} + rc_{\text{m}}s)^{1/2}l] \tag{6-35}$$

Thus, once $I_{\text{syn}}(t)$ is known, one in principle could transform it to $I_{\text{syn}}(s)$ [perhaps aided by tables of integral transforms, such as Erdélyi *et al.* (1954)], substitute the transform into equation (6-35), and then take the inverse transform (perhaps again with the aid of tables of integral transforms) to determine $i(l, t)$. For example, if $I_{\text{syn}}(t)$ is an impulse delivered at $t = 0$, then

$$I_{\text{syn}}(s) = Q = \text{charge delivered by current impulse} \tag{6-36}$$

and, from Table 6-1, which shows a few useful transforms,

$$\begin{aligned} i(l, t) = {} & Q(rc_{\text{m}}l^2/\pi t^3)^{1/2} \sum_{n=1}^{\infty} (-1)^{n-1}(2n-1)\exp[-(2n-1)^2 rc_{\text{m}}l^2/4t] \\ & \times \exp[-(g_{\text{m}}/c_{\text{m}})t] \end{aligned} \tag{6-37}$$

the first few terms of which are

$$\begin{aligned} i(l, t) = {} & Q(rc_{\text{m}}l^2/\pi t^3)^{1/2} e^{-(g_{\text{m}}/c_{\text{m}})t}[e^{-rc_{\text{m}}l^2/4t} - 3e^{-9rc_{\text{m}}l^2/4t} \\ & + 5e^{-25rc_{\text{m}}l^2/4t} - \cdots] \end{aligned} \tag{6-38}$$

Clearly, unless l is sufficiently large to allow the factor $e^{-(g_{\text{m}}/c_{\text{m}})t}$ to dominate at small values of time, many terms of the series will be required for convergence to a reasonably accurate solution. This is always the case with solutions derived by Laplace transforms for distributed systems. In fact, the only truly simple form of the solution is for $l \to \infty$. In that case, we have the boundary conditions

$$\lim_{x\to\infty} v_{\text{m}}(x, s) = 0; \qquad \lim_{x\to\infty} i(x, s) = 0 \tag{6-39}$$

which require that

$$v_{\text{m}}(0, s) = [r/(g_{\text{m}} + c_{\text{m}}s)]^{1/2} i(0, s) \tag{6-40}$$

and leads to the solutions

$$v_{\text{m}}(x, s) = v_{\text{m}}(0, s)\exp[-(rg_{\text{m}} + rc_{\text{m}}s)^{1/2}x] \tag{6-41}$$

$$i(x,s) = i(0,s)\exp[-(rg_{\mathrm{m}} + rc_{\mathrm{m}}s)^{1/2}x] \tag{6-42}$$

For a current impulse $I_{\mathrm{syn}}(s) = Q$, at $x = 0$, these solutions can be transformed with the aid of Table 6-1:

$$i(x,t) = [Q(rc_{\mathrm{m}}x^2)^{1/2}/2(\pi t^3)^{1/2}]e^{-rc_{\mathrm{m}}x^2/4t}e^{-(g_{\mathrm{m}}/c_{\mathrm{m}})t} \tag{6-43}$$

$$v_{\mathrm{m}}(x,t) = [Qr/(rc_{\mathrm{m}}x^2\pi t)^{1/2}]e^{-rc_{\mathrm{m}}x^2/4t}e^{-(g_{\mathrm{m}}/c_{\mathrm{m}})t} \tag{6-44}$$

Note that the current and voltage responses both diminish extremely rapidly with distance (x) from the source, the dominant spatial attenuation term in each case being

$$e^{-rc_{\mathrm{m}}x^2/4t}$$

It also is interesting to compare the two response functions with the normal, or Gaussian, probability density:

$$p(x) = \frac{1}{\sigma(2\pi)^{1/2}}e^{-x^2/2\sigma^2} \tag{6-45}$$

where σ is the standard deviation. At any instant in time, the distribution of the response along the fiber is very close to being Gaussian with standard

TABLE 6-1. A Few Laplace-Transform Pairs Useful for Cable Analysis

Function of s	Function of time
$F(s)$	$f(t)$
$G(s)$	$g(t)$
$F(s + G(s)$	$f(t) + g(t)$
$dF(s)$	$af(t)$
$F(s + a)$	$e^{-at}f(t)$
$F(as)$	$(1/a)f(t/a)$
$dF(s)/ds$	$-tf(t)$
1	unit impulse at $t = 0+$
$1/s$	unit step at $t = 0+$
$1/s^{1/2}$	$1/(\pi t)^{1/2}$
$\exp[-s^{1/2}]$	$(\exp[-1/4t])/2\pi^{1/2}t^{3/2}$
$(\exp[-s^{1/2}])/s$	$\mathrm{Erfc}\,[1/(4t)^{1/2}]$
$\mathrm{csch}\,[s^{1/2}]$	$[2/\pi^{1/2}t^{3/2}]\sum_{n=0}^{\infty}(n+\frac{1}{2})\exp[-(n+\frac{1}{2})^2/t]$
$\mathrm{sech}\,[s^{1/2}]$	$[1/(\pi t^3)^{1/2}]\sum_{n=0}^{\infty}(-1)^n(2n+1)\exp[-(2n+1)^2/4t]$
$s^{1/2}\exp[-s^{1/2}]$	$(1-2t)(\exp[-1/4t])/(4\pi^{1/2}t^{5/2})$

deviation proportional to the square root of time elapsed since the application of the impulse. The current and voltage are distributed according to the normal function multiplied and divided, respectively, by the distance from the point of impulse application.

The impulse is the most rapid transient stimulus that can be applied to the system (in fact, its application can only be approximated). It is interesting to contrast the responses to this transient with those to the slowest stimulus, a steady current. Returning to equations (6-7) and (6-10), we have $\partial v_m/\partial t = 0$, and thus

$$\partial^2 v_m/\partial t^2 = rg_m v_m \tag{6-46}$$

whose solution for an infinite dendrite is

$$v_m(x) = v_m(0)\exp[-(rg_m)^{1/2}x] \tag{6-47}$$

$$i(x) = i(0)\exp[-(rg_m)^{1/2}x] \tag{6-48}$$

In this case the response diminishes exponentially with the distance from the source. One can define a "space constant," $1/(rg_m)^{1/2}$, which is the distance over which the response decreases by a factor of $1/e$. Similarly, all the solutions derived by Laplace transform for time-varying sources contain the factor

$$e^{-(g_m/c_m)t} \tag{6-49}$$

In this case we have the signal or part of the signal diminishing exponentially with time after the beginning of the source stimulus. One can define a "time constant," c_m/g_m. This is the so-called "membrane time constant" and is the longest time constant appearing in the solutions obtained by separation of variables.

It was mentioned at the beginning of this section that the analysis herein was originally applied to electrical cables. The modern extension of "cable theory" is the theory of transmission lines, and many of the shortcuts and tools developed in that field are applicable to dendrites. One interesting notion from transmission lines is that of reflected waves. For example, one can interpret the various terms of equation (6-38) as wavefronts that have made $2n - 1$ passes over the length of the dendrite. The term

$$e^{-rc_m l^2/4t} \tag{6-50}$$

represents a wave that has made a single pass from $x = 0$ to $x = l$; according to (6-43), this term gives the spatial attenuation expected for a waveform that has passed over this distance. The term

$$e^{-9rc_m l^2/4t} \tag{6-51}$$

represents a wavefront that has made three passes along the dendrite, beginning at $x = 0$ and ending at $x = l$; once again, the attenuation is appropriate according to equation (6-43). One can perfectly reasonably conceive of this wavefront as having been initiated at $x = 0$ by the stimulus at that point, having passed along the dendrite to $x = l$, having been reflected back down the dendrite from that point and back again from the point $x = 0$, and finally returning to $x = l$ after having made three passes over the length of the dendrite, two coming toward $x = l$ separated by one return to $x = 0$. All the remaining terms in equation (6-38) can be given similar interpretation.

6.5. Continuous Analysis of Branching Dendritic Trees: Rall's Equivalent Cylinder Model

The abrupt dendritic termination at the synapse in Figure 6-7 represents a discontinuity as far as signals spreading down the dendrite from the soma are concerned. Similarly, for signals spreading in the other direction, the soma itself represents a discontinuity. When the cable equations were applied to that situation, the two discontinuities imposed important constraints on the distributions of v_m and i along the dendrite. In the convenient terms of transmission-line theory, the discontinuities can be conceived as generating reflections of the waves or signals spreading toward them. Thus, at any point along a fiber, v_m and i each can be considered to comprise the sum of a wave moving forward from the source (e.g., from a receptor site) and waves reflected back and forth from various discontinuities along the fiber, and in this manner the constraints imposed by those discontinuities can be met.

In the case of a uniform, infinitely long fiber with a source connected at one end, there are no discontinuities in the path of the signal spreading down the fiber from the source, so there are no reflected waves. On the other hand, if a uniform fiber is terminated abruptly, as in the example of Figure 6-7, reflections arise, and they complicate the analytical solutions quite drastically [e.g., compare equation (6-38) with equations (6-43) and (6-44)]. Signals spreading down the dendrite from the soma generate reflections at the end of the dendrite which spread back toward the soma. At the soma, these reflected waves in turn are reflected back in the other direction. Thus, waves bounce back and forth indefinitely between the soma and the other end of the dendrite.

In this way, even a single discontinuity in the path of a wave from the source will generate reflections back toward the source, and these in turn will bring into play any discontinuities at the source or behind it.

Clearly, all dendrites have finite lengths and thus have discontinuities at either end which in themselves may create significant analytical difficulties. In addition, many dendritic structures consist of multiply branching fibers, and each branch point is a possible discontinuity and source of reflections for waves spreading in either direction past it. In fact, only under very special circumstances will a branch point not generate reflected waves. Transmission-line designers have known about these circumstances for many years, and they have developed design criteria to meet them. To see how a nonreflecting branch point can be designed, consider the infinitely long, uniform fiber, which exhibits no reflections and which we therefore wish to emulate. Over its entire length, the infinite, uniform fiber exhibits a constant, intrinsic relationship between v_{m} and i. One can find the Laplace transform of this relationship quite easily by combining equations (6-40), (6-41), and (6-42):

$$v_{\mathrm{m}}(x,s) = [r/(g_{\mathrm{m}} + c_{\mathrm{m}}s)]^{1/2} i(x,s) \tag{6-52}$$

In this case, the intrinsic relationship is embodied in the term $[r/(g_{\mathrm{m}} + c_{\mathrm{m}}s)]^{1/2}$, which often is called the *characteristic impedance*. If one were to transect an infinitely long uniform fiber at any point along its length and connect an electrical source to the cut end of the remaining infinite segment (as depicted in Figure 6-8a), the electrical impedance presented by the fiber to the source simply would be the characteristic impedance of the fiber. If an equal impedance, regardless of how it was constructed, were connected to the end of a finite segment of the fiber (as depicted in Figure 6-8b), then as far as signals spreading over the fiber segment toward the impedance are concerned, the fiber would appear exactly as though it were infinitely long. In other words, when the fiber segment is terminated in its own characteristic impedance, no reflections are generated at that termination by signals spreading toward it along the fiber.

Suppose that several uniform, infinitely long fibers join a single, finite, uniform fiber segment at a branch point (Figure 6-8c). If the impedance looking simultaneously into all of those infinite fibers is equal to the characteristic impedance of the fiber segment, then the branch point itself is intrinsically nonreflective with respect to signals spreading toward it along the finite segment; it will not generate reflections of those signals. Since the infinite branches represent parallel loads on the fiber segment, the total impedance that they represent will be

$$1/\sum (1/Z_{oi}) \tag{6-53}$$

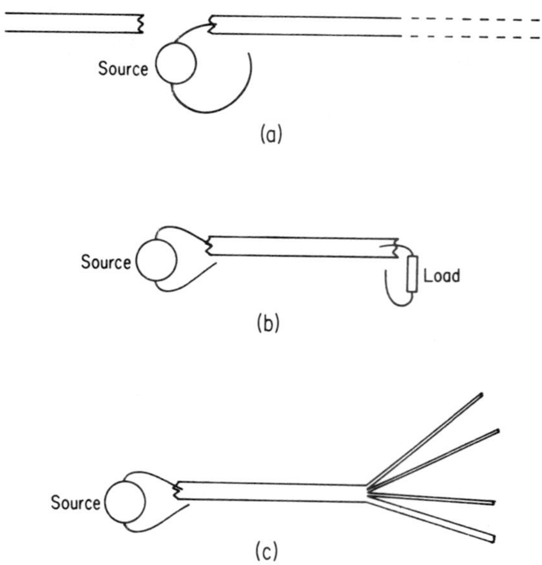

Figure 6-8. Idealization of the effects of branching in dendritic trees. (a) An electrical source connected to the broken end of an infinitely long dendrite. (b) The same system, but with most of the dendrite replaced by a simple, lumped-load impedance; if the impedance were equal to that looking into the infinite dendrite, no reflections would occur at the load. (c) The same system, but with four infinitely long branches replacing the lumped load.

where Z_{oi} is the characteristic impedance of the ith branch. Therefore, we can write the following criterion for a branch point that is intrinsically nonreflective of signals spreading toward it along a given fiber:

$$[r/(g_m + c_m s)]^{1/2} = 1/\sum \{1/[r_{oi}/(g_{moi} + c_{moi} s)]^{1/2}\} \tag{6-54}$$

where r, g_m, and c_m are the parameters of the fiber along which the signal is approaching the branch point; r_{oi}, g_{moi}, and c_{moi} are the parameters of the ith branch through which the signal will leave the branch point; and the summation is carried out over all such branches.

It is clear from equation (6-54) that if a branch point is intrinsically nonreflective with respect to signals spreading along one of the fibers leading into it, it certainly will be reflective of signals spreading toward it along any one of the other fibers. In other words, the branch point can be intrinsically nonreflective with respect to only one of the fiber segments connected to it. Signals spreading toward the branch point along the other paths will automatically generate reflections. Furthermore, since the ultimate dendritic branches are not infinite in length, but end blindly at distances from the soma that are quite finite, they will generate reflections of signals spreading into them from the soma. These reflections in turn will converge back on the branch points in the dendritic tree, generating further reflections back toward the periphery, and so on. Since such reflections cannot be avoided, the next best thing, with respect to making models of dendritic trees analytically tractable, is to design the models in such a way that the reflections are simplified. One way to do this is to make them coherent and apparently generated by a single discontinuity at the extreme periphery of the tree. In that case, the entire dendritic tree can be represented as a single (unbranched) equivalent dendrite, with one end connected to the soma and the other end sealed.

Rall was the first to point out the equivalent-dendrite model and the geometric constraints it imposes. By considering situations in which the extracellular components of r and r_{oi} in equation (6-54) are negligible compared to the intracellular components, and by assuming that the fibers under consideration were cylindrical, he reduced the criterion of an intrinsically nonreflective branch point to a purely geometric relationship:

$$d^{3/2} = \sum d_{oi}^{3/2} \tag{6-55}$$

where d is the diameter of the fiber leading into the branch point from the direction of the soma, and d_{oi} is the diameter of the ith fiber leading out of the branch point, away from the soma. He also pointed out that to ensure coherent reflections from the periphery of the tree, all blind terminations must be at the same "electrotonic distance" from the soma, where the electrotonic distance is given in number of space constants [see discussion surrounding equation (6-47)]. Once again, for cylindrical fibers, this criterion can be reduced to geometry. The sum

$$\sum L_i/\sqrt{d_i} \tag{6-56}$$

must be the same for all direct paths from the soma to the periphery, where L_i is the length of the ith fiber segment in the path, d_i is the diameter of that segment, and the summation is taken over all segments in a given path, beginning at the soma, always moving toward the periphery, and ending with the inclusion of the blindly terminated peripheral fiber. Originally, Rall demonstrated that the constraints of equations (6-55) and (6-56) led to coherence of all reflections of waves originating at the soma or simultaneously at current sources distributed over the dendritic tree in a symmetrical manner with respect to the soma (i.e., on every branch of the tree at the same electrotonic distance from the soma). Later, with a clever superposition argument, Rall and Rinzel (1973) were able to demonstrate that as far as reflections to the soma are concerned, the constraints of equations (6-55) and (6-66) guarantee coherence for waves initiated at a single current source located anywhere in the dendritic tree or at the soma. Thus, the equivalent dendrite (usually called the *equivalent cylinder*) can be used to deduce waveforms at the soma in response to a current source in the dendritic tree.

Although the equivalent-cylinder model itself is an analytical convenience, the constraints it imposes conform reasonably well to observed geometries in dendritic trees, and the multiple reflections generated when those constraints are relaxed have a much smaller effect on the deduced dynamic behavior of the tree than they do on the analysis required to deduce the behavior.

In the 1950's, electrophysiologists had begun to penetrate neuronal somata and to stimulate them intracellularly. Several of them soon observed what they considered to be anomalous transients in the observed membrane potentials. As long as the dendritic tree connected to the soma is reasonably symmetrical, and as long as the three-halves power law of equation (5-55) is not grossly disobeyed, the equivalent cylinder model is a reasonable analytical tool for this experimental situation. Rall (1960) applied it and by doing so demonstrated that the observed transients were indeed to be expected. The apparent anomaly resulted simply from the fact that both redistribution and annihilation of dipoles were present. The surplus of dipoles established in the soma by the stimulus was redistributed rather rapidly over the dendritic tree (represented analytically by the equivalent cylinder model), while the annihilation of surplus dipoles by transmembrane ion flow took place more slowly. Thus, the stimulated voltage or current did not exhibit a simple exponential decay.

6.6. Spatially Discrete Analysis of Passively Conducting Fibers and Fiber Trees

Once the geometry becomes sufficiently complicated to make spatially continuous analysis intractable, one can turn to discrete representations of fibers and fiber trees. Each segment of a dendritic tree can be represented as a finite ladder network of the form illustrated in Figure 6-6. One then might proceed with continuous-time analysis, employing standard ladder-network or matrix methods; one might construct a lumped analog simulation with resistors and capacitors; or he might go directly to discrete-time simulation on a digital computer. The latter two approaches are particularly attractive for nonlinear situations where discrete network analysis becomes virtually as intractable as spatially continuous analysis.

The basic equations of the ladder network are derived very easily from Figure 6-6. For the signal (v_i) at the ith stage, we have

$$\frac{dv_i}{dt} = -v_i\left(\frac{g_i}{c_i} + \frac{1}{r_{i+1}c_i} + \frac{1}{r_i c_i}\right) + v_{i-1}\left(\frac{1}{r_i c_i}\right) + v_{i+1}\left(\frac{1}{r_{i+1}c_i}\right) \quad (6\text{-}57)$$

This expression can be simplified by the use of Rall's (1964) notation,

$$\mu_{ij} = \frac{1}{r_{ij} c_i} \tag{6-58}$$

$$\mu_{ii} = -\sum_{\substack{j=1 \\ j \neq i}}^{m} \frac{1}{r_{ij} c_i} \tag{6-59}$$

where r_{ij} is the sum of the intracellular and extracellular resistances between stages i and j, and r_{ij} between nonadjacent stages is taken to be infinite, making μ_{ij} zero for nonadjacent stages. Equation (6-57) now becomes

$$\frac{dv_i}{dt} = \sum_{j=i}^{N} \mu_{ij} v_j - \frac{g_i}{c_i} v_i \tag{6-60}$$

where N is the total number of stages in the model (but the only nonzero terms in the sum are those for stages adjacent to the ith stage). In Figure 6-6 the ith stage is depicted as having only two neighbors, i.e., $i + 1$ and $i - 1$. On the other hand, the ith stage could represent a branch point in the dendritic tree, in which case it would have more than two immediately adjacent stages. Equation (6-60) covers this more general case.

If g_i remains constant for each of the n stages in the model (i.e., if synaptic or sensory modulation of dendritic membrane conductance is negligible or is ignored), then continuous-time analysis using Laplace transforms might be feasible. In carrying out such an analysis, one can follow the established algorithms for ladder networks, beginning at the stages in the model most remote from the signal source and working gradually, branch by branch, back toward the source. If more than one source is considered, solutions are obtained independently and then summed (since the model is linear, the principle of superposition applies). For each branch in the model, the following algorithms and definitions are useful [see Lewis (1974) for derivations]:

$$Y_i = sc_i + g_i \tag{6-61}$$

$$Z_i = r_{i,i+1} \tag{6-62}$$

$$i_i = \text{net flow of dipoles through } r_{i,i+1}$$

where increased value of the stage index indicates increased remoteness from the source, and the nth stage is most remote from the source,

$$
\begin{aligned}
v_i/v_n &= 1/(D_i)_n \\
i_i/v_n &= 1/(N_i)_n \\
v_i/v_j &= (D_i)_n/(D_j)_n \\
v_i/i_j &= (D_i)_n/(N_j)_n \\
i_i/i_j &= (N_i)_n/(N_j)_n
\end{aligned}
\tag{6-63}
$$

where N and D are polynomials in s, obtainable as follows: $(D_i)_n$ is formed by taking the sum of all possible products of alternating Y's and Z's between and including stage n and stage i, beginning with a Y, always moving toward stage i, and ending with a Z. For example, two of the products in this sum might be

$$Y_n Z_n Y_{n-1} Z_{n-4} Y_{n-5} \cdots Z_i$$

and

$$Y_{n-3} Z_{n-3} Y_{n-5} Z_{n-5}$$

whereas

$$Y_{n-2} Z_{n-3} Y_{n-3} Z_i$$

would not be included, since Y_{n-3} is on the wrong side of Z_{n-3}, and we have not moved consistently toward stage i in forming the product. $(N_i)_n$ is formed by taking the sum of all possible products of alternating Y's and Z's between and including stages n and i, beginning with a Y, always moving toward stage i, and ending with a Y.

If the fiber under consideration is uniform, then all the Y's and Z's in the model of that fiber are the same. In that case, $(D_i)_n$ and $(N_i)_n$ take on very simple forms:

FORMULAS FOR LADDER NETWORKS WITH n IDENTICAL STAGES

$$
\begin{aligned}
(N_i)_n &= \sum_{k=0}^{n-i} \binom{2n-2i-k+1}{k} Y(ZY)^{n-i-k} \\
(D_i)_n &= 1 + \sum_{k=1}^{n-i} \binom{n-i+k}{2k} (ZY)^k
\end{aligned}
\tag{6-64}
$$

where $\binom{a}{b}$ is the binomial coefficient given by $a!/(a-b)!\,b!$ These expressions provide rather simple alternatives to the results of analysis of the spatially continuous, uniform fiber of finite length.

It is clear from the preceding algorithms that standard ladder-network methods will rapidly become intractable as the geometry of the model becomes complicated. However, as Rall (1964) recognized, the basic equation (6-60) of the model is the same as that for a linear, multicompartmental system; and it can be solved in the same manner, namely, by matrix manipulation. In that case, the various v_i can be treated as elements of a state column vector **V**; and if μ_{ii} is redefined to include the term $-g_i/c_i$, the various μ_{ij} can be treated as the elements of a square matrix **M** relating the rate of change of **V** to **V** itself:

$$\frac{d\mathbf{V}}{dt} = \mathbf{MV} \tag{6-65}$$

Since the model basically is a resistive-capacitive electrical network, it must possess the well-known properties of such networks, and among these is the property that the eigenvalues of the matrix **M** are distinct, negative real numbers. The dynamic behavior of any of the state variables (the various v_i) of the model will be given by a linear combination of exponentials $e^{m_i t}$, where the m_i are the eigenvalues of **M**. The coefficients of the exponential terms are determined by the boundary conditions (i.e., by the description of the signal source), and the eigenvalues of **M** can be determined by well-established methods that are easily implemented on a digital computer.

If the model is to include the effects of conductance changes occurring at various receptor sites distributed over a fiber or fiber tree, then linear analysis cannot be employed. For such situations, equation (6-60) can be modified to include receptor currents and to include g_i as a function of time for all stages representing receptor sites. Generally, formal analysis of the modified equation in such cases will not be possible. However, analog simulation, employing multipliers to form the products of the time functions v_i and g_i, can be carried out if the total number of stages is not too great for realization on available facilities. If analog simulation is not feasible, then time can be treated as a discrete variable and the model can be simulated on a digital computer. Although simulation results often are very specific in nature and lack the generality usually present in formal analytical results, they can nonetheless be quite revealing and are generally considerably better than no results at all. Rall (1964) was the first to carry out such simulations, and he based them on a variation of equation (6-60) similar to the following:

$$\frac{dv_i}{dt} = \sum_{j=1}^{N} \mu_{ij} v_j - \left(\frac{g_{i0} + \sum_{r=1}^{k} g_{ir}(t)}{c_i} \right) v_i + \sum_{r=1}^{k} g_{ir}(t) V_r \tag{6-66}$$

where g_{i0} is the membrane conductance of the ith stage in the absence of receptor modulation, $g_{ir}(t)$ is the component of membrane conductance induced by stimulation of the rth receptor site associated with the ith stage, and there are k such sites at the ith stage.

In his simulation, Rall considered primarily a model with 10 stages arranged in a row, with the first stage representing the soma. A few of his

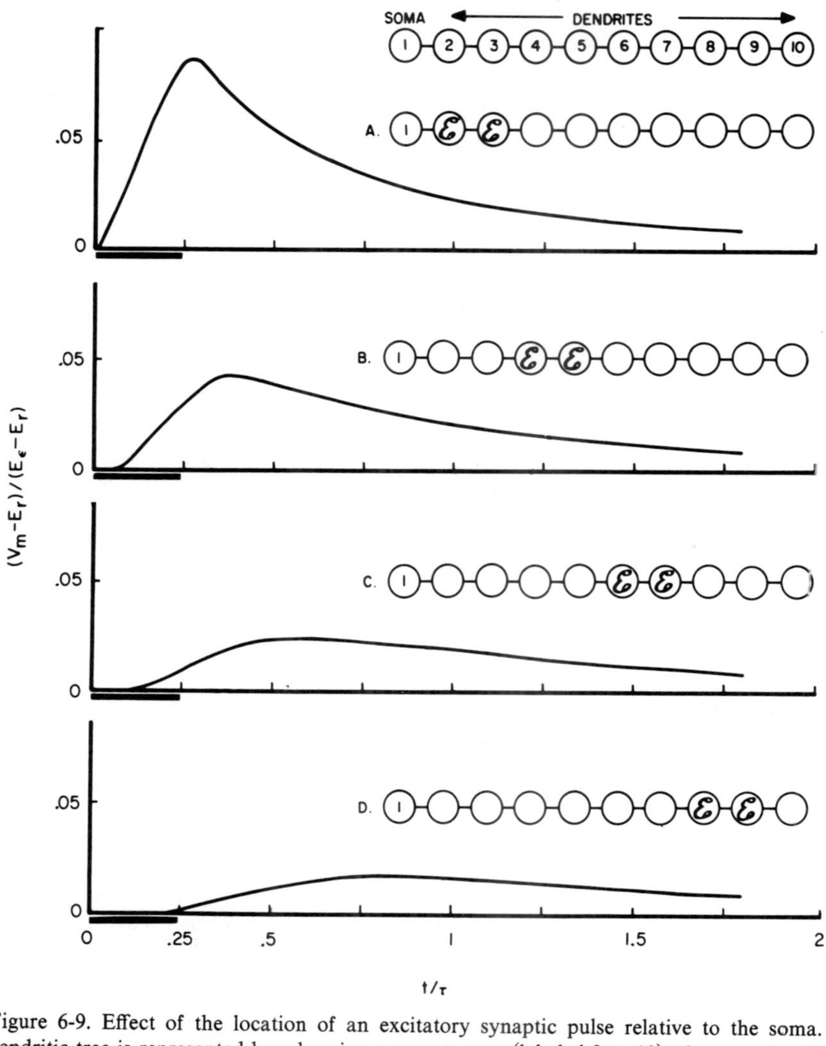

Figure 6-9. Effect of the location of an excitatory synaptic pulse relative to the soma. A dendritic tree is represented here by nine compartments (labeled 2 to 10); the soma to which the tree is connected is represented by a single compartment, labeled 1. The location of the synaptic pulse is indicated by $\mathcal{E}$. The time (horizontal) axis is normalized with respect to the membrane time constant. The vertical axis is the deduced amplitude of the synaptic response observed at the soma. [From Rall (1964).]

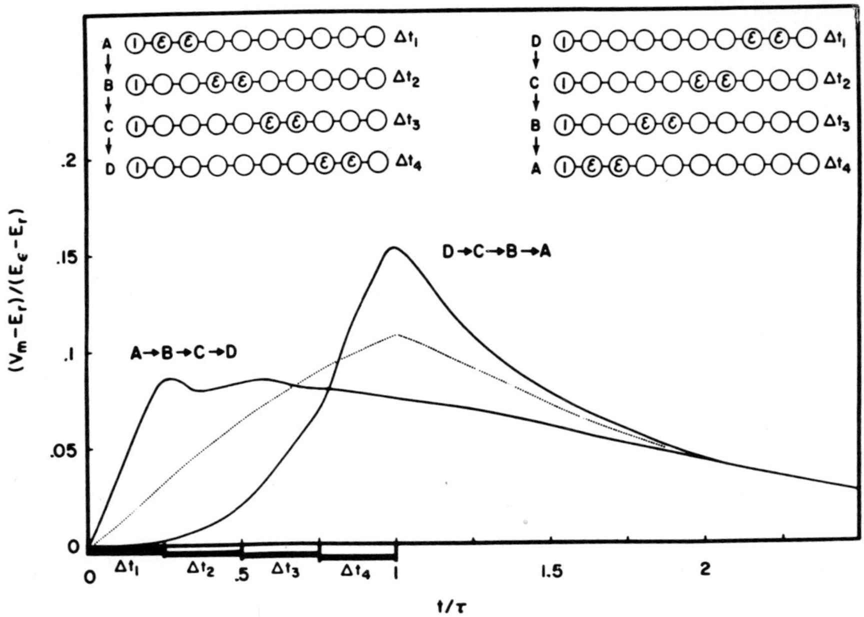

Figure 6-10. Effects of two spatiotemporal sequences of excitatory synaptic pulses over the dendritic-tree model of Figure 6-9. Clearly, according to the model, the sequence involving excitation successively closer to the soma generates a response of larger amplitude than does the sequence involving successively more remote excitations. The dotted line shows the response deduced for the same average excitation, uniformly distributed over the compartments 2 to 9. This too produces a response of lower amplitude than that of the sequence moving toward the soma. [From Rall (1964).]

results are summarized in Figures 6-9 to 6-11. In the course of his research, Rall came to two rather significant, qualitative conclusions. First, he found that, in general, the nonlinear effects of receptor-modulated conductances were important with respect to interactions at a given stage of the model (i.e., when more than one receptor site associated with a given stage was stimulated), but were relatively unimportant for interactions of remotely stimulated responses (i.e., when stimulated receptor sites were associated with different stages). In the latter case, the linear model provided a reasonably good approximation. His second conclusion relates to the possible roles of receptor sites on dendrites as opposed to those on somata. In his own words (Rall, 1964):

> A simplified treatment of synaptic excitation and inhibition leads to the theoretical prediction of a functional distinction between somatic and dendritic synapses: Dendritic synaptic activity would be expected to dominate slow adjustments of the

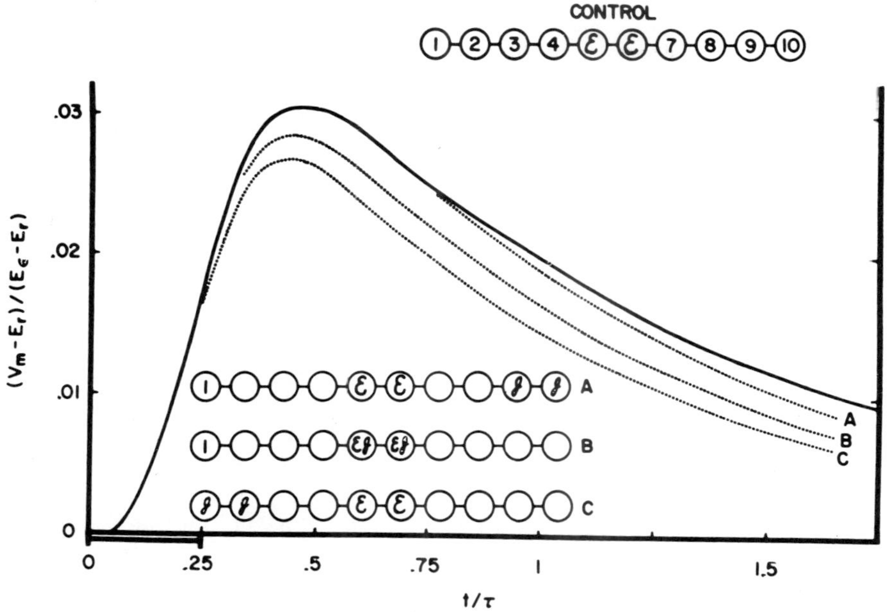

Figure 6-11. Spatiotemporal interactions of an excitatory synaptic pulse and a steady inhibitory synaptic background. The solid curve is the control response, excitatory pulse alone (in compartments 5 and 6). Dotted curves show the effects on this response of steady, low-level inhibitory synaptic conductance at three different locations, as indicated by $\mathcal{I}$.

background excitation level, while somatic synapses would be best suited for triggering of impulses.

The theoretical results that I am about to sketch for you indicate that the dendrites can indeed make an important contribution to the receptive and integrative functions of the neuron. In fact, the dendritic tree can be credited with a dominant role in determining the background level of the neuronal excitatory state that waxes and wanes with the amount of (spatiotemporally) scattered synaptic excitatory and inhibitory background activity delivered to the neuron's receptive surface. On the other hand, a brief increment of synchronous synaptic activity would be most effective when delivered to the soma and proximal dendritic surface.

The theoretical results can be seen to suggest a general functional distinction between dendritic and somatic synaptic activity. In contrast with the relatively sluggish and finely graded depolarization that would spread to the trigger zone from the dendritic periphery, brief synchronous activity of a few synapses at or near the soma could produce precisely timed and relatively sharp depolarizations. These would be well suited to precisely timed triggering of axonal impulses; their amplitudes might well be too small to reach threshold without the help of earlier dendritic synaptic activity. Thus the dendritic activity could determine success or failure in the onward transmission of temporal synaptic patterns delivered at the soma. A different mode of behavior would result when sustained dendritic synaptic activity was sufficient to cause a rhythmic discharge of impulses (similar to the action of a generator potential in a sensory neuron); in such a case, synaptic inhibition at the soma could block the discharge of impulses over specific periods, or it could reduce the frequency of rhythmic discharge.

6.7. Shapes of Passively Conducted Signals

In response to a stepwise change in current at a source, the passively conducted signal will exhibit an inflected, or sigmoid, rise toward a final steady value. The degree of inflection and the attenuation of the response both increase with distance along a fiber from the source. In response to a current impulse, or a current pulse that is very brief relative to the membrane time constant, the passively conducted signal will exhibit an inflected rise followed by a falling phase that is more complicated than a simple, decaying exponential. Proportionately, the initial part of the falling phase is steepest, and later parts become progressively less steep. The response is always skewed toward longer times, the falling phase always being more prolonged than the rising phase. This skewing is most pronounced at or near the source and gradually diminishes with increasing distance from the source. In the limit, as the distance from the source becomes very large, the response approaches (but never achieves) symmetry, with the duration of the rising phase approaching that of the falling phase. At these same distances, of course, attenuation becomes extreme. Figure 6-12 shows examples of these various shape changes along a uniform, infinitely long fiber. Recently, Jack and Redman (1971) and others have attempted to catalog the various shape changes that one might expect in synaptic responses passively conducted over various distances along dendrites. In doing so, they have made extensive use of the Laplace-transform methods.

6.8. Conduction of Signals to Very Remote Sites

Apparently, dendrites seldom are longer than one or two length constants. Therefore, they are not modeled especially well by an infinitely long cable, and equations (6-44) and (6-47) will not be appropriate. On the other hand, many axons are sufficiently long and sufficiently uniform that the model should be quite reasonable for signals conducted along them passively. In that case, equation (6-44) should provide a good description of the fate of a very brief signal as it spreads passively along the axon, and equation (6-47) should provide a reasonably good description of the fate of a spreading dc, or prolonged, signal. Clearly, in either case the signal

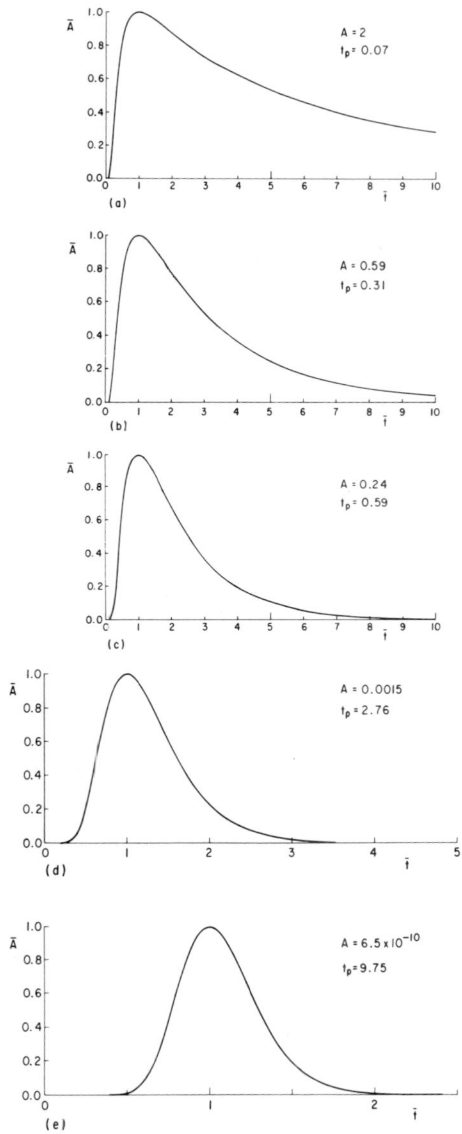

Figure 6-12. Shape of a passively conducted waveform as a function of distance from source along an infinite, uniform cable. The waveforms are presented on normalized axes so that their shapes can be compared easily. The normalization factor A for the vertical axis $\overline{A}$ is given in each case; the ratio of the amplitudes of any two responses is simply given by the ratio of the corresponding A's. The normalization factor for time in each case is t_p, which is the time from the stimulus (a current impulse) to the peak of the response, given in number of membrane time constants. The distances from the source are as follows: (a) 0.2 space constants; (b) 0.5 space constants; (c) 0.8 space constants; (d) 3.0 space constants; (e) 10 space constants. Curves were computed from equation (6-44).

diminishes rather rapidly as it spreads from its point of origin (i.e., either exponentially as x^2 or exponentially as x).

Considering a cylindrical fiber with typical parameters as an example, one can very quickly determine approximately the space constants to be expected in axons. Cytoplasm resistivity is typically close to 100 Ω-cm, so even if we neglect extracellular current paths, we have

$$r = 100/(\pi d^2/4)\,\Omega/\text{cm} \tag{6-67}$$

where d is the diameter of the fiber in centimeters. Neuronal membranes typically have resting conductances close to 0.25 mmho/cm^2, which leads to

$$g_{\text{m}} = 0.00025 \times 2\pi d \text{ mho/cm} \tag{6-68}$$

and

Typical space constant

$$1/(rg_{\text{m}})^{1/2} = (5d)^{1/2}\,\text{cm} \tag{6-69}$$

For a huge fiber, such as the squid giant axon, d typically is close to 0.1 cm, in which case the space constant should be close to 0.7 cm. Thus, a prolonged signal spreading passively along that giant would decay by a factor of $1/e$ in every 0.7 cm or so. Clearly, the signal would be vanishingly small by the time it reached the end of such an axon even a few centimeters long, let alone an axon spanning the half-meter tail section of a moderately small squid. Most axons are even smaller, of the order of a micrometer in diameter. In such cases, the space constant would be a few hundredths of a centimeter, and passively spreading signals would not travel very far at all.

Of course, conduction of signals in axons is not passive. The signals are continually reinforced as they progress along the fiber. Models of the mechanisms underlying this reinforcement are presented in the next chapter. As we conclude that chapter, we shall return to the basic formulations presented in Chapter 6, but they will be modified considerably.

Suggested Further Readings

For further readings we refer the reader to Jack, Nobel, and Tsien (1973); Rall (1969); Taylor (1963); Erdélyi *et al.* (1954); and Roberts and Kaufman (1966).

7

Models of Spike Generation and Conduction

The previous chapters in this section have discussed the generation and passive conduction of signals in neurons, noting that passively conducted signals are attenuated very severely even over short distances. When such signals are excitatory, they may lead to the generation of large-amplitude spikes, which in turn are continually provided with energy as they propagate and thus are conducted over essentially unlimited distances without attenuation. The low-amplitude signals that are excitatory and thus tend to trigger spikes are often called *generator potentials*. In this chapter, we shall discuss several models related to the triggering of spikes by generator potentials and the subsequent attenuationless propagation of those spikes along neuronal fibers. We will progress from the iron-wire analogy of the 1920s, through the threshold/accommodation models of the 1930s, to the Hodgkin–Huxley model of the early 1950s.

The first model was based on observed analogous behavior of chemically treated iron and nerve fibers capable of active conduction (i.e., conduction of spikes). Models of the second series were mathematical abstractions, with little or no basis in underlying mechanism, but with remarkable ability to predict the efficacy of almost any signal with respect to the triggering of a spike. The final model, that of Hodgkin and Huxley, was based on observed, dynamic changes in chord conductances of the membrane of the squid giant axon prior to, during, and following a spike.

7.1. The Iron Wire (or Heathcote–Lillie) Model

In 1922, Lillie pointed out that in at least 14 ways, the generation and conduction of spikes was essentially the same as the generation and conduction of a phenomenon known as *activation* over iron wire that had been specially chemically treated (*passivated*):

1. Both systems (nerve and passivated iron wire) consist of a conductive core inside an insulating or semiinsulating sheath surrounded by an electrolytic solution.
2. Local chemical, mechanical, or electrical alterations may initiate a propagated disturbance in either system.
3. Once initiated, the propagated disturbance in either system has a form and velocity determined by the system itself and independent of the characteristics of the stimulus.
4. Local regions in both systems recover following the passage of a disturbance.
5. In both systems propagation is associated with local variations of electrical potential and local current flow.
6. The propagation velocities in both systems are of the same order of magnitude and exhibit similar, strong temperature dependences.
7. In both systems propagation may be blocked by certain chemical or electrical alterations.
8. Both systems exhibit stimulus thresholds for initiation of the propagated disturbance.
9. In both systems effects of subthreshold stimuli can accumulate and trigger a propagated disturbance.
10. Following the passage of a propagated disturbance, both systems exhibit temporary refractory periods during which their thresholds are higher than normal.
11. In either system an electric stimulus may excite or inhibit, depending on its polarity.
12. Inhibitory electric stimuli increase mechanical and chemical thresholds in both systems and may block passage of a disturbance initiated elsewhere.
13. In either system an electric current must be above a certain amplitude in order to trigger, no matter how long it is maintained; and a current sufficient for triggering must flow for more than a certain time in order to be effective.
14. A current rising too gradually from a subthreshold value to a normally sufficient intensity will not trigger either system; both systems thus exhibit accommodation.

Impressed by this long list of similarities between propagation in nerve and propagation in a passivated iron wire, Lillie inferred that the underlying mechanisms were basically the same, and for the next 14 years he studied iron-wire systems in the hope of learning more about the mechanisms of spike generation and conduction in nerve fibers. Lillie's last paper on the subject appeared in 1936, but the use of passivated iron wire as a model of nerve fibers and neurons has continued almost to the present day.

The history of this endeavor is rather interesting and illuminating. Physical chemists and metallurgists had long been interested in the surface properties of iron, hoping to discover a treatment that would prevent corrosion of iron structures exposed to the atmosphere or to water. In 1790, Keir reported that after iron was treated with strong nitric acid or silver nitrate it behaved more like a noble metal than like iron. Ordinary iron, for example, would precipitate silver from a silver nitrate solution and copper from copper sulfate or copper nitrate solutions, whereas the treated iron would not. On the other hand, Keir noted that if the surface of the treated iron was scratched, its chemical properties immediately returned to those of ordinary iron. Several decades later, Wetzlar (1827) and Herschel (1833) observed that the visible surface changes during the chemical treatment of the iron and the visible reversal of those changes induced by scratching both tended to be propagated over the surface of the iron. During the next 170 years, at least 200 papers were published on the treatment and its reversal. Many of those were attempts to explain the effects of the treatment, which had come to be called the *passivation* of iron (the return of iron to its normal state being called *activation*). Faraday proposed that the surface of passivated iron was oxidized; Helmholtz believed that it had a thin coat of iron nitrate, while Andrews thought that it simply was ultraclean. Other explanations for the passivity included mechanical alteration of the iron and protective films of oxygen, nitric oxide, nitrogen, or cyanogen.

Whatever its cause might have been, passivation definitely inhibited the solubility of iron and markedly reduced its chemical activity in comparison to that of normal iron; so it stirred up considerable interest. By 1900, most of the properties of passivated iron were known, but the physical and chemical bases of those properties were not. Two questions remained prominent: What was the composition of the passivated surface, and how were activation and passivation propagated over the surface of the iron? Mousson (1843) and others had suggested that the propagation of passivity and activity depended on local electric currents. Noting that local electric currents had been observed in nerves during propagation of signals, Ostwald suggested that propagation in iron might be analogous to that in nerve; and he convinced Heathcote (1907) to look into this analogy in the hope of gaining a better understanding of passivation in iron.

During the course of his studies, Heathcote performed experiments that would be repeated time after time by neural modelers in the years that followed. He found, for example, that if he touched a piece of zinc to one end of a passivated iron wire immersed in nitric acid, a visible state of activity would propagate along the entire length of the wire. Because of the presence of the nitric acid, however, the wire would gradually regain its passivation, only to have another wave of activation begin. These alternations between passivity and activity would continue until the wire was dissolved or the zinc was removed. The velocity of propagation decreased as the volume of the nitric acid surrounding the wire was decreased. Furthermore, if the wire was sufficiently long, the activation did not engulf the entire wire at one time, but propagated as a zone of activity, with passivated iron in front and behind. On the basis of his experiments, Heathcote concluded that passivation and activation were electrolytic processes, and he was able to develop a large part of the list of properties it had in common with nerve propagation, but the analogy itself provided no deeper understanding of the iron-wire phenomena.

Lillie became interested in electrolytic phenomena associated with nerve as a result of his fascination with Bernstein's ionic hypothesis (an early predecessor of the Hodgkin–Huxley model). Between 1909 and 1916, he pointed out a number of electrochemical analogies that he believed gave support to Bernstein's hypothesis. Among these were a mercury–hydrogen peroxide system that oscillated, the ordinary electrolytic battery which provides a reasonably permanent electrical potential in an ionic system, and a system comprising a galvanized iron wire in dilute sulfuric acid which provided a phenomenon similar to the injury potentials observed in nerve. In 1922, he finally settled upon the iron-wire model, which he studied extensively. Many of his observations are discussed in a review by Lewis (1968). Perhaps his most significant results were obtained when he attempted to model myelinated nerve fibers with the iron wire. To simulate the segmented myelin sheath, he used segments of glass tubing, through which he threaded the passivated iron wire. Short lengths of exposed wire between the glass segments represented nodes of Ranvier. With this configuration, he noted that activation propagated by jumping from exposed surface to exposed surface, apparently skipping over the intervening, glass-covered wire. Furthermore, the velocity of propagation was considerably increased over that of the same wire without the glass tubes. On the basis of these findings, Lillie proposed that the segmental structure of myelinated nerve fibers might be responsible for the observed high velocity of propagation in those fibers, a view still prevalent today.

These experiments by Lillie were repeated many years later by Yamagiwa (1951) and by Franck (1956), both of whom reasserted Lillie's hypothesis concerning propagation in myelinated fibers. Yamagiwa also repeated and extended other aspects of Lillie's research, concentrating on decremental conduction over insulated regions of iron wires and on the interactions of iron-wire models in a single solution. As a result of his observations, Yamagiwa proposed (but could not demonstrate conclusively) that threshold in the iron-wire system was related to area (i.e., when the area of activated surface reached a fixed threshold, the activation would then spread over the entire wire). Matumoto (1955) and his colleagues at Gunma University were later able to confirm this proposal by the use of a marker dye with which area of activated surface could be measured quite accurately. However, the active area they observed was not continuous, but consisted of a large number of discrete spots. As a result of their observations of squid axon under voltage clamp, where they noted numerous discrete current pulses in response to a single voltage step, Tasaki and Spyropoulos (1958) proposed that at any level of depolarization, the nerve membrane is a mosaic of active and inactive patches, with the ratio of active to inactive area determining the gross membrane conductance. Knowing of the work of Matumoto, Tasaki and Bak (1959) set out to determine whether or not the iron-wire system was analogous to the squid axon. In performing their experiments, they inverted the usual iron-wire analogy, taking the wire core to represent the fluid bathing the squid axon and the nitric acid to represent the axoplasm. With this convention, the potentials and currents in the model were of the same polarity as those in the axon. Applying voltage clamps to the iron wire, they found that its transient and steady-state voltage–current characteristics bore extremely close resemblences to those of the squid axon under clamp, and that the same discrete current pulses were present. Coupling these observations with those of Matumoto, Tasaki and Bak offered the iron-wire model as support for their mosaic-membrane hypothesis.

Whether or not it provides valid support for the mosaic-membrane hypothesis, the iron-wire model does appear to behave in a manner very similar to that of an axon. A local stimulus that exceeds threshold for activation causes a local change in the surface of the passivated iron. This in turn produces a current that tends to carry the change in the surface further, which in turn causes an even greater current, which carries the surface change even further, and so on until the surface change is complete and the surface is locally activated. The surface provides an electrical capacitance between the iron core and the surrounding nitric acid solution, and the local current discharges this capacitance and reverses the polarity of the voltage across it. As a consequence of the resulting voltage gradients (almost all of which are in the surrounding solution), currents flow to neighboring regions of the wire surface and trigger the same sequence of events there. Thus, the local activation is propagated over the entire wire. As in an axon, the velocity of propagation will depend on the local electrical capacitance (decreasing as the capacitance per unit length of wire increases) and on the electrical resistance faced by currents flowing along the inside and the outside of the wire (decreasing as the resistance increases).

The inverted, or inside-out, analogy used by Tasaki and Bak is quite reasonable not only from the point of view of voltage and current polarity, but also on the basis of the relative electrical resistances of the iron core and the surrounding nitric acid solution. In his elaboration of Weber's core-conductor theory, Weinberg (1941) concludes that it applies very well to the iron-wire model, with its very high core conductance, but that its applicability to axons with their relatively low axoplasmic conductance is questionable. As we have shown in Chapter 6, one can apply perfectly well the analysis of cable theory without assuming that the core of the cable is a perfect conductor, as Weber had. On the other hand, Weinberg's point is well taken. In fact, the current paths outside of the axon apparently offer considerably less resistance than do the paths through the axoplasm, whereas the resistance of the iron core is negligible compared to the resistance offered by the nitric acid. As a result of this difference, the velocity of propagation exhibits completely different relationships to geometric parameters in the two systems. In the iron wire, the velocity decreases with increasing wire diameter, whereas in the axon, velocity increases with increasing diameter. In the iron wire, velocity decreases with decreasing volume of the surrounding nitric acid, whereas in the axon, the velocity is relatively independent of the volume of extracellular fluid. The reasons for these differences are quite clear. Increasing the wire diameter serves only to increase the area of the interface between the iron core and the surrounding nitric acid, thus increasing the local capacitance but not significantly changing the resistance to the flow of currents along the outside of the wire (compared with which the resistance of the wire itself is negligible). Thus, the propagated wave should be slowed. On the other hand, increasing the diameter of an axon decreases the dominant axoplasmic resistance more quickly than it increases the electrical capacitance per unit length (the former being proportional to the reciprocal of diameter squared, the latter to diameter directly); so the velocity of propagation is increased.

Lillie and Yamagiwa both recognized this difference between the iron wire and the nerve fiber, and both proposed remedies. Lillie proposed a model comprising a glass tube filled with nitric acid, with the wire running axially down the center. Yamagiwa proposed an iron tube filled with nitric acid. Both models would give a qualitatively correct relationship between propagation velocity and the diameter of the tube. Many of the more recent investigators, on the other hand, have not concerned themselves with the inside-out nature of the iron-wire model but have accepted it as it is and have gone on to construct model neural networks with it. This idea was proposed independently by Aldrich *et al.* (1962) and by Stewart (1962). Stewart in fact has proposed that an adaptive system could be constructed with iron and glass spheres in a nitric acid solution. Even more recently, Nagumo (1975, personal communication) has found two-dimensional wave propagation over iron screens in nitric acid.

Two strong critics of the iron-wire analogy were Rosenblueth and Wiener (1945), who pointed out that experiments on the iron wire were at least as difficult to perform as they were on nerve fibers and that the phenomena of passivated metals were not better understood than those of nerve fibers and involved quite as much physical conjecture. They suggested that the useful model in the pair might be the axon rather than the wire, which brings us full-circle back to Heathcote. Regardless of whether or not it is an inside-out analogy and regardless of its present dubious scientific value, the iron-wire model did serve one important role. It demonstrated that nervelike, decrementless conduction could occur in a purely electrochemical system and thus fulfilled the purpose for which it was originally taken up by Lillie; it bolstered the foundations of membrane theory and of the ionic hypothesis against vitalistic arguments.

7.2. Threshold and Accommodation (or the Hill–Rashevsky–Monnier Model)

Before biologists were able to insert microelectrodes into neurons, they were forced to draw all their conclusions about neural properties from external measurements. Generally, they would place two electrodes against the surface of a nerve and pass a current from one electrode to the other as indicated in Figure 7-1. Then with another pair of electrodes they would determine whether or not the current had been strong enough to generate spikes. Since the nerve is composed of many axons running along together in parallel, the voltage observed at the voltmeter would reflect combined spikes from many different cells. In spite of this complication, such experiments on whole nerves provided considerable information about the properties of individual axons.

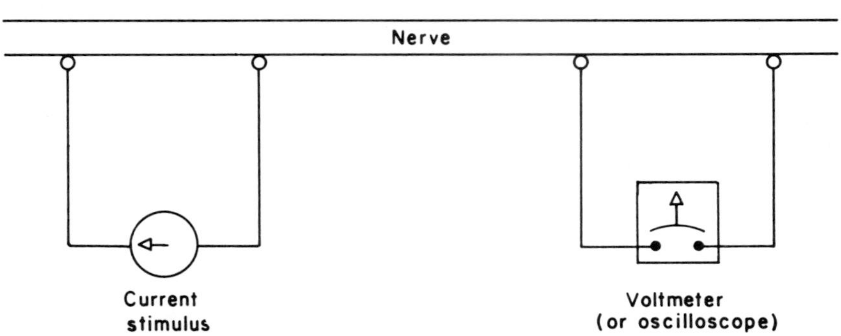

Figure 7-1. Typical experimental arrangement for observing axon properties before microelectrodes were available. Usually, observations were made on whole nerves or bundles of fibers teased from whole nerves. Thus, several or many axons were stimulated at one time, and the responses of many axons were observed simultaneously.

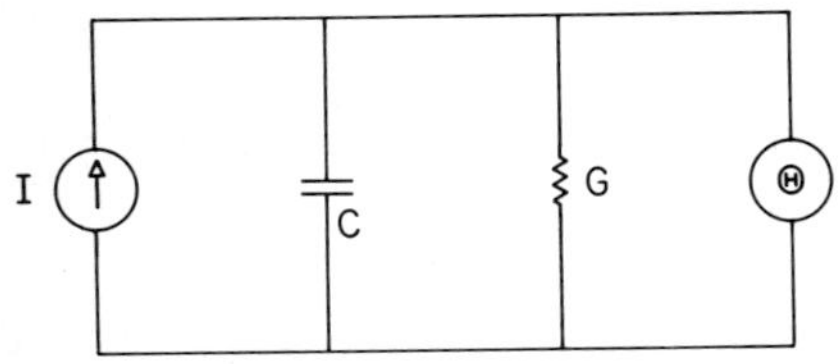

Figure 7-2. A simple model that could account for many of the observed properties of nerve fibers in the configuration of Figure 7-1. The stimulus current is applied across a simple resistor–capacitor circuit in parallel with a threshold device that produces a spike whenever the voltage across the circuit reaches a certain fixed threshold.

After nearly a decade of such measurements, it was found that a large proportion of the results could be explained by the extremely simple model of Figure 7-2. The threshold device θ produces a spike whenever the voltage across it reaches a certain, fixed threshold value. Otherwise, no current flows through θ. This model follows directly from the membrane equivalent circuit of Figure 4-2 if one assumes that the membrane chord conductance and its reversal potential are essentially constant for subthreshold potentials and that in addition to its passive electrical properties the nerve-cell membrane exhibits a fixed voltage threshold. Figure 7-3 depicts the passive

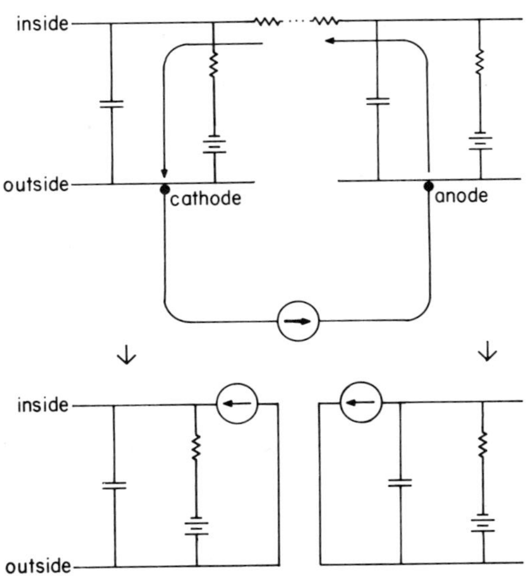

Figure 7-3. Model representing the local membrane (of a single fiber) under the electrodes delivering the stimulating current. The upper diagram depicts the actual experimental arrangement; the lower diagrams show the local equivalent current sources under the anode and the cathode. Although the anode is delivering current to the outside of the fiber, its net effect is to drive current away from the membrane on the inside of the fiber, thus hyperpolarizing the local membrane. Similarly, the cathode tends to depolarize the local membrane. Therefore, it is the cathode that tends to excite a fiber when current is applied.

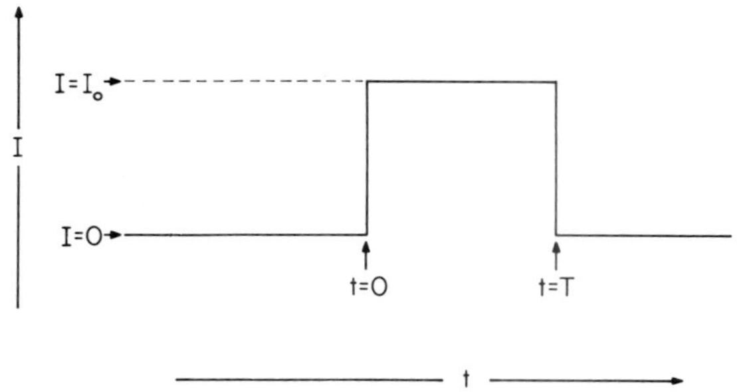

Figure 7-4. Typical current stimulus applied to nerves in the configuration of Figure 7-1. The current suddenly is turned on to a value I_0 at $t = 0$, then turned off again at $t = T$.

membrane under the two electrodes of the stimulus source (the anode and the cathode). Since it is the current at the cathode that tends to depolarize the membrane, excitation tends to occur there. The anode, on the other hand, would tend to hyperpolarize the membrane and thus suppress excitation at its location.

The model of Figure 7-2 can be analyzed very easily with the aid of Laplace transforms. The rate of change of the voltage across the threshold device is given by equation (7-1a), the transform of which is given in (7-1b):

$$\frac{dV(t)}{dt} = bI(t) - \frac{V(t)}{k} \tag{7-1a}$$

$$V(s) = \frac{V_{t=0} + bI(s)}{(s + 1/k)} \tag{7-1b}$$

where $k = C/G$, the membrane time constant; and $b = 1/C$. Probably the most common form of stimulus applied to nerves was the square pulse of current, as shown in Figure 7-4. Since the observable output of the nerve was a discrete, all-or-nothing spike, the interpretations of square-current-pulse experiments were limited to the parameters of the well-known "strength-duration curves." In other words one could plot the relationships between the two parameters of the stimulus (its amplitude and its duration) when threshold was just reached. The prediction of the model in this case is quite simple. The voltage during the pulse follows a time course given by

$$\begin{aligned} V(s) &= \frac{bI_0/s + V_{t=0}}{(s + 1/k)} \\ V(t) &= V_{t=0}e^{-t/k} + bkI_0[1 - e^{-t/k}] \end{aligned} \tag{7-2}$$

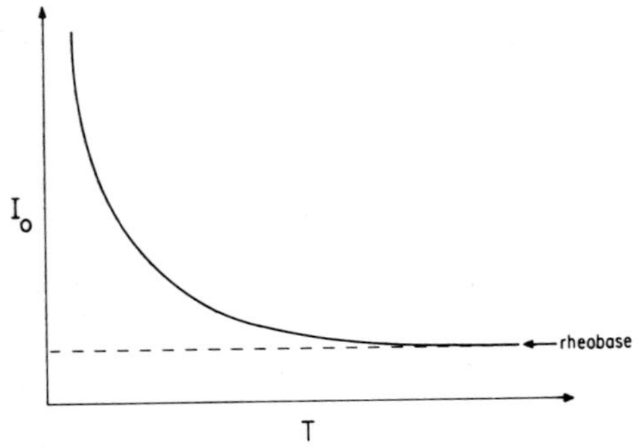

Figure 7-5. Strength-duration curve predicted by the model of Figure 7-2. For a current pulse of duration T, this curve gives the value of I_0 (the current strength) just required to produce a spike. When I_0 is sufficiently small (i.e., at the "rheobase" level or below), no spike would occur for any pulse duration.

where I_0 is the amplitude (or strength) of the current pulse, and $I(s) = I_0/s$ is the Laplace transform of a current step of amplitude I_0. If one assumes that the system initially was at rest ($V_{t=0} = 0$), then the maximum voltage occurs at the end of the pulse and is given by

$$V(T) = bkI_0[1 - e^{-T/k}] \tag{7-3}$$

where T is the duration of the pulse. By adjusting I_0 or T until $V(T)$ is equal to the threshold voltage V_θ, we have the following relationship between I_0 and T:

$$I_0 = V_\theta/bk[1 - e^{-T/k}] \tag{7-4}$$

which is plotted in Figure 7-5. The strength-duration curves actually measured in nerves were always extremely similar to this; in fact, one could match them almost precisely with proper choices of G, C, and V_θ.

One difficulty arose, however. The model predicts that as one approaches the minimum current required to reach threshold, the corresponding time required to reach threshold should increase without limit. In other words, the strength-duration curve should continue along the "rheobase" asymptote indefinitely. In actual measurements, on the other hand, a "Hauptnutzzeit" or maximum effective duration always was found, suggesting that the magnitude of the threshold voltage was not constant. This suggestion was supported further by the fact that while it did depress the nerve while current was flowing, as predicted by the model, the anode also tended to excite the nerve and even produce spikes when the current was turned off at the end of the pulse.

To account for these phenomena, several modelers independently proposed the same modification of the model. The modified model, which we shall call the Hill–Rashevsky–Monnier model, is depicted in Figure 7-6. Here the threshold element is floating so that the effective threshold is not fixed but tends to follow (or "accommodate" to) the changes in membrane potential induced by the stimulus.

We will follow the presentation of Hill (1936) in the following description of this model. The presentations of Rashevsky (1933) and Monnier (1934) are similar. Hill's main assumption is that the threshold for excitation rises according to the level of the local excitation of the nerve. More precisely, his model consists of the following concepts: There exist two state variables, V, the local potential, and U, the threshold, such that excitation ensues if and only if V is greater than U; an applied current I brings about an increase in V, and an increase in V brings about an increase in U; moreover, in fully accommodated nerve, the extra current required to excite the nerve is the same as in the original unaccommodated case. In Hill's own words (Hill, 1936):

> All the facts of excitation lead to the idea of a "threshold," of a level at which a discontinuity of some kind sets in. We suppose that excitation occurs when V reaches a certain level U; that for very short currents this level U is constant; that for longer currents it rises; that its rate of rise at any moment is determined by the value of V at that moment; and that when V is restored to its original value, U returns to its original level also in a simple exponential manner with time constant λ. We have made no assumption of the physical nature either of U or of λ; but those we have made as to their behavior express the facts simply, and further hypotheses are at the moment unnecessary.

Either from Hill's description or from the circuit of Figure 7-6, one can develop the following equations and their transforms for the model:

$$\frac{dv(t)}{dt} = bI(t) - \frac{v(t)}{k}$$
$$\frac{du(t)}{dt} = \frac{v(t)}{\lambda} - \frac{u(t)}{\lambda} \qquad (7\text{-}5a)$$

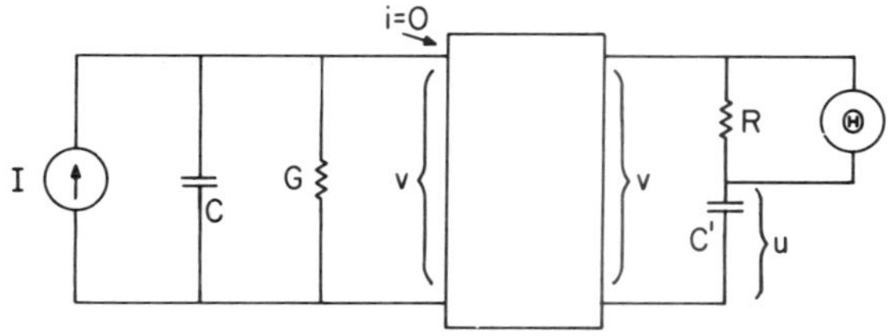

Figure 7-6. An electrical equivalent of the Hill–Rashevsky–Monnier model (see Hill, 1936, for an interesting hydraulic equivalent). The current stimulus again is applied to a resistor–capacitor circuit, but the threshold element is no longer directly in parallel with that circuit. Instead, it is connected in such a manner that the voltage across it is equal to $v - u$. When $v - u$ reaches a certain fixed value, a spike is generated; but as v increases, u tends to increase as well. Therefore, threshold is more difficult to reach if the stimulus is weak or gradually increasing.

$$v(s) = \frac{bI(s) + v_{t=0}}{(s + 1/k)}$$
$$u(s) = \frac{v(s)/\lambda + u_{t=0}}{(s + 1/\lambda)} \tag{7-5b}$$

where $v(t)$ is the difference between the local potential V and its original or resting value; $u(t)$ is the difference between the threshold U and its original or resting value; $v_{t=0}$ and $u_{t=0}$ are the values of those variables when the stimulus $I(t)$ is first applied; $k = C/G$ and $b = 1/C$, as before; and λ is Hill's time constant of accommodation. Note from the second equation that in the steady state (i.e., when $du/dt = 0$), $u - v$ is equal to zero. This corresponds to Hill's assumption that the nerve in the fully accommodated state is exactly as excitable as it is in the resting state. The remainder of Hill's work consists of spelling out some of the implications of this equation system, with particular reference to strength-duration curves for step and ramp currents.

If one begins with the model at rest (i.e., with $v = 0$ and $u = 0$) and applies a current pulse of amplitude I_0, he would have the following responses during the pulse:

$$v(s) = \frac{bI_0/s}{(s + 1/k)}; \qquad v(t) = bkI_0(1 - e^{-t/k})$$
$$u(s) = \frac{v(s)/\lambda}{(s + 1/\lambda)}; \qquad u(t) = bkI_0\left[1 + \frac{ke^{-t/k} - \lambda e^{-t/\lambda}}{\lambda - k}\right] \tag{7-6}$$

When the difference between $v(t)$ and $u(t)$ reaches or exceeds the resting threshold, $U_0 - V_0$, a spike will be triggered. Although $v(t)$ and $u(t)$ both increase monotonically during the current pulse, the difference $v(t) - u(t)$ does not. Instead, it passes through a peak

$$v(t) - u(t) = bk\lambda I_0 \frac{e^{-t/\lambda} - e^{-t/k}}{\lambda - k} \tag{7-7}$$

at the time when the term $e^{-t/\lambda} - e^{-t/k}$ is maximum:

$$t_{\max} = \frac{k\lambda}{\lambda - k} \ln\left(\frac{\lambda}{k}\right) \tag{7-8}$$

Thus, when I_0 is the minimum current strength required for excitation, the time at which threshold will be reached is $t_{\max}$, and that will be the maximum effective stimulus duration, or Hauptnutzzeit ("maximum utilization time"). The minimum (rheobase) current I_r is found by substituting the right side of equation (7-8) for t in equation (7-7):

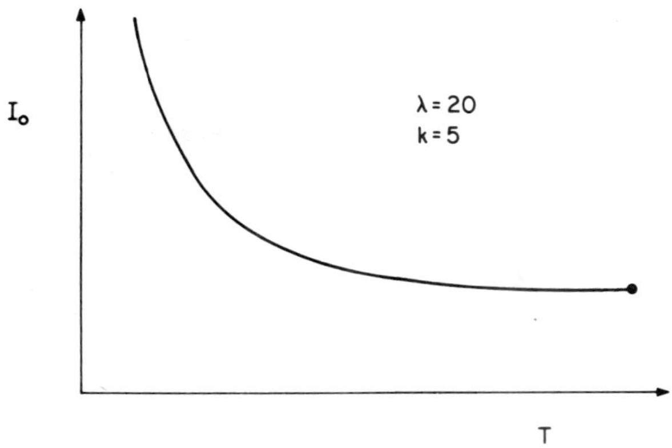

Figure 7-7. Strength-duration curve predicted by the Hill–Rashevsky–Monnier model. In this case, the parameters were set to relative values similar to those determined by Hill (1936) for frog nerve fibers. The principal distinction between this curve and that in Figure 7-5 is the presence of a "Hauptnutzzeit," or maximum latency, between the onset of the current stimulus and the occurrence of the spike.

$$I_r = \frac{U_0 - V_0}{bk}\left(\frac{\lambda}{k}\right)^{1/[(\lambda/k)-1]} \tag{7-9}$$

For currents greater than this value, threshold is exceeded prior to t_{max}; for currents less than this value, threshold is never exceeded. Equations (7-7), (7-8), and (7-9) together describe the strength-duration curve of the Hill–Rashevsky–Monnier model, which is displayed in Figure 7-7.

In addition to providing a Hauptnutzzeit, the Hill–Rashevsky–Monnier model can also exhibit a threshold response to the cessation or "break" of anodal current. For example, consider an anodal current pulse that has a very long duration relative to k and λ. By the end of the pulse, v and u will have reached steady-state values

$$u(T) = v(T) = -bkI_0 \tag{7-10}$$

After the pulse is turned off, we have

$$v(s) = \frac{-bkI_0}{s + 1/k} \tag{7-11a}$$

$$v(t) = (-bkI_0)e^{-t/k} \tag{7-11b}$$

$$u(s) = \frac{-bkI_0/\lambda(s + 1/k) - bkI_0}{s + 1/\lambda} \tag{7-11c}$$

$$u(t) = (-bkI_0)[k\lambda/(k - \lambda)][(1/\lambda)e^{-t/k} - (1/k)e^{-t/\lambda}] \tag{7-11d}$$

and

$$v(t) - u(t) = (bkI_0)[\lambda/(\lambda - k)][e^{-t/\lambda} - e^{-t/k}] \tag{7-12}$$

which can exceed U_0 if I_0 is sufficiently large and *if* $\lambda > k$. In other words, the Hill–Rashevsky–Monnier model predicts the excitatory effect of "anodal break" if and only if the time constant of accommodation is greater than the time constant of excitation.

If I_0 were sufficiently strong to allow $v - u$ to reach U_0 in a time that was very short compared to λ and k, then equation (7-7) reduces to the extremely simple form

$$v(T) - u(T) = bI_0 T \tag{7-13}$$

Thus, the parameter b in the model could be matched to the corresponding parameter in the nerve simply by observation of the vertical asymptote of the strength-duration curve. The parameters k and λ could be determined by observation of the Hauptnutzzeit and the rheobase current. Once these three parameters were known for the nerve, the model could be tested by determining whether or not it predicted the current required to reach threshold on anode break, or the current dependence of the time required to reach threshold after anode break, or any of a very large number of other observable relationships. Hill, in particular, tested the model for a very wide variety of situations, including "make" and "break" excitation, current ramps, exponentially increasing currents, sinusoidal currents, and repetitive pulses. In all cases, the model faithfully predicted the observations in the nerve with respect to the triggering of the first spike in response to the stimulus.

For example, the response of the model to a ramp current $I = mt$ is given by

$$\begin{aligned} v(t) &= mbk[t - k(1 - e^{-t/k})] \\ u(t) &= mbk\left[t - (\lambda + k) + \frac{\lambda^2}{\lambda - k}e^{-t/\lambda} - \frac{k^2}{\lambda - k}e^{-t/k}\right] \end{aligned} \tag{7-14}$$

Perhaps the most significant aspect of this result is the fact that $v(t) - u(t)$ approaches its maximum value as time approaches infinity:

$$\lim_{t\to\infty} [v(t) - u(t)] = \lambda mbk = [v(t) - u(t)]_{\max} \tag{7-15}$$

Thus, there is a minimum slope m of the current ramp that will allow threshold even to be reached:

$$m|_{\min} = (U_0 - V_0)/\lambda bk \tag{7-16}$$

Such "critical slopes" were found in nerve, and their values were predicted correctly by equation (7-16).

Hill also explored the frequency response of his model. In particular, he showed that the response to a sinusoidal input is given by

$$\begin{aligned} v(t) &= \frac{bkI \sin \omega t}{[1 + (\omega k)^2]^{1/2}} \\ u(t) &= \frac{bkI \sin(\omega t + \phi)}{\sqrt{[1 + (\omega k)^2][1 + (\omega\lambda)^2]}} \end{aligned} \tag{7-17}$$

To find the minimum current for excitation, I^*, for a given frequency ω, Hill demanded that $v - u = U_0 - V_0$ and that the slopes of u and v be equal as well. This leads to the relationship between I^* and ω given by

$$I^* = \{[1 + (\omega k)^2][1 + 1/(\omega\lambda)^2]\}^{1/2} \frac{U_0 - V_0}{bk} \tag{7-18}$$

This expression implied that there exists a minimum value for I^* and a corresponding resonant frequency $\omega^*{}_{\mathrm{m}}$ at which it occurs, as given by

$$\begin{aligned} I^*_{\min} &= (+k/\lambda) \frac{U_0 - V_0}{bk} \\ \omega^*_{\mathrm{m}} &\simeq 1/(k\lambda)^{1/2} \end{aligned} \tag{7-19}$$

Note that the curve of I^* versus ω rises at the low frequencies because slow changes imply a considerable amount of accommodation in this model, whereas excitation is also difficult at high values of ω because the membrane capacitance prevents the membrane potential from rising very much in response to rapidly oscillating stimulation. These factors combine to determine the resonant frequency as given by equation (7-19). The relationship is symmetric on a log scale. As ω gets very large, I^* is approximately equal to ωk.

At first impression, the Hill–Rashevsky–Monnier model is severely limited by virtue of the fact that it was not designed to provide threshold changes after the first spike in response to a given stimulus. In other words, it does not deal with the postspike refractoriness. The squid-axon data of Hodgkin and Huxley, which we will take up momentarily, suggest that mechanisms underlying accommodation to depolarizations are essentially the same as the mechanisms underlying postspike refractoriness, namely, increased potassium conductance and increased inactivation of the sodium conductance. The increased potassium conductance should have two effects: By giving the potassium flux a headstart in its rising phase, it should increase the actual threshold depolarization; and, by its tendency to draw the membrane potential more strongly toward the potassium reversal potential, it should make the stimulus current less effective in eliciting depolarization.

If "accommodation" is defined strictly as a change in voltage threshold, then the latter effect is not, strictly speaking, an accommodative mechanism; it does act in the same way, however, with respect to the current stimulus, and it undoubtedly played an important role in the observations of Hill and others. The increased sodium-conductance inactivation increases the threshold depolarization but should have negligible influence on the capabilities of current stimuli to elicit depolarizations. The difference between accommodation to subthreshold depolarization and refractoriness following a spike is very likely one of degree (i.e., the extent to which these conductance mechanisms come into play). The dynamics, on the other hand, may be quite similar in the two cases. The detailed dynamics of the onset of refractoriness are not especially important in the model; the onset is abrupt and essentially concurrent with the spike. The dynamics of recovery from refractoriness, on the other hand, are important. If they are essentially the same as the dynamics of accommodation (i.e., have the same time constant), then the Hill–Rashevsky–Monnier model can be modified in a very simple way to provide the refractoriness that follows a single spike. Since Hodgkin and Huxley were interested chiefly in events leading to and during the spike, their data are not sufficient to describe completely the postspike recovery phase. Specifically, they reported only one set of data pertaining to the recovery from inactivation of the sodium conductance (Hodgkin and Huxley, 1952, p. 504). Therefore, we cannot be sure that the time constant of this recovery is not dependent on previous levels of depolarization (even though no such dependence is included in the Hodgkin–Huxley equations). The data are more clear in the case of the return of the potassium conductance to its resting state (Hodgkin and Huxley, 1952, p. 491). The time constant of this return is slightly (but systematically smaller for recovery from large depolarizations than it is from small depolarizations. This might be reflected in a slight, but systematic, difference in the dynamics of refractoriness and the dynamics of accommodation.

Another problem in the design of a modified Hill–Rashevsky–Monnier model is the question of whether or not refractoriness is cumulative from spike to spike. Again looking ahead to the Hodgkin–Huxley data, one finds that the underlying phenomena (increased potassium conductance and inactivation of sodium conductance) are cumulative, but that their accumulation is extremely nonlinear, exhibiting marked saturation at large depolarization. This aspect of refractoriness might be difficult to embody simply in the modified model. On the other hand, the best approach would very likely be an empirical one, involving only the nerve fiber to be modeled and neglecting any constraints that might be implied by the squid-axon data. As far as we know, no one has attempted to develop a modified Hill–Rashevsky–Monnier model on this basis. One possible modification of the model shown in Figure 7-6 would be to allow the threshold element simply to dump a predetermined amount of charge on C' each time it is triggered, thus providing a transitory increase in u.

Since we have considered the effects of depolarization on the Hill–Rashevsky–Monnier model in the light of the Hodgkin–Huxley data (which were obtained more than a decade later), we might as well complete the comparison by considering the effects of hyperpolarization, specifically, the phenomenon of spike triggering on anodal break. According to the Hill–Rashevsky–Monnier model, the threshold potential tends to follow the membrane potential, so that in steady state the difference between the two potentials tends to be constant. The Hodgkin–Huxley data concerning sodium conductance can be interpreted in at least two ways, one of which is consistent with the Hill–Rashevsky–Monnier model, the other completely inconsistent. The Hodgkin–Huxley equations are an embodiment of the inconsistent interpretation. This does not imply that the Hodgkin–Huxley equations represent an incorrect interpretation; it does imply, however, that the Hill–Rashevsky–Monnier model and the Hodgkin–Huxley model cannot both be correct for the same fiber.

7.3. The Hodgkin–Huxley Model

Perhaps the best way to begin a discussion of the Hodgkin–Huxley model is to describe very briefly the experiments that immediately preceded it. Using the voltage-clamp techniques that had been developed recently by Cole and others, Hodgkin and Huxley shifed the membrane potential of the squid giant axon and observed the resulting currents. Because of the large diameter of that axon, they were actually able to thrust a pair of fine, independent silver electrodes through the cut end of the axon and into its axoplasm for a considerable distance. Thus, they were able to effectively short-circuit a segment of axoplasm, making it essentially isopotential over its length. Employing a clever guard system and cylindrical electrodes, they were able to make the adjacent extracellular solution (seawater) isopotential as well, eliminating all axial components of current and leaving only the possibility of radial components in the portion of the axon being voltage clamped. Thus, in effect, they were applying the clamp to a fairly large patch of membrane. This technique generally is called the space clamp.

In response to a stepwise depolarization of the membrane from its resting potential or some voltage close to the resting potential, they observed four phases of current. First, there was a very brief current spike, discharging the membrane capacitance and thus allowing the membrane potential to shift to its new level. As soon as this spike subsided (which required only a few microseconds), they observed an increase in the net outward current (just as one would expect in response to a depolarization in a passive-membrane system). Within a fraction of a millisecond, however, this outward current gave way to a net inward current, which, if the membrane were not clamped by a nearly ideal voltage source, would bring about further depolarization. The time course and magnitude of this inward current phase depended markedly on the magnitude of the depolarizing step and on the voltage from which the step was made. Generally, within 1 or 2 msec, the inward phase was overwhelmed by a second outward phase, whose dynamics and magnitude also depended upon the size of the voltage step and the membrane potential prior to it. This outward phase persisted for many tens of milliseconds if the clamped depolarization were maintained, but subsided rather quickly if the membrane were repolarized.

Substitution of choline ions for the sodium ions of the seawater bath abolished the inward phase of the current. Under these circumstances, the growth of the late outward phase could be observed to begin almost immediately after the clamped-voltage step, being continuous with the early outward phase. Thus, the picture emerged of a single outward phase, interrupted by a transitory inward phase when sodium ions were present in the extracellular solution. The reversal potential for the outward phase was very close to the Nernst potential for potassium ions, while that of the inward current was very close to the Nernst potential for sodium ions. This is what one would expect on the basis of equation (4-4b) if the outward current were carried predominantly by potassium ions and the inward current were carried predominantly by sodium ions, just as Hodgkin and Huxley had already assumed.

To model the system they were observing, Hodgkin and Huxley selected a mixed equivalent circuit with chord conductances and fixed reversal (Nernst) potentials for potassium and sodium ions and a combined chord conductance and fixed reversal potential for all other ion species. They assumed that the combined chord conductance for the other ions was constant and estimated its magnitude on the basis of data from the membrane near its resting state. The membrane capacitance was found to

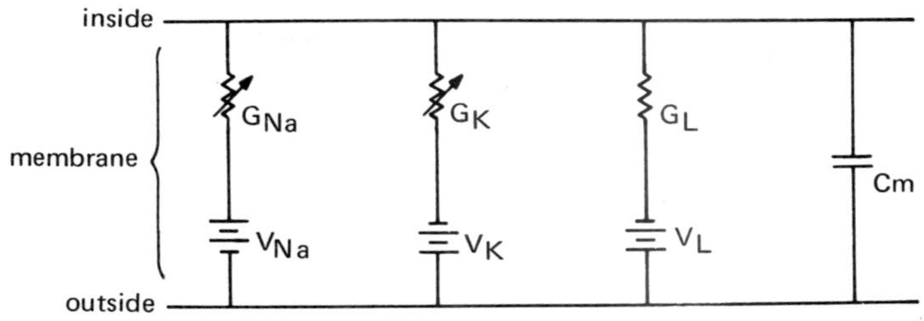

Figure 7-8. Chord-conductance model (mixed equivalent circuit) employed by Hodgkin and Huxley in their analysis of the squid giant axon.

be quite constant over the entire voltage range of interest; so the only variable parameters in their model were the sodium and potassium chord conductances. The assumed potassium and sodium currents were distinguished experimentally on the basis of the choline experiments, and the corresponding conductance values were found, consistent with the equivalent circuit (Figure 7-8).

In response to a stepwise depolarization, the potassium chord conductance increased gradually, with a deep inflection, then approached a final magnitude which was maintained. In response to a stepwise repolarization or hyperpolarization, the potassium chord conductance decreased in a noninflected manner toward a final, lower magnitude. The rate at which the potassium chord conductance rose and the final magnitude it attained both increased monotonically as the amplitude of the stepwise depolarization was increased. In order to describe the dynamics of this chord conductance for their model, Hodgkin and Huxley sought a mathematical function that would exhibit both an inflected rise and a noninflected fall. The function that they selected was

$$f(t) = (a + be^{-ct})^{\beta} \tag{7-20}$$

where the initial value is $(a + b)^{\beta}$ and the final value is a^{β}. If the final value is greater than the initial value, then b must be negative, and the transition from $a + b$ to a will be inflected (a fact that the reader can easily verify for himself). On the other hand, if the final value is less than the initial value, then b must be positive and the transition will be noninflected. Thus, the function of equation (7-20) was a reasonable candidate for describing the inflected rise and noninflected fall of the potassium chord conductance. What remained was to select a, b, c, and β to provide the best fit to the data. Noting that the term within the parentheses on the right side of equation (7-20) is the solution of a simple linear differential equation, Hodgkin and Huxley described their observed potassium chord conductance as follows:

$$G_K = \bar{g}_K n^4 \qquad (7\text{-}21a)$$

$$\frac{dn}{dt} = \frac{n_{ss} - n}{\tau_n} \qquad (7\text{-}21b)$$

where $\bar{g}_K$ is a constant and n_{ss} and τ_n depend on membrane potential. All of the voltage dependence of the dynamic chord conductance is thus associated with the two parameters n_{ss} and τ_n, while all of the innate time dependence is embodied in the differential equation. Hodgkin and Huxley used curve-fitting methods to develop equations to describe the dependences of τ_n and n_{ss} on membrane potential. However, these equations themselves are not nearly as important as the curves they describe, which are shown (redrawn from Cole, 1971) in Figure 7-9.

In response to a stepwise depolarization, the dynamic chord conductance for sodium ions rose with an inflection, passed through a peak, then fell again to a low magnitude. If the membrane suddenly were repolarized, the sodium chord conductance fell much more rapidly and in a distinctly noninflected manner. To describe the inflected rise and noninflected, rapid fall of the sodium chord conductance, Hodgkin and Huxley again used the function of equation (7-20). To describe the slower fall, which took place when the depolarization was maintained, they used the same basic function, but with β set equal to unity. Again, parameters (τ_m, m_{ss}, τ_h, and h_{ss}) were matched to the data (see Figure 7-9), and the conductance dynamics were embodied in linear differential equations:

$$G_{Na} = \bar{g}_{Na} m^3 h \qquad (7\text{-}22a)$$

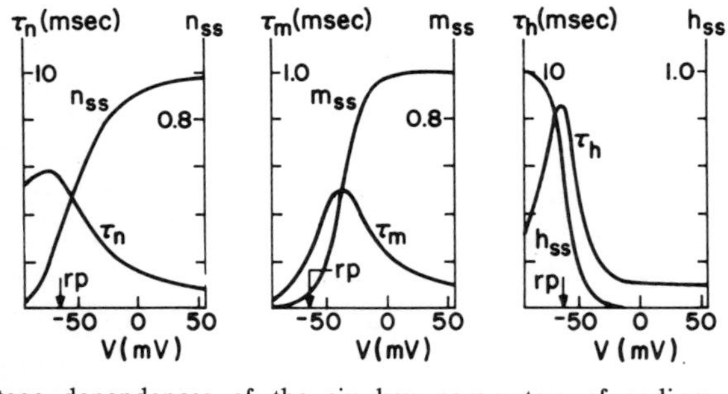

Figure 7-9. Voltage dependences of the six key parameters of sodium and potassium conductances in the Hodgkin–Huxley model. (Redrawn from Cole, 1971.)

$$\frac{dm}{dt} = \frac{m_{ss} - m}{\tau_m} \tag{7-22b}$$

$$\frac{dh}{dt} = \frac{h_{ss} - h}{\tau_h} \tag{7-22c}$$

The dynamic equation for the mixed equivalent circuit employed by Hodgkin and Huxley is simply

$$\frac{dV}{dt} = -\frac{1}{C_m}[G_K(V - V_K) + G_{Na}(V - V_{Na}) + G_L(V - V_L)] \tag{7-23}$$

where V is the membrane potential. By combining equations (7-21), (7-22), and (7-23), Hodgkin and Huxley had a complete dynamic description of the space-clamped patch of squid-axon membrane, derived under the rather severe constraint of voltage clamp. The next obvious step was to determine whether or not this same set of equations correctly described the dynamics of the unconstrained axon membrane prior to, during, and following a spike. Incorporating an impulse-like depolarizing current in their equations, Hodgkin and Huxley calculated the corresponding responses of the two dynamic chord conductances and the membrane potential and found excellent correspondence with the observations on the actual axon. In other words, they had, in their equations, an excellent dynamic description of the unconstrained axon as well as the clamped axon. The equations themselves are now known collectively as the Hodgkin–Huxley model. They apparently apply specifically only to the squid giant axon, but with rather minor modifications they appear to be applicable to other invertebrate axons (e.g., see Dalton and FitzHugh, 1960) and to myelinated axons in several vertebrates (see Dodge, 1963; Hille, 1971). The general phenomena that these equations represent are apparently almost universal in spike-generating nervous tissue.

Followng a depolarization that exceeds threshold, the sodium chord conductance rises with inflection but rather rapidly, allowing sodium ions to flow into the axon and thus leading to further depolarization. As a result of its voltage dependence, the sodium chord conductance responds to the increased depolarization by increasing even more and even faster. Thus, the rising phase of the spike consists of a regenerative feedback loop involving the sodium chord conductance and the membrane potential. The resulting sodium current falls only as the membrane potential approaches the sodium reversal potential. By this time, however, two things have begun to happen. The sodium conductance has begun its gradual decline (represented by a declining h in the Hodgkin–Huxley equations), and the slower-rising potassium chord conductance has begun to overtake the sodium chord conductance. As soon as the efflux of potassium ions exceeds the influx of sodium ions, the falling phase of the spike begins. This too has a regenerative aspect, brought about by the rapid decline in sodium chord conductance in response to repolarization. In the squid giant axon, but apparently not in many other invertebrate axons, the potassium reversal potential is well below the resting membrane potential. Since the potassium chord conductance is rather slow to fall in response to repolarization, it remains fairly large when the membrane potential reaches its original resting value, and the potential undershoots and approaches the potassium reversal potential. Gradually, as the potassium chord conductance falls toward its resting value, the membrane potential settles back to its resting level. However, because the potassium chord conductance of the hyperpolarized axon is somewhat less than that of the resting axon, a slight rebound depolarization is quite likely.

Although most nerve fibers are far too small to permit the direct space clamp employed by Hodgkin and Huxley, an indirect (or virtual) clamp can be imposed on myelinated fibers by means of a scheme perfected by Dodge and Frankenhaeuser (1959). Later, Julian, Moore, and Goldman (1962) adapted the technique to small, nonmyelinated fibers [see Moore (1971) and Hille (1971) for evolution of the technique]. Dodge (1963) developed equations similar to those of Hodgkin and Huxley to describe the ionic currents in myelinated axons from five organisms. In essence, Dodge's model took exactly the same form as the Hodgkin–Huxley model and is given by equations (7-21) to (7-23). There is, however, an interesing and probably important difference in the parameters $\bar{g}_{\mathrm{Na}}$ and $\bar{g}_{\mathrm{K}}$ measured in the squid giant axon and those measured by Dodge and Frankenhaeuser. Hodgkin and Huxley found that the sodium and potassium channels in the squid axon bathed in seawater were instantaneously ohmic (i.e., $\bar{g}_{\mathrm{K}}$ and $\bar{g}_{\mathrm{Na}}$ in the squid were constants). In other words, the change in sodium current or potassium current in response to an applied instantaneous (stepwise) change in membrane potential was directly proportional to the magnitude of the voltage change. Although this was true or nearly so for some current components in the myelinated fibers, it was not true for all of them. For example, in the grass frog (*Rana pipiens*), the potassium channel appeared to be nearly instantaneously ohmic, but the sodium channel was markedly nonohmic; in the clawed toad (*Xenopus laevis*), both channels were nonohmic. Interestingly, the nonohmic channels exhibited the imperfect-diodelike nonlinearities of the Goldman model, and the $\bar{g}$ parameters were given quite well by that model [i.e., by equation (3-18)].

The parameters m_{ss}, h_{ss}, n_{ss}, τ_m, τ_h, and τ_n have very similar dependences on membrane potential in all the axons that have been observed with

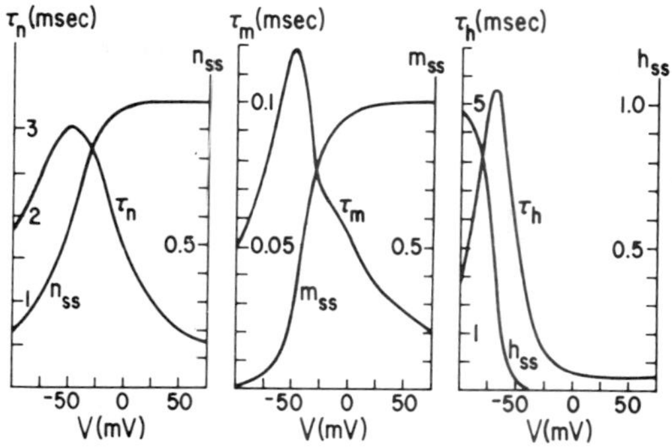

Figure 7-10. Voltage dependences of the sodium-conductance and potassium-conductance parameters deduced by Dodge (1963) for a grass frog axon.

voltage clamps and modeled (e.g., see Figure 7-10). Noble and Tsien (1969) attributed this to gating mechanisms that are common to all the axons in question, perhaps to axons in general (see Noble, 1972). They proposed the view that the $\bar{g}$ parameters reflect the properties of the individual channels, and that the factor m^3h represents the proportion of sodium channels actually conducting, while n^4 represents the proportion of potassium channels actually conducting, the remaining channels being closed or blocked in each case. Thus, the dynamics of n, m, and h would reflect the unblocking (or gating) of the individual channels. According to this view, then, the individual potassium and sodium channels in the squid axon in seawater are ohmic, but conduct only when they are unblocked, or gated. Similarly, the potassium channels in *Rana pipiens* are nearly ohmic, but the sodium channels in that animal and both the potassium and sodium channels in *Xenopus* exhibit Goldman-like chord conductances.

Accommodation is represented in the Hodgkin–Huxley model by the variables n and h. Threshold can be exceeded only if the sodium influx can maintain dominance over the potassium efflux. If the membrane is depolarized sufficiently slowly, the slowly rising potassium chord conductance (represented by the rising magnitude of n) tends to keep the potassium efflux abreast of the sodium influx. Furthermore, during slow depolarizations, the slow fall of the sodium chord conductance, represented by a falling magnitude of h and often called the sodium inactivation, tends to dominate and prevent the sodium conductance from reaching a large magnitude. In other words, the decline of h (sodium inactivation) tends to stay abreast of the rise of m (sodium "activation"). Thus, with n rising and h falling, the threshold tends to accommodate to gradually depolarizing stimuli. In order to be completely consistent with the Hill–Rashevsky–

Monnier model, however, the Hodgkin–Huxley model would require some device whereby the threshold could accommodate to strongly hyperpolarizing stimuli as well. One such device could be a dependence of m on the magnitude of the depolarizing step (i.e., on the magnitude of the *change* of voltage) rather than on the voltage itself. Such an interpretation would not be inconsistent with the Hodgkin–Huxley data (Lewis, 1965).

Other slow, or long-term, phenomena that were not included in the original Hodgkin–Huxley model are the modification of the potassium reversal potential as a result of potassium-ion flow across the membrane (Frankenhauser and Hodgkin, 1956) and a slow inactivation of the potassium current, possibly the result of a gating off of the potassium channels [see Frankenhaeuser (1963) and Hille (1967)].

Although the sodium–potassium antagonism as described by the basic Hodgkin–Huxley equations appears to be an extremely common source of excitability in neurons, there are exceptions emerging. For example, Frankenhaeuser (see Hille, 1971) found it necessary to invoke a third component of current to describe the dynamics of the *Xenopus* axon, and the spike in the soma of *Aplysia* cells apparently involves calcium rather than sodium, whereas the axon spikes in the same cells involve sodium (Kado, 1973). A model of excitable membrane in *Aplysia* neurons was proposed by Connor and Stevens (1971) and again is very similar to the basic Hodgkin–Huxley model.

Hodgkin and Huxley partially spelled out the properties of their model. Most notably, it describes the shape of the observed axon spike very well. It also describes reasonably well the total conductance changes as well as the specific changes of potassium and sodium chord conductances.

The model does relatively well in describing the net fluxes of sodium and potassium ions during an action potential; however, it predicts somewhat too little exchange of sodium and too much exchange of potassium during an impulse. Refractoriness in the model depends essentially on two mechanisms: first, sodium inactivation which reduces the level to which the sodium conductance can be raised by a given depolarization, and second, a delayed rise in potassium conductance which tends to maintain the potential near the equilibrium potential for potassium and thereby oppose excitatory influences. With regard to basic processes of excitation, the model exhibits threshold-like behavior even though such a threshold is only implicitly defined and is hard to identify explicitly. The model displays "anode break" behavior upon the release from high-intensity hyperpolarizing stimuli. Hodgkin and Huxley did not explore accommodation in the model, but indicate (as we have) that the model should exhibit some form of accommodation. Finally, they noted that low-level oscillations were more marked in both their space-clamped experimental preparations and in their model than in the normal animal, primarily because in the normal animal oscillations are damped by the leakage of current away from the local region.

Various investigators since Hodgkin and Huxley have used digital or analog computer techniques in attempts to spell out the various properties of the Hodgkin–Huxley system. These include Cole *et al.* (1955), FitzHugh (1960), Cooley and Dodge (1966), and many others. One very interesting observation has to do with the response of the isolated patch model to a step current. For suprathreshold currents the isolated Hodgkin–Huxley patch responds with steady firing of spikes as long as the stimulus is applied. Moreover, as the intensity of the applied current is increased, the firing rate increases and the amplitude of the individual pulses decreases. These relations, as given by Sabah and Spangler (1970), for example, are shown in Figure 7-11.

Investigators have also employed computers in attempts to represent the spatial context of the Hodgkin–Huxley system. For example, Huxley (1959), using a result of cable theory, showed that the equation system exhibits three possible solutions for an assumed wave-form solution: a normal train of spikes which agree in shape and velocity with observed data, and two subthreshold phenomena including a single subthreshold wave and an indefinitely long train of subthreshold oscillations which propagate much more slowly than the normal impulse. As far as we know, there is no evidence that such waves occur in real tissue. Cooley and Dodge (1966) employed a partial differential version of the Hodgkin–Huxley equation to represent a continuous single axon and obtained strength-duration curves for a step current in both the space-clamped and the continuous case, which agree reasonably well with Hill's theoretical curves. However, they found that, in response to a step current, the continuous axon fired repetitively at a rate that was almost independent of stimulus current intensity. This contrasts quite sharply with the frequency dependence for the isolated patch shown in Figure 7-11 and underlines again the importance of the spatial context of the model. Further emphasis is provided by Zeevi (1972), who incorporated partial-differential forms of the Hodgkin–Huxley equations into models of branching axonal trees and models of dendritic trees with spike-initiating loci (see Llinas and Nicholson, 1969).

An interesting question that has been attacked by several investigators concerns the range of behavior available from the Hodgkin–Huxley model if one is given the freedom to vary its parameters at will. Among those who have indulged in this endeavor are Frankenhaeuser and Valbo (1965), Noble (1966), and Lewis (1968). It is clear from the results of these efforts that the range of obtainable behavior is extremely large, and, in fact, almost any observed neuroelectric phenomena can be reproduced by the extended model. It is important to note, however, that the fact that the extended

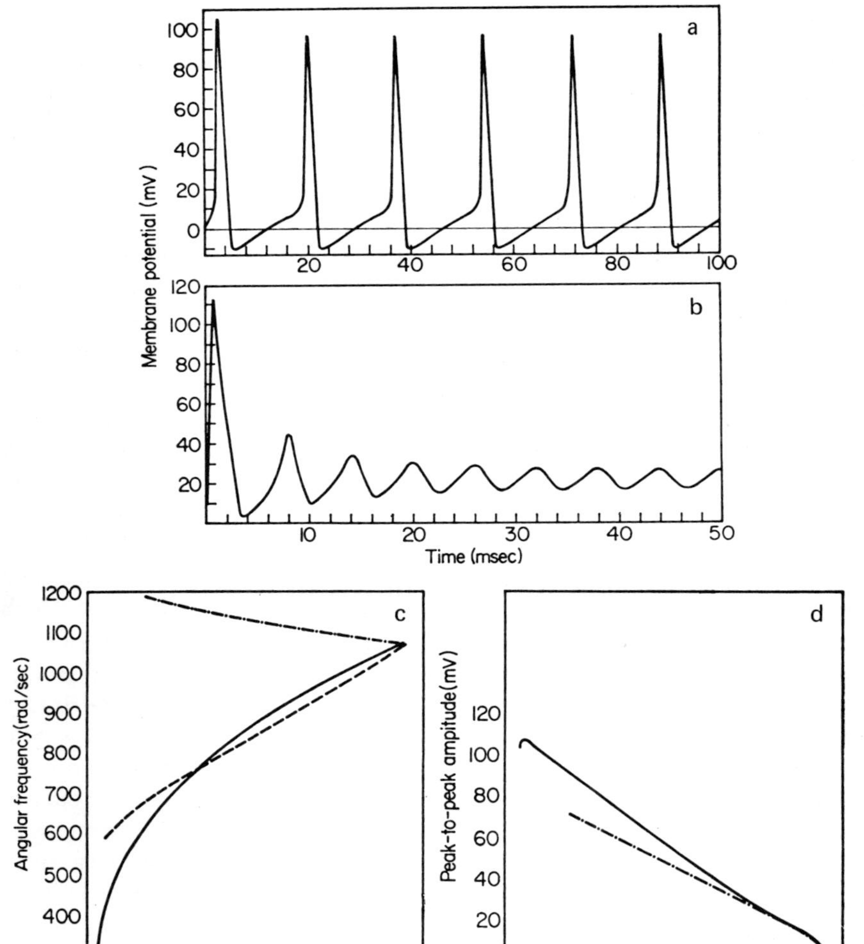

Figure 7-11. Repetitive firing characteristics of the Hodgkin–Huxley equations. Membrane potential for a lower (a) and higher (b) applied step current. Repetitive firing rate (c) and peak-to-peak amplitude (d) as functions of amplitude of applied step currents. (From Sabah and Springer, 1970.)

model can account for a particular phenomenon does not in itself imply that the mechanisms underlying the phenomenon actually are those underlying the Hodgkin-Huxley model. An example of this situation is provided by Lewis (1965) who found that the extended Hodgkin–Huxley model has inherent capability of synaptic facilitation and antifacilitation; yet these phenomena are now generally acknowledged to be the result of mechanisms essentially unrelated to those represented in the model (see Chapter 5). On the other hand, some slow oscillatory phenomena, rebound phenomena, and many others may indeed have their roots in Hodgkin-Huxley-like mechanisms, and the extended model certainly can produce them (see Lewis, 1968).

7.4. Abstractions of the Hodgkin–Huxley Model

Because of the complexity of the Hodgkin–Huxley equations with their four state variables (V, n, m, and h), several investigators have attempted to develop simplified or generalized abstractions of the equations that would retain their essential properties and provide conclusions that transcend detailed specification. Perhaps the most notable among these investigators is FitzHugh (1959, 1961) who developed the Bonhoeffer–Van der Pol ("BVP") models as an abstraction of the Hodgkin–Huxley model. Rather than moving directly to that model, however, we shall begin by examining a simpler abstraction proposed by Noble (1966) and illustrated in Figure 7-12. In response to a sudden change in membrane potential, the variable m in the Hodgkin–Huxley model changes much more rapidly than either n or h. Noble (1966) derived quasi-instantaneous voltage–current relations from the Hodgkin–Huxley model by assuming that m reached its new value m_{ss} instantaneously in response to a sudden depolarization, while n and h remained constant at their original values. According to these abstracted voltage–current relations, the instantaneous threshold voltage (i.e., the threshold for a sudden voltage shift such as that imposed by a current impulse) should be the value of membrane potential at which the instantaneous inward sodium current is just large enough to balance the instantaneous outward potassium current. On one side of that voltage, the potassium current will dominate and a spike will not occur; on the other side, the sodium current will dominate, leading to a spike.

Extending this abstraction, Noble represented the slowly varying state variables n and h in terms of shifts in the quasi-instantaneous voltage–current relations. Curves C and D in Figure 7-12 represent situations in which there is no threshold whatsoever (i.e., the system is in an absolute refractory state), and curve B represents a situation of elevated, but finite, threshold (relative refractoriness or accommodation). One can use this scheme to envision the effects of slowly varying subthreshold stimuli as well as the effects of a spike. During slowly increasing depolarization, for example, n will gradually increase and h will gradually decrease, causing a

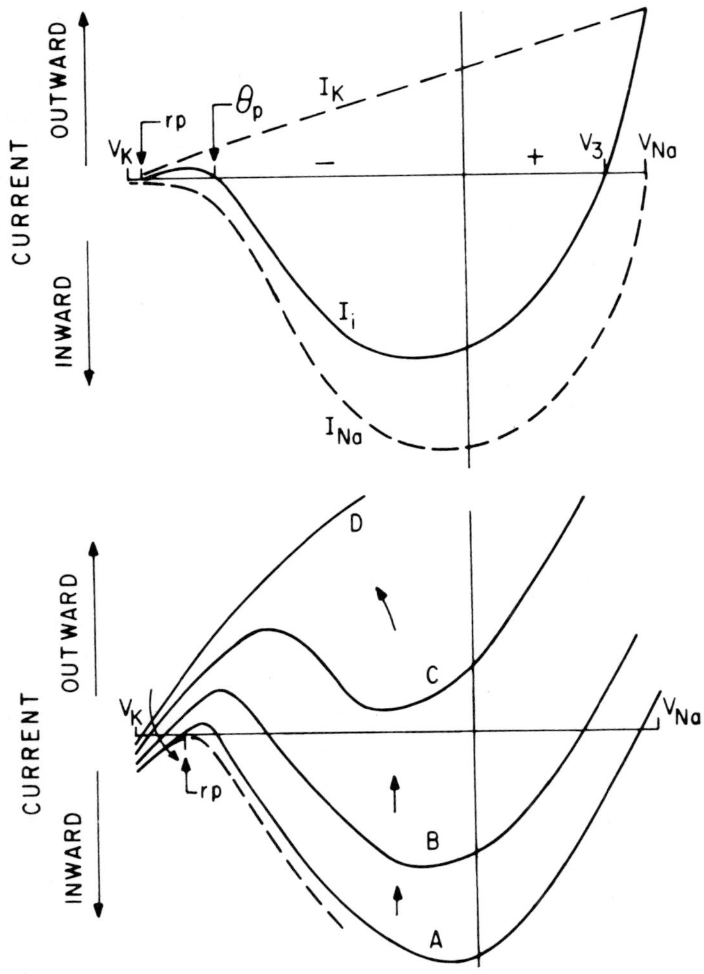

Figure 7-12. Noble's representation of threshold in a patch of squid-axon membrane (based on the Hodgkin–Huxley model). The solid curve (labeled I_i) represents the "instantaneous" ionic current in response to a sudden shift in membrane potential (represented by the abscissa). When the current is outward (above the horizontal axis in the figure), the potential tends to shift toward the left (hyperpolarization); when the current is inward (below the horizontal axis), the voltage tends to shift toward the right (depolarization). Thus, small excursions from the resting potential (r.p.) tend to be resisted, and the membrane tends to return to that point. The current flow reverses at the threshold (θ_p), and, beyond that point, the membrane tends toward V_3 (a potential close to the sodium-ion reversal potential). The dashed lines (labeled I_K and I_{Na}) depict the instantaneous potassium and sodium currents calculated by Noble from the Hodgkin–Huxley equations and summed to yield I_i. The lower diagram in the figure depicts the shifts in I_i in response to the changes in the parameters n and h as the membrane undergoes depolarization (e.g., the early phases of a spike). The line labeled A depicts the resting state, while B, C, and D depict states of progressively increasing depolarization. (Redrawn from Noble, 1966.)

gradual upward tilting of the current–voltage relationship. This tilting would represent a gradual shift of threshold to increasingly larger depolarizations. Following a spike, n would be very large and h would be very small, leading to a current–voltage relation similar to curve D. As the membrane recovered from the spike, the current–voltage relation would gradually tilt downward, back toward the position of curve A. Thus, this model very graphically depicts the phenomena of accommodation and absolute and relative refractoriness.

Extending his simplified model to a spatially distributed situation, Noble developed the picture presented in Figure 7-13. Here, the membrane potential is taken to be depolarized by a current impulse at $x = 0$ on a cable, and the resulting depolarization gradually diminishes with distance along the cable from the source. The corresponding current density at each point along the cable was estimated from the instantaneous voltage–current relations of Figure 7-12a. When the area under the current-density curve below the horizontal axis (i.e., the net inward current) is less than the area under the current-density curve above the axis (i.e., the net outward current), then the initial stimulus is not sufficient to trigger a spike; the net outward current over regions of cable remote from the stimulus source is more than sufficient to compensate for the net inward current close to the stimulus source. Thus, the local voltage threshold for initiation of a spike in a cable should be somewhat greater than that for initiation of a spike in a membrane patch. Threshold in this model corresponds to the situation in which the area under the current-density curve is the same on both sides of the horizontal axis.

One might abstract this system further by considering the operation of a single patch of membrane projected on a single axis representing membrane potential, as depicted in Figure 7-14. The system normally resides at the resting potential (r.p.), and if the membrane potential is displaced suddenly a short distance from r.p., it will return, as indicated by the arrows in the diagram. If the potential is displaced suddenly to the right (corresponding to membrane depolarization) and beyond threshold (θ_p if the membrane patch is isolated, θ_c if the patch is part of a cable), then it will move on to the right, almost to the sodium-ion reversal potential, before it finally returns to the resting potential. Of course, as we already know, the threshold may be relatively fixed with respect to sudden shifts of membrane potential (i.e., with respect to current-impulse stimuli), but it will definitely shift in response to slowly varying membrane potentials, and it will shift drastically during and following a spike.

To depict the shifting threshold along with the shifting membrane potential, one might consider the operation of the system projected onto a plane, with one axis representing membrane potential and the other

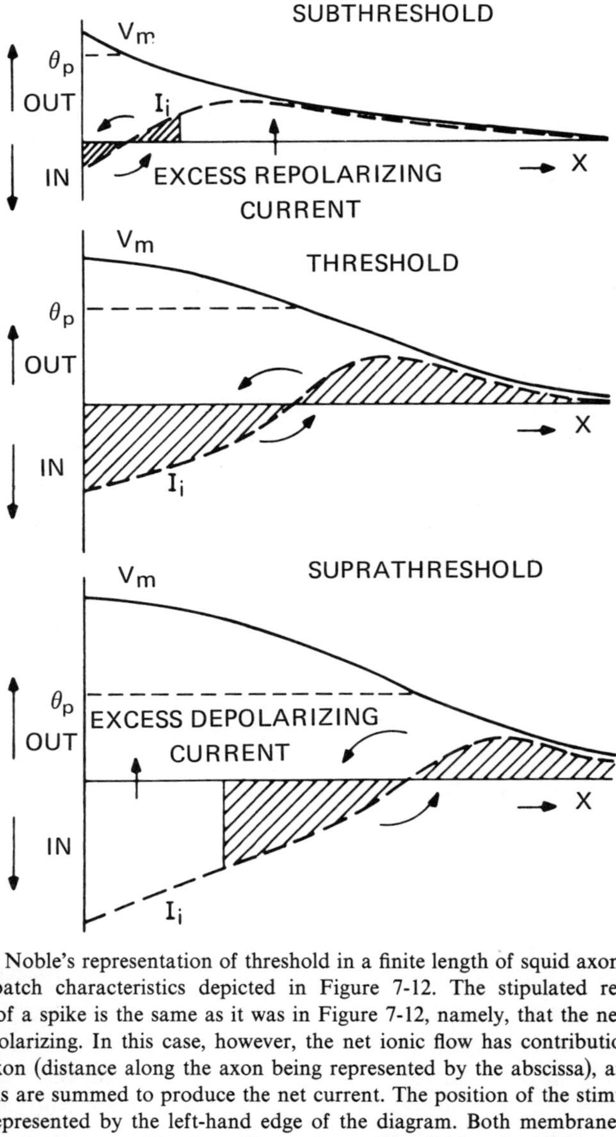

Figure 7-13. Noble's representation of threshold in a finite length of squid axon, based on the membrane-patch characteristics depicted in Figure 7-12. The stipulated requirement for production of a spike is the same as it was in Figure 7-12, namely, that the net ionic current flow be depolarizing. In this case, however, the net ionic flow has contributions distributed along the axon (distance along the axon being represented by the abscissa), and all of these contributions are summed to produce the net current. The position of the stimulus (a current source) is represented by the left-hand edge of the diagram. Both membrane potential and ionic current are represented by the ordinate, the upper, solid line depicting potential, and the lower, dashed line depicting current. Where the potential exceeds the local patch threshold (θ_p from Figure 7-12), there is a net inward flow, and elsewhere there is a net outward current. When the overall inward current exceeds the overall outward current, as in the lower diagram, the effective threshold of the fiber is exceeded. Note that the potential under the stimulating electrode in that case will be greater than θ_p. (Redrawn from Noble, 1966.)

Figure 7-14. One-dimensional state diagram for a squid axon, abstracted from Noble's representations in Figures 7-12 and 7-13. The horizontal axis represents membrane potential; the arrows indicate the direction of change of membrane potential for each segment of the line, given that the potential was shifted instantaneously into that segment (i.e., to a potential corresponding to a point in that segment). The quantity θ_p is the threshold of a patch of membrane; θ_c is the threshold of a finite length of axon (i.e., many patches together forming a cable).

representing threshold. Such a plane is depicted in Figure 7-15. Here, the horizontal axis represents membrane potential, and a few landmarks are plotted along it for reference. The vertical axis represents the difference between the actual threshold and the instantaneous threshold of the resting system (i.e., the threshold relative to the instantaneous resting threshold). The plane can probably be explained most easily in terms of a few trajectories (paths) along which the system might proceed after a current-impulse stimulus. The current impulse, or very-brief current pulse, will transfer charge very quickly to the membrane capacitance and cause the membrane potential to shift very quickly before any of the other state variables (e.g., m, n, and h, all represented here by the threshold) can begin to respond. Thus, the current impulse will shift the system along the voltage axis. A hyperpolarizing impulse will shift the potential from the resting potential to the left (e.g., to the point labeled 1). From there, the potential will gradually shift back along path A toward the resting potential; but as it does so, the threshold will have time to react and will decrease slightly. Before its return to r.p., the potential may actually overshoot slightly. A small depolarizing impulse will shift the potential to the right (e.g., to point 2). From there, it will return along path B to r.p., but with a slight increase in threshold and perhaps a slight undershoot. A slightly stronger, but still subthreshold, impulse might shift the potential to point 3, from which it might continue to move (after a brief delay) in a positive direction for a short time before beginning its return toward r.p. This continued positive motion would be brought about by the sodium current excited by the stimulus. The brief delay in the positive excursion is the result of the fact that the excited sodium conductance does not develop instantaneously. Clearly, path C must never exceed threshold, otherwise a full spike would have developed. If the scale of the vertical axis is the same as the scale of the horizontal axis (i.e., the same number of millivolts represented by each centimeter of axis length), then we could pass a 45° line through θ_c, and all trajectories that fail to exceed threshold would be confined to the left of that line, while early phases of all trajectories that lead to spikes would be confined to the right of that line. In nonlinear mechanics, such a line is known as a *separatrix* (it separates two distinct classes of trajectories). If the stimulus impulse were strong enough to drive the membrane potential past the resting threshold, to point 4, then the system would continue to move to the right (along path D), and a full spike would develop. Thus, path D would extend close to the sodium reversal potential, and it would extend markedly upward as the threshold increased. The threshold (according to Noble's representation) would approach a finite limit and then abruptly become infinite as the system became absolutely refractory. A strong stimulus might drive the membrane potential farther along the potential axis (e.g., to point 5), from which it would move on to the right even more quickly (along path E). All trajectories leading to spikes would tend to converge close to the sodium reversal potential and might tend to become indistinguishable. The phase of relative refractoriness following a spike would begin as the threshold once again attained its finite maximum (returning abruptly to that point from infinity), and the various trajectories in this phase would probably be very difficult to distinguish, all being extremely close to a single path such as F.

Bonhoeffer (1948) employed a graphical model similar to that of Figure 7-15 to represent both nerve fiber and iron wire in nitric acid, attempting to demonstrate the common features of the two. FitzHugh (1961) used the same scheme but quantified the model by combining it with a modified version of the Van der Pol equation for a simple nonlinear oscillator, which provided detailed specification of the trajectories. The resulting Bonhoeffer–Van der Pol model, as FitzHugh labeled it, is quite interesting and instructive to contemplate. The BVP model consists of the following two equations:

$$\begin{aligned} dx/dt &= c(y + x - x^3/3 + z) \\ dy/dt &= -(x - a + by)/c \end{aligned} \tag{7-24}$$

FitzHugh suggested that x in these equations might represent the membrane potential and the Hodgkin–Huxley variable m, both of which have rather fast dynamic responses, that y could represent the Hodgkin–Huxley variables n and h, both of which have a slow dynamic responses, and that z could represent any current stimulus applied to the system.

Figure 7-16 depicts the BVP model in the x,y plane. Two lines of special significance in this plane are those corresponding to $dx/dt = 0$ and $dy/dt = 0$. Crossing the former, all trajectories are vertical since the horizontal (x) component of motion is zero; similarly, crossing the latter, all trajectories are horizontal. These two lines (usually labeled *nullclines*) divide the x,y plane into four regions, in each of which the trajectories move in a

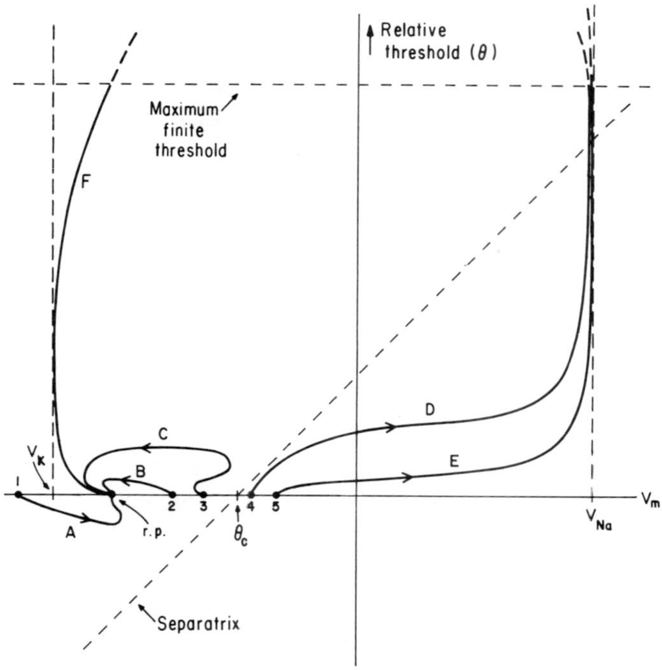

Figure 7-15. Two-dimensional state diagram for a squid axon. An ordinate, representing threshold, has been added to the abscissa of Figure 7-14, which represented membrane potential. Thus, changes in threshold can be depicted along with changes in potential. See text for details.

specific direction (i.e., upward and to the right, upward and to the left, downward and right, downward and left). Furthermore, the intersection of the two nullclines is a point at which there is no movement whatsoever. In the absence of stimuli (i.e., $z = 0$), all trajectories tend to converge on this point, and this corresponds to a stable equilibrium state of the model system, which FitzHugh identifies with the resting state of the membrane. An impulse of z (corresponding to a very brief current stimulus pulse) applied to the resting system will shift it horizontally from this equilibrium or resting point, effecting an instantaneous shift of the state variable x but no immediate change of y. Once again, there is a separatrix. If the state of the system is shifted to the left of the separatrix, a spike occurs. If it is shifted to some point on the right of the separatrix, the system returns to the resting point without having produced a spike. An important and simplifying feature of the model is the fact that (with proper parameter adjustments) dx/dt is comparable to dy/dt only close to the x-nullcline (where dx/dt is close to zero), otherwise dx/dt is much greater than dy/dt. For this reason, the trajectories tend to be horizontal when they are remote from the x-nullcline, and they tend to follow that nullcline when they are close to it and being driven toward it by dx/dt.

FitzHugh pointed out that, in principle, if stimuli were sufficiently well controlled and sufficiently close to the separatrix, a continuous gradation of responses should be observable. However, with the parameter values that he had selected for his model, he was unable to follow the separatrix very far into the region labeled "no-man's-land" in Figure 7-16. This simply was due to the fact that his analog simulation contained a modicum of noise (as would any real axon), and the probability was very small that the system could follow a trajectory close to the separatrix for very long without being bumped off of it by noise. Thus, the presence of noise in the system would tend to guarantee an early commitment to a trajectory on one side of the separatrix or the other and moving rapidly away from it. In this way, the noise guaranteed essentially all-or-none response of the model system to graded stimuli, and "no-man's-land" represents an essentially unreachable region of the plane. Similar all-or-none response, of course, had long been observed in axons; FitzHugh (1955, 1961) deduced that it also owed its presence to noise. Recently, Cole *et al.* (1970) have observed completely

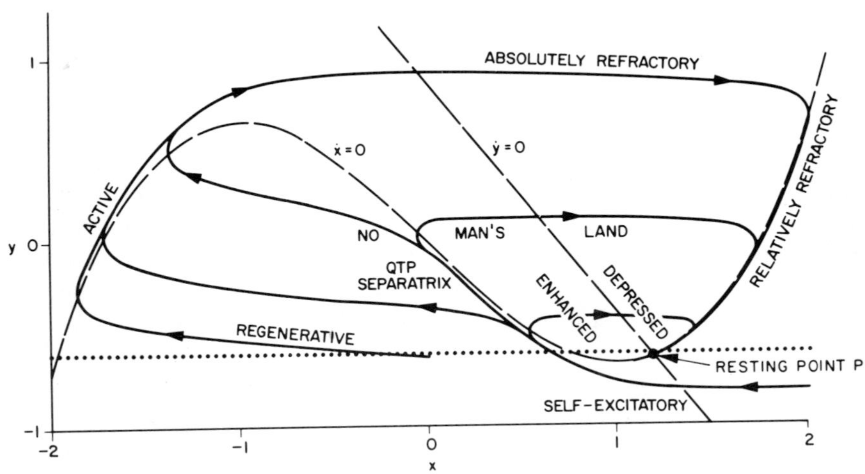

Figure 7-16. Two-dimensional state diagram for FitzHugh's BVP model. The dashed lines are the $x(\dot{x} = 0)$ and $y(\dot{y} = 0)$ nullclines. Solid lines with arrows represent trajectories along which the system progresses. As they rise upward and to the left along the x nullcline, the trajectories tend to merge into a single separatrix (labeled "QTP Separatrix" by FitzHugh). See text for details. (Redrawn from FitzHugh, 1961.)

graded responses in axons at elevated temperatures. Formally analyzing the effects of noise on the BVP model, Clay (1975) has shown that this is precisely what one would expect if FitzHugh's hypothesis were correct. Adjusting the parameters of the model to correspond to the slowly responding, cold axon, Clay found that effects of noise were profound and the probability of penetrating "no-man's-land" was indeed negligible. On the other hand, when the parameters were adjusted to correspond to the rapidly responding, warm axon, the effects of noise were markedly reduced and the probability of penetrating "no-man's-land" became quite high.

Like the Hodgkin–Huxley model, but unlike the squid axon upon which it was based, the BVP model will provide a continuous train of spikes in response to a constant stimulus of sufficient amplitude. The effect is illustrated in Figure 7-17. Here, a steadily applied z has shifted the x-nullcline upward so that trajectories no longer converge on the intersection of that nullcline with the y-nullcline. Thus, there is no equilibrium, or resting point, and all trajectories converge on a single cyclical path (the stable limit cycle).

7.5. Conduction of Spikes

So far, our discussion has been concerned primarily with the generation of spikes; little has been said about their conduction from one place to another in a neuron. We have already mentioned that Hodgkin and Huxley themselves (1952), Cooley and Dodge (1966), Zeevi (1972), and others have examined the Hodgkin–Huxley equations in the context of spatially distributed membrane systems. In doing so, they found that indeed the model accounts for the propagation of spikes without attenuation. Cooley, Dodge, and Zeevi solved partial-differential forms of the equations with the aid of digital computers These solutions were difficult to achieve, especially when the geometry of the distributed system being represented was complicated. Zeevi, for example, examined the effects of tapering, such as one might find at an axon hillock, and the effects of branching. The chief difficulty in doing so was the establishment of efficient algorithms to handle the equations. For a discussion of this topic, the reader is referred to the original papers. In this section, we shall concentrate instead on the approach used by Hodgkin and Huxley in their original modeling study.

Returning to the cable equation (6-29), we can rewrite it in terms of the membrane current as follows:

$$\frac{\partial^2 v_\mathrm{m}}{\partial x^2} = rI_\mathrm{m} \tag{7-25}$$

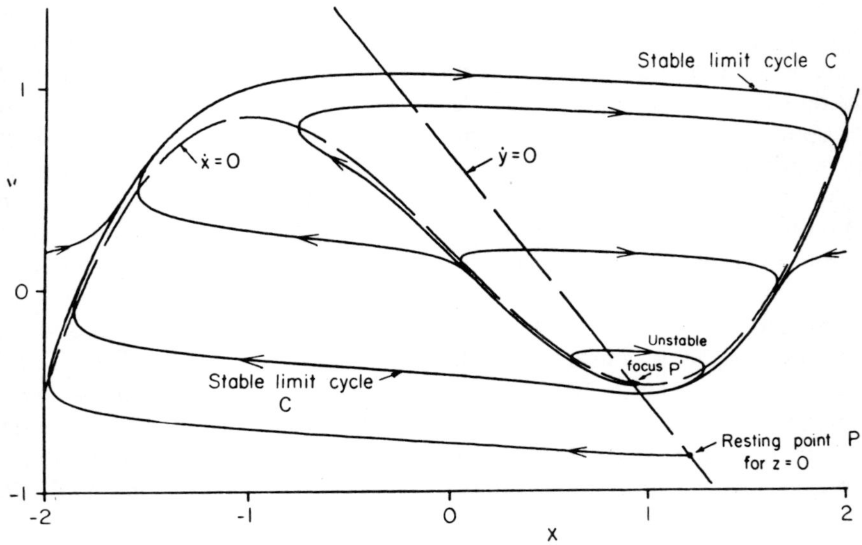

Figure 7-17. BVP state diagram showing the effects of a steadily applied excitatory stimulus, namely, cyclical progression about a limiting trajectory. (From FitzHugh, 1961.)

where v_m is the local membrane potential, r is the sum of the intracellular and extracellular resistances, I_m is the membrane current per unit length of the fiber, and x represents position along the fiber. Now, if instead of being passive the membrane of the fiber exhibits the properties of the membrane of the squid giant axon, then the Hodgkin–Huxley model can be used to determine I_m. The total current supplied to the local membrane, according to the Hodgkin–Huxley model, comprises four parts: that supplied to generate or annihilate dipoles across the local membrane capacitance, that carried through the membrane by sodium ions, that carried by the potassium ions, and that carried by all the other ions:

$$I_m = c_m \frac{\partial v_m}{\partial t} + I_{Na} + I_K + I_L \tag{7-26}$$

or

$$I_m = c_m \frac{\partial v_m}{\partial t} + g_{Na}(v_m - V_{Na}) + g_K(v_m - V_K) + g_L(v_m - V_L)$$

where c_m is the capacitance of the membrane per unit length of the fiber, g_{Na} is the time- and voltage-dependent sodium conductance per unit length of fiber, g_K is the time- and voltage-dependent potassium conductance per unit length of fiber, and g_L is the remaining membrane conductance per unit length of fiber. The quantities G_{Na}, G_K, and G_L as derived by Hodgkin and Huxley are conductances per square centimeter of membrane area. They can be converted to g_{Na}, g_K, and g_L (mho/cm) by multiplying them by the fiber circumference C_F (cm). Similarly, the membrane capacitance per unit area (C_m) can be converted to c_m. For the squid axon in seawater and for many other axons *in situ*, the extracellular component of r is negligible compared to the intracellular component, in which case r is given by the resistivity ρ_F of the cytoplasm of the fiber divided by the cross-sectional area A_F of the cytoplasm (i.e., of the fiber).

Thus, we are led to

$$\frac{\partial^2 v_m}{\partial x^2} = \frac{\rho_F C_F}{A_F}\left[C_m \frac{\partial v_m}{\partial t} + \bar{g}_{Na} m^3 h(v_m - V_{Na}) + \bar{g}_K n^4 (v_m - V_K) + G_L(v_m - V_L)\right] \tag{7-27}$$

which can be combined with the dynamic descriptions of n, m, and h to give the nonlinear partial differential form of the Hodgkin–Huxley equations. Imagining that C_F and A_F are functions of x, we have the problem undertaken by Zeevi and others. Hodgkin and Huxley, on the other hand, took a different approach. They assumed that the geometry of the fiber was uniform (a very good assumption for many axons, including that with which they were working), and that the conducted spike was well under way and reasonably distant from the stimulus source that triggered it. In this manner, they could treat the spike as a traveling wave with constant velocity. Such a wave, if it is continuous, must obey the wave equation

$$\frac{\partial^2 v_m}{\partial t^2} = \phi^2 \frac{\partial^2 v_m}{\partial x^2} \tag{7-28}$$

where ϕ is the velocity of the wave. Combining equations (7-27) and (7-28), Hodgkin and Huxley derived an ordinary differential equation to describe spike propagation:

$$\frac{d^2 v_m}{dt^2} = (\phi^2 \rho_F C_F/A_F) \times \left[C_m \frac{dv_m}{dt} + \bar{g}_{Na} m^3 h(v_m - V_{Na}) + \bar{g}_K n^4 (v_m - V_K) + G_L(v_m - V_L)\right] \tag{7-29}$$

which they attempted to solve numerically. The procedure consisted of the assumption of a value for conduction velocity followed by step-by-step numerical integration of the equation. Each attempt led to a numerical instability (i.e., the solution for v_m moved off toward infinity in the positive or negative direction), after which they would refine their estimate of velocity and try again. Gradually, they were able to approach a single value of velocity that would provide numerical stability and a reasonable solution for v_m. That value was given by

$$\phi = (168,000 A_F/\rho_F C_F C_m)^{1/2} \tag{7-30}$$

from which they calculated the velocity for specific fibers and compared the results with the observed velocities in those fibers. The results were exceedingly close. In one case, for example, they calculated a velocity of 18.8 m/sec and observed a velocity of 21.2 m/sec.

The functional form of equation (7-30) is very nearly correct for the squid axon (as well as many other fibers), making the velocity directly proportional to the square root of the ratio of cross-sectional area to the circumference. In a cylindrical fiber, this ratio would be one-fourth of the fiber diameter. Thus, according to (7-30), the velocity of conduction should increase roughly as the square root of fiber diameter. In the squid giant axon, the observed relationship was given approximately by

$$\phi \propto \text{diameter}^{0.6} \tag{7-31}$$

The difference between this and the calculated relationship, which is rather small, could be the result of the Frankenhaeuser–Hodgkin space discussed in Chapter 4.

Spike conduction velocity is limited by several factors whose relative importance varies with the ambient temperature. To derive a qualitative picture, one can consider two patches of Hodgkin–Huxley membrane connected by a resistive current path. The conduction velocity is determined by the same factors that determine how quickly a spike will be triggered in one of the patches in response to a spike in the other: (1) how quickly the membrane capacitance can be discharged to threshold, (2) how quickly the sodium conductance can respond to superthreshold potentials, and (3) how quickly the membrane capacitance can be discharged by the sodium current. The speed of the first process is inversely proportional to the product of the resistance between the patches (i.e., the series resistance per unit length of fiber) and the capacitance of a patch (i.e., the membrane capacitance per unit length of fiber). The speed of the second process depends on the speed of the first and also on temperature, becoming markedly faster as ambient temperature is increased. The speed of the third process depends upon the speed of the other two and also on the maximum value attained by the sodium conductance, being directly proportional to the ratio of that conductance to the local membrane capacitance. The latter apparently is essentially independent of temperature. Thus, spike conduction velocity can be increased by increased ambient temperature up to the point at which the second process no longer is rate limiting. Beyond that point, it can be increased by decreasing the product rc_m or increasing the ratio $\bar{g}_{Na}/C_m$. In vertebrate evolution, the rc_m product was effectively reduced by the development of myelin sheaths, which drastically reduce c_m. The effects of myelin sheath have been modeled by many investigators, including Lillie (1936), Franck (1956), and FitzHugh (1962), and, as we already have mentioned, it was Lillie's modeling study that led to the proposition that the myelin sheath indeed enhanced conduction velocity.

7.6. Concluding Remarks

In Part II, we have discussed models of passive and active neuronal membrane, typically representing extremely simple cellular geometries (e.g., patches of membrane, simple fibers, branching fiber trees with strong restrictions, and the like). Before moving on to models of neurons and networks of neurons, we would very much wish to complete this basic treatment with a reasonably thorough discussion of the modeling of membrane systems with geometries approaching the complexity of real neurons. We cannot. The appropriate investigations simply have not been carried out. Therefore, happily, we can end this section by saying that here is a field ripe for exploration and as yet virtually untouched. It is clear, however, from the works of Zeevi (1972), Goldstein and Rall (1974), and others who have penetrated the frontiers of this area that a very large obstacle will be the design of efficient computer algorithms for managing these complicated systems in reasonable times and at reasonable costs.

PART III

NEURAL CODING: MODELS OF ELECTRICAL SIGNAL PROCESSING

In this section we discuss quantitative techniques for describing and modeling the ongoing processing of electrical signals in neurons and groups of neurons. The techniques are best conceived as tools to help discover and predict the overall significance of neuronal electrical signals with respect to the networks and systems they are recorded from and, often, in terms of their underlying physiological mechanisms. These activities, together with certain experimental electrophysiological investigations (primarily at the single unit level) comprise the area known as *neural coding*. Such activities are also clearly included in an only slightly expanded interpretation of the term *neural modeling*. The first three chapters of the section focus primarily on the generation and characteristics of spike trains in individual or small numbers of neurons. Chapter 8 discusses various electronic, mathematical, and computer devices for simulating the ongoing electrical activity of patches of neural membrane—devices which we call *neuromimes*. The next two chapters consider the application of the theory of stochastic processes to characterize and model time series of action potentials. The last four chapters of the section consider models for describing ongoing activity in large numbers of neurons interconnected in various types of networks. Chapter 11 considers models intended to describe or simulate activity of hundreds or thousands of interconnected neurons, while Chapters 12 and 13 consider analytic and computer-oriented approaches, respectively, to describing the activity of networks containing vast numbers (tens of thousands to millions) of interconnected neurons. Chapter 14 reviews quantitative models for field potentials, slow waves, and EEG signals, while Chapter 15 discusses models of specific neural networks.

8

Neuromimes

A neuromime is a device for simulating the electrical activity of a patch of neural membrane. The interest in neuromimes derives mainly from their application to simulating input–output activity of entire neurons according to mechanisms for generating trains of action potentials by postsynaptic potentials. This interpretation is approximately valid as long as the action potentials of the neuron in question can be considered to originate from a single main "triggering" section. A group of neuromimes may be interconnected with appropriate time delay elements to simulate the activity of a small neural network. Alternatively, several neuromimes may be used to represent the activity of a single neuron. Here, each neuromime represents a patch of membrane and is interconnected with the others by way of intervening resistances according to the morphology of the cell to be simulated.

Typically, neuromimes are constructed with electrical circuitry, mathematical equations, or digital computer algorithms. (Less commonly, various analogous physical processes have been used.)

Neuromime studies may be evaluated according to the dynamic properties of the model itself, and also according to the way or ways in which they are used. Thus, one may characterize a neuromime according to: (a) the way in which it accepts input and how various input events interact. Is there, for example, nonlinear interaction of PSPs according to simulated conductance changes, or are positive and negative currents merely added? (b) the way in which output spike trains are triggered and the manner in which accomodation is included or ignored; (c) the way and extent to which refractory properties are included in the model. Moreover, in recent years it has become possible to take a critical look at the way in which neuromimes are used. In earlier, more optimistic days neuromimes were built with the idea that if one built them according to the right rules one had simply then to turn them on and thereby find out how "standard" neurons behaved. One supposed that small numbers of neuromimes could be hooked together in various canonical arrangements to spell out patterns of activity characteristic of various arrangements of cells which ould then be identified in various nervous systems. It has been increasingly realized, however, that neuromime studies, to be most fruitful, should be directed to specific systems where model behavior and predictions can be compared directly to experimental behavior. Under these conditions, parameter values can be explicitly fixed, and the theoretical system can be tested in penetrating detail. The chore of describing the operating characteristics of a particular neuron model is not very instructive if that model is not related to any real situations.

In the last 15 years or so, accompanying the intensive development of microelectrode studies of single neurons, many neuromimes of various degrees of complexity have been designed. In this chapter we will describe a selected few of these that are representative of the class as a whole. Our apologies to researchers whose models have been necessarily, but with no disparagement intended, omitted from the discussion. The approach will be essentially chronological.

One of the earliest and most representative of neuromimes was presented by Harmon in 1961. Harmon's model was an electronic circuit representing a two-state-variable simulation of neural excitable processes. The circuit diagram for this model, along with a picture of its realization, is shown in Figure 8-1.

One state variable represents the transmembrane potential; the other variable represents a threshold. The model includes nonlinear interaction of simulated synaptic inputs, and triggers spikes according to a threshold rule. Accommodation is approximated by supposing that excitation of the model leads to two processes: The usual depolarizing process is followed by a hyperpolarizing process with a larger time constant. Absolute and relative refractoriness are included by elevating the threshold for excitation above its resting level when an output spike occurs, and subsequently letting the threshold decay exponentially back to its resting level. In the absence of output action potentials the threshold remains at its resting value.

Harmon (1961*a,b*) has obtained various elemental input–output relationships from this model. These include the occurrence of a pronounced burst output from the model at the onset of a step input when the accommodative mechanism is pronounced. Harmon showed that under these conditions the number of spikes in a burst is proportional to the time constant of the accommodative process. He also found that for the case of a single cell being driven by a single regular series of PSPs, the output firing rate depends on the input firing rate in a staircase manner, as shown in

a

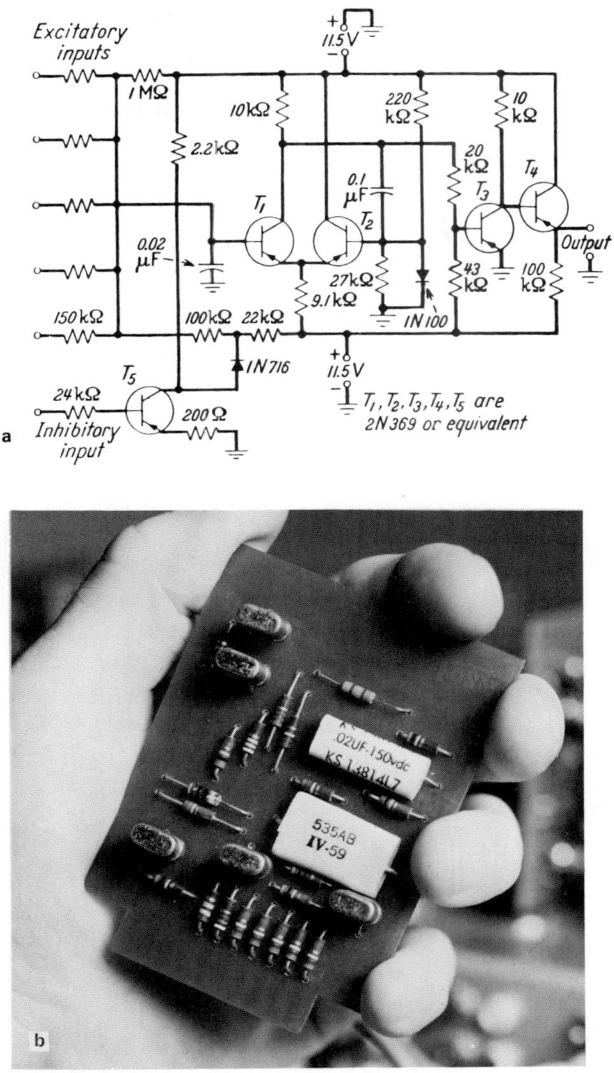

b

Figure 8-1. Early electronic neuromime built by L. D. Harmon of Bell Laboratories. (a) Circuit diagram of artifical neuron; (b) photograph of its realization. (Photograph courtesy of L. D. Harmon.)

Figure 8-2; i.e., the overall firing rate transfer relationship consists of a series of plateaus, each of which is characterized by a constant ratio of input PSPs per output spike. As the input firing rate is increased, this ratio decreases from one plateau to the next. Between plateaus various other

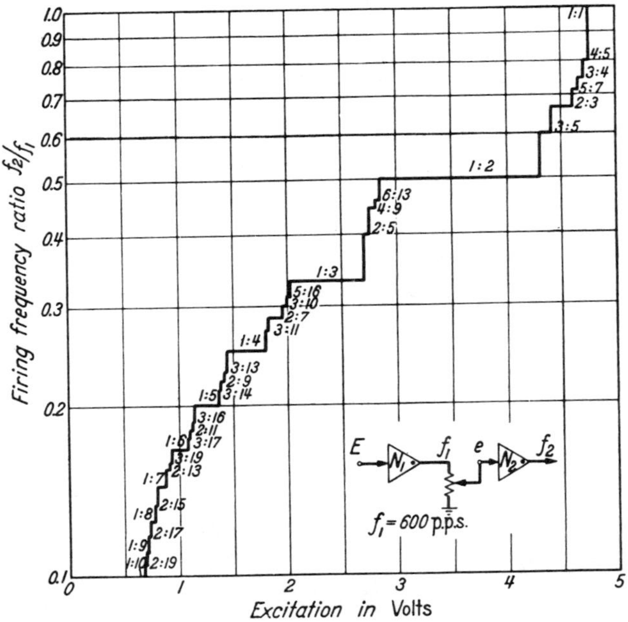

Figure 8-2. Firing response of one unit being driven by another. "Following" is precise at high levels of excitation. As the driven unit receives progressively smaller stimuli, it begins to drop pulses. The 1:2 and 1:3 steps result from ordinary temporal summation. Other ratios of firing frequencies stem from another phenomenon, caused by asynchrony between driving pulses and the refractory period of the driven unit.

regularly repetitive patterns occur. This effect is a characteristic feature of activation of spike potentials according to the threshold rule by summating PSPs.

Harmon (1964) examined the properties of a pair of mutually inhibiting neuromimes. This arrangement was thought to reflect a basic mode of neuronal organization, particularly in motor control. The arrangement had been modeled earlier by Reiss (1962). Reiss used a three-state-variable neuromime realized on a digital computer. The three state variables in Reiss' model are a membrane potential S, a threshold θ, and a "fatigue" mechanism. A variable E corresponding to the stored energy of the neuron was defined such that E was a maximum when the cell had not fired in a very long time and decreased an incremental amount according to each output spike of the cell. The behavior of E is determined by

$$E' = E - Q\frac{E}{E_{\max}}$$
$$E = E' + (E_{\max} - E')e^{-1/Pt} \tag{8-1}$$

The value of E is reduced to E' immediately after an output spike. When the unit is not spiking, E returns to E_{max} according to the second of equations (8-1). The quantities Q, E_{max}, and P are parameters which determine the degree of fatigue and the recuperative and fatigue rates. Reiss' model includes refractoriness by elevating the threshold and letting it decay to its resting level according to equation (8-2), immediately after each output spike:

$$\theta = \frac{1/E + t^2}{t + t^2} \tag{8-2}$$

This form was chosen to allow for a supernormal period following refractoriness, as had been observed in mammalian axon membranes. Fatigue is then effected by having the decay rate of the threshold increase with E. Thus, a cell which has fired a large number of spikes in a given time period will have a smaller value of E than normal, and, therefore, the threshold will decay to its normal level from its postspike elevated level more slowly than normal. With this mechanism in his units, Reiss found what he called the "multivibrator effect" for mutually inhibiting pairs of neuromimes driven by a common source. This arrangement and effect are shown in Figue 8-3. The effect may be readily interpreted in terms of a struggle for dominance. Thus, one neuron momentarily gains the upper hand and is able to keep the other neuron turned off with the inhibition it imposes upon it. Subsequently, however, it fatigues, whereupon the second neuron is able to fire spikes and, being refreshed, gains the upper hand and is able to hold the first neuron down by its inhibition. Subsequently, this neuron fatigues and the process repeats. Two kinds of "switchover" transition between the firing of cell a_2 and that of cell a_3 are seen in Figure

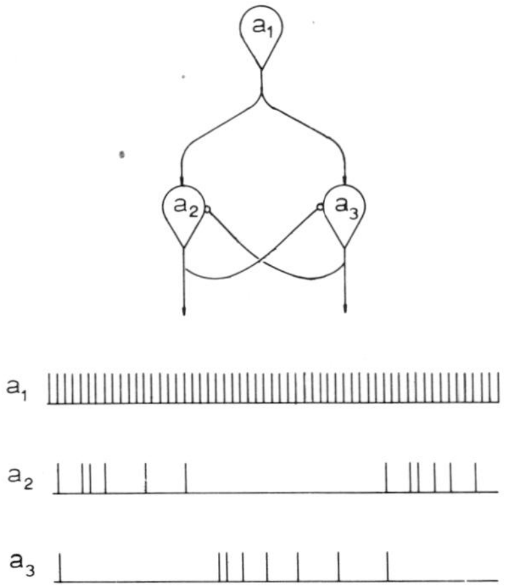

Figure 8-3. Two-unit reciprocal inhibition. Top: Driving unit a_1 delivers excitatory impulses in common to cross-coupled pair a_2, a_3. Bottom: Units a_2 and a_3 fire in alternating bursts.

8-3. The switchover from a_2 to a_3 does not involve synchronous firing of the two units; an appreciable gap appears between the end of the a_2 burst and the beginning of the a_3 burst. A second type of switchover appears when a_3 loses dominance; both units fire simultaneously at the transition. The first kind is associated with takeover by a unit having less sensitivity; synchronous switchover, on the other hand, appears to reflect the accession to dominance of the intrinsically more sensitive unit. This alternate bursting of spikes prevails over a wide range of moderate input rates. At rates less than about 20/sec, one unit dominates completely. At rates higher than about 100/sec, either one unit dominates or there is rapid, mixed firing in both units.

These response patterns are very similar to those observed in reciprocally innervated flexor and extensor muscles in the legs of locusts. As the insect walks progressively faster, the alternation rate increases. With very rapid walking, the extensor is continuously excited (dominant neural control) and provides an elastic band against which the flexor works differentially (via rhythmic burst input).

Harmon also used his neuromimes to investigate a mutually inhibiting pair of neurons, as indicated in Figure 8-4. His time constants were considerably shorter than in Reiss' model. Because of this, alternating dominance took the form of alternations and interchanges of combinations of a few spikes, as may be seen in Figure 8-4, rather than alternations of long bursts of spikes. Harmon showed that these various patterns changed in discrete steps with changes in stimulus rates and, moreover, exhibited hysteresis. This circuit was originally arranged to simulate the locust flight motor systems. Although accounting in a simple way for alternating spike-pair patterns seen in the locust thoracic ganglion, the observed behavior differed from the model's behavior in showing long pauses between bursts and acceleration of spikes within bursts. This work was taken up by Wilson (1968) and is discussed later in this chapter.

Harmon's work is representative of a large number of electronic neuromimes which were constructed and used in the early 1960s. Other similar neuromimes were presented by Jenik (1965) and by Kornmuller [see Reiss (1964)]. This modeling activity is admirably summarized in the proceedings of a conference held at Ojai, California, in 1964 (Reiss, 1964).

Two investigators at the Rand Corporation performed digital computer neuromime studies in the mid-1960s. Perkel (1964*a,b*) presented a two-state-variable neuromime with transmembrane potential and threshold. The main rules for this model are shown as

$$\begin{aligned} \theta(t) &= \theta_R + [\theta(0) - \theta_R]\exp(-\mu t) \\ p(t) &= p_R + [p(0) - p_R]\exp(-\lambda t) \\ p_{new} &= fp_R + (1-f)p_{old} + D \end{aligned} \tag{8-3}$$

PSPs are effected by a sudden quantal jump in membrane potential triggered by an input event, followed by an exponential decay. The model includes an option for diminution of PSP amplitude in proportion to the prior potential and thereby includes nonlinear interaction of the PSPs according to the conventional model of synaptic conductance changes. Spikes are triggered according to a threshold rule. No accommodation is included in the model. Absolute and relative refractory periods are included by elevating the threshold when an output spike occurs and subsequently letting it decay exponentially to resting level. The cell level is reset at zero

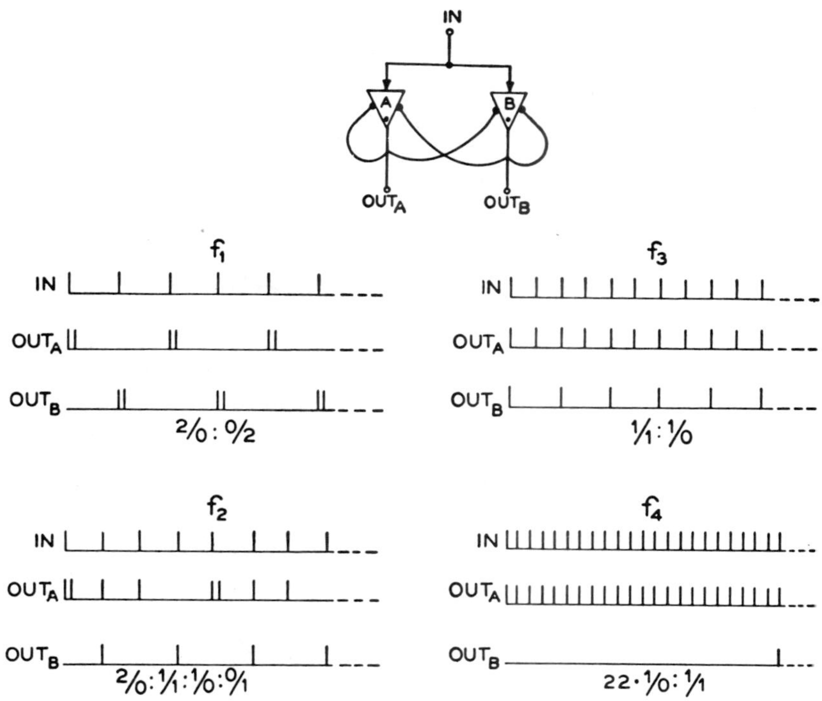

Figure 8-4. Four typical firing patterns derived from a pair of reciprocally inhibiting neuromimes driven in common by a constant-frequency spike train. All patterns repeat indefinitely. f_1: alternate double firings similar to neuromuscular patterns in insect flight motor systems. The name given to the reponse pattern is A/B: A/B: . . . , each colon marking one stimulus period, and the entire name representing the response period: thus, 2/0: 0/2 indicates that, for the first stimulus impulse, unit A fires twice and unit B not at all, while for the second stimulus, A is silent as B fires twice, and then this pattern repeats. f_2: at a slightly higher common input frequency the pattern of A becomes more complex, and *B* fires on alternate input stimuli. f_3: for still higher input frequency, unit A fires with each stimulus spike while B fires on alternate ones. f_1: at a relatively high frequency, A fires 22 times while B remains silent; B then fires once (as does A), and the cycle repeats.

following the occurrence of an output spike. Perkel (1964*a,b*) looked at many input–output transfer relations, particularly examining the influence of randomness in input patterns on the shapes of output interval histograms, autocorrelations, and the like.

A simple, interesting application of Perkel's neuromime to the interaction of small numbers of pacemakers—or regularly firing cells—was presented in 1964. The model was thought to be relevant to various invertebrates which exhibit many pacemakerlike cells. Perkel simulated a single endogenously active pacemaker unit activated directly by excitatory or inhibitory pacemakers. The main results are shown in Figure 8-5 which displays the firing rate of the activated pacemaker as a function of the firing rate of the excitatory input pacemaker (to the right of the origin) or that of the inhibiting input pacemaker (to the left of the origin). The most salient point is the stepwise character of the curve. This curve consists of a series of slanted plateaus each of which corresponds to a constant integral ratio of input PSPs to output spikes. As in Harmon's staircase effect, this phenomenon reflects basic rules of interacting PSPs. Perkel *et al.* (1964) showed that data from a thoracic stretch receptor of *procambarus clarkii* are

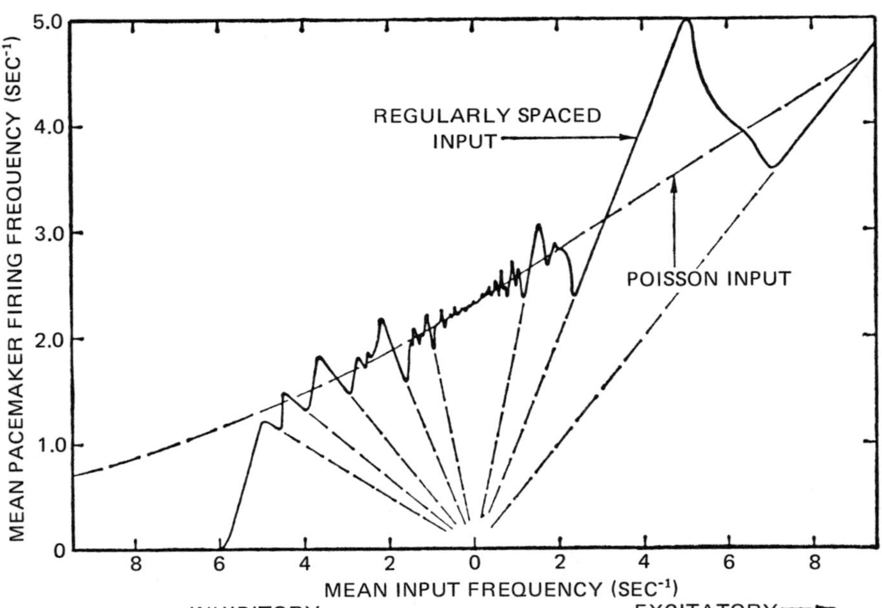

Figure 8-5. Mean firing frequency of computer-simulated neuron as a function of mean input frequency. Inhibitory frequencies are plotted to the left of the origin, and excitatory frequencies to the right.

matched by this model very well. The agreement includes the apparent paradoxical effect that increasing IPSP rate often leads to an increase in overall firing rate rather than a decrease as one would naively suspect. Finally, Perkel showed that activation of an endogenously active pacemaker by excitatory or inhibitory inputs with interspike intervals distributed according to a Poisson rule gave rise to the firing-rate curve shown as the dashed line in Figure (8-5). Clearly, the plateaulike behavior for the regularly spaced input appears as deviations around the smooth curve for the Poisson input.

MacGregor (1966*a*, 1967*b*) also investigated various input–output relationships using a digital computer neuromime. The main relationships for this neuromime are shown here as

$$
\begin{aligned}
E(t) &= \mathrm{ER} + \sum \mathrm{EPSP}_i - \sum \mathrm{IPSP}_i \\
\mathrm{ER}(t) &= A[e^{-c(t-\tau)} - ke^{-d(t-\tau)}] \qquad t > \tau \\
\mathrm{EPSP}_i &= \xi_j[e^{-\alpha(t-\tau_i)} - e^{-\beta(t-\tau_i)}] \qquad t > \tau \\
\mathrm{IPSP}_i &= \delta_j[e^{-\eta(t-\tau_i)} - e^{-\chi(t-\tau_i)}] \qquad t > \tau_i \\
\alpha &= \alpha_\infty[1 + re^{-n(t-\tau)}] \\
\eta &= \eta_\infty[1 + re^{-n(t-\tau)}] \\
V_c(t) &= V_{c\infty}[1 + ae^{-b(t-\tau)}] \qquad t > \tau
\end{aligned}
\tag{8-4}
$$

An output spike is generated when the membrane potential E exceeds the threshold V_c. The quantity τ is the time since the previous output spike. This model included nonlinear interaction of PSPs by letting the amplitudes of PSPs, ξ_j and δ_j, in equations (8-4) be dependent on the potential at the time of their initiation. No accommodative behavior was included. Refractoriness was determined by three mechanisms. The threshold was elevated upon firing and decayed to resting level exponentially. Also, upon firing, the membrane potential went through an appropriate after-hyperpolarization phase as given by ER. Finally, PSPs were flattened and their time courses altered according to the expressions given for the constants α and η. The latter expressions were determined by matching appropriate data on the influence of action potentials on the shape of PSPs from the publications of Eccles (1957, 1964). (See Figure 2-8 in Chapter 2.)

This model produced the staircase frequency-transfer relationships for regular input which had been found by Harmon (1961*b*) and by Perkel (1964). The frequency transfer for various irregular temporal patterns of input is illustrated in Figure 8-6. The smooth line in this figure represents the staircase relationship for regular input. No simple rule was found to

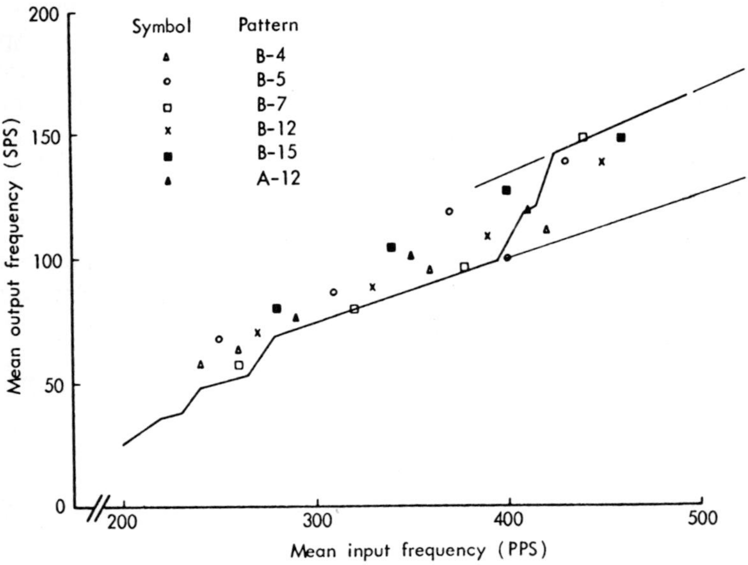

Figure 8-6. Frequency transfer with irregular input.

relate the effects of temporal pattern to frequency transfer; the coefficient of variance of the input was not systematically related to the transfer ratio. A general finding was that the coefficient of variance of output patterns was consistently smaller than that of the corresponding input pattern.

MacGregor also indicated in broad terms how the probability that a given input event will be associated in time with an output spike depends on the level and character of background activity. One can say that the probability that a given input event E will produce an output spike in the presence of other input activity O is an increasing function of $[e + o - \text{th}(o)]$, where e and o are the amplitudes of the graded event associated with E and O, and th is the threshold. The threshold is an increasing function of o because of refractoriness; i.e., increasing o increases the probability that the cell will have fired in a short time interval before E, and that E will therefore arrive during a refractory period. Any other input O then which tends to produce output spikes will, on the one hand, facilitate the effectiveness of E by summation of graded responses and, on the other hand, tend to decrease the effectiveness of E by increasing the likelihood that E encounters a refractory state. The facilitating effect should predominate for low values of o, whereas at high values of o the latter effect should predominate. The probability that E is associated with an output spike, then, should generally exhibit a maximum at an intermediate value of o for the period. This general mapping of the influence of background activity on the probability of spike elicitation by a given input event is shown in Figure 8-7. Computer simulation with this neuromime corroborated this mapping. This model was also used to explore the influence of inhibitory input patterns on frequency transfer, and also to show that frequency transfer relationships for axodendritic input patterns tend to saturate if one assumes that the individual PSPs take on very large values at their point of origin (MacGregor, 1968*b*).

In the mid-1960s, Lewis (1965, 1968*a*) constructed several elaborate electronic neuromimes. In the first phase of this work, Lewis was concerned with evaluating what he called the unifying ionic hypothesis: that "the Hodgkin–Huxley descriptions of electrically excitable conductances can be combined with the Eccles descriptions of synaptically induced conductances to account for most of the neuroelectric phenomena exhibited by single neurons." Lewis constructed neuromimes to simulate the basic membrane circuit shown in Figure 8-8, using the block diagram also shown there. Electronic circuits were specially designed to match the voltage- and time-dependent behavior of the potassium, chloride, and sodium conductances according to the Hodgkin–Huxley equations. With these circuits, Lewis made his model exhibit dynamic potential and conductance behavior equivalent to that of the Hodgkin–Huxley model. Moreover, Lewis used potentiometers so that one could systemically vary crucial parameters in the Hodgkin–Huxley formulation and thereby create various different kinds of artificial neurons. These variable parameters included the time constants of rise and decay of potassium and sodium conductance, their magnitudes, and the voltage sensitivity of the sodium conductance. With this model,

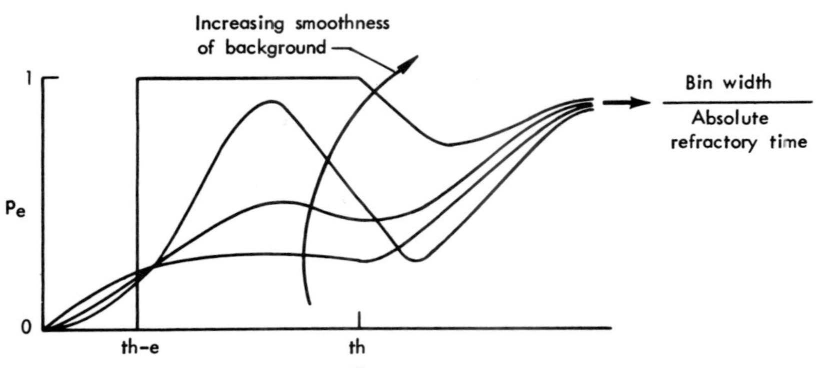

Figure 8-7. Influence of character and magnitude of background activity on the effectiveness of a given input event.

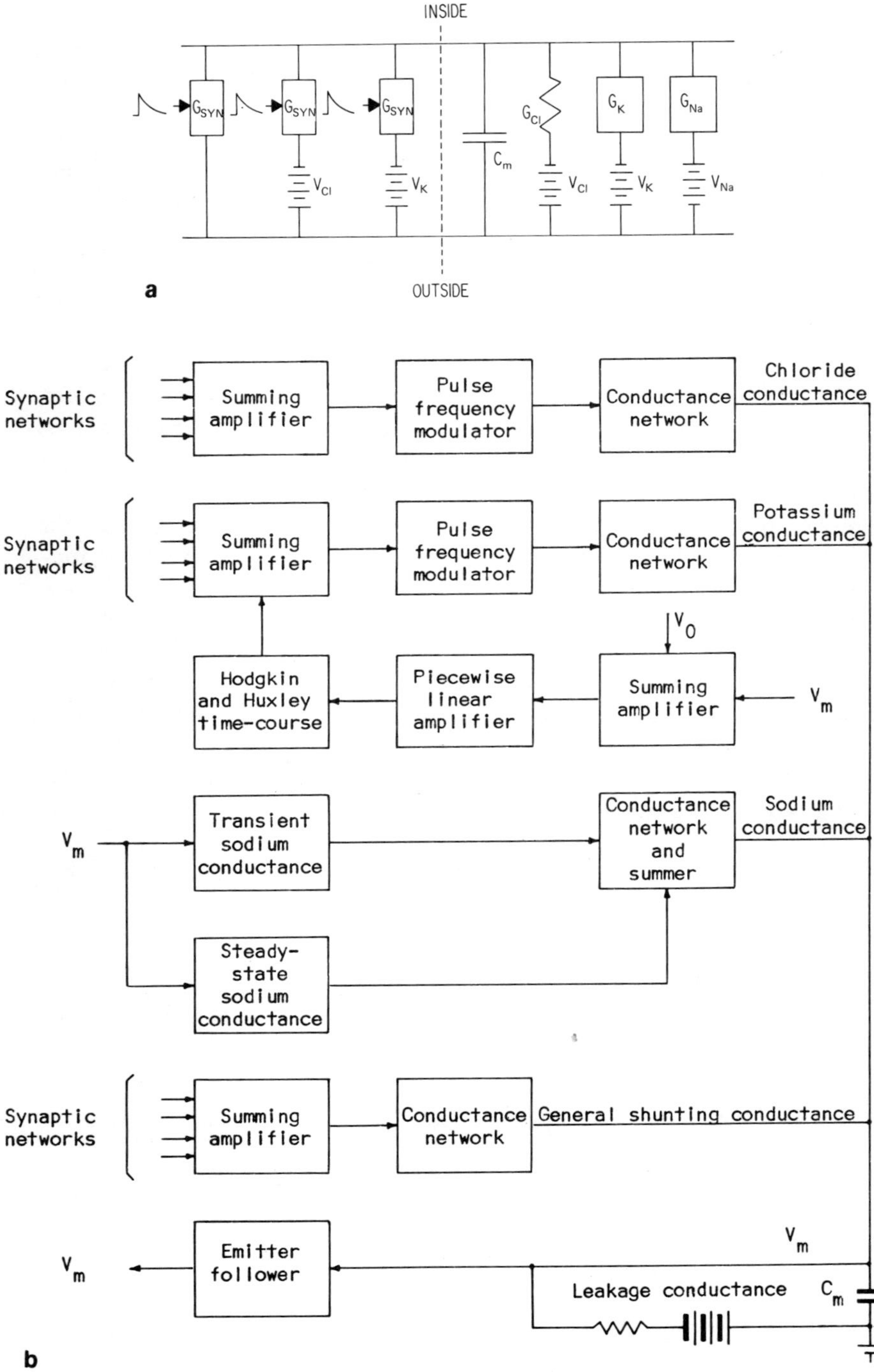
INSIDE
OUTSIDE
G_{SYN}
G_{Cl}
G_K
G_{Na}
C_m
V_{Cl}
V_K
V_{Na}
a
Synaptic networks
Summing amplifier
Pulse frequency modulator
Conductance network
Chloride conductance
Potassium conductance
V_0
V_m
Hodgkin and Huxley time-course
Piecewise linear amplifier
Transient sodium conductance
Steady-state sodium conductance
Conductance network and summer
Sodium conductance
General shunting conductance
Emitter follower
Leakage conductance
C_m
b

Lewis was able to account for most of the neuroelectric phenomena exhibited by single neurons, including almost all the categories enumerated by Bullock for synaptic responses (excitatory or inhibitory, and facilitating or antifacilitating) and spontaneous oscillations. However, Lewis noted that his model could not readily produce the one-cycle-per-second subthreshold oscillation observed in certain neurons of the lobster cardiac ganglion, nor would it produce the dc current–voltage characteristics of that ganglion.

In the second phase of this research, Lewis systematically simplified the circuitry of this extensive model, at each step ascertaining that the behavior of the simplified version still faithfully represented that of the full Hodgkin–Huxley model. Lewis then had 10 of these slightly simplified circuits built. Figure 8-9 shows a picture of one completed module together with circuit diagrams for the main conductance elements. The potentiometers on the front of the module allow one to adjust the indicated parameters g_{Na}, g_K, g_{Cl}, V_0 (the voltage sensitivity of g_{Na}), and C_m (the membrane capacitance). These 10 neuromimes were then connected by resistors to provide lumped approximations to an axon; in pairs to provide a two-patch neural model for an initial segment—somadendritic configuration; and with time delay elements to simulate small networks of interacting neurons.

To culminate this work, Lewis designed a simplified but highly flexible neuromine retaining many of the subtle properties of the more elaborate earlier versions. Its circuit diagram is shown in Figure 8-10. These units are very conveniently used to simulate small numbers of interacting neurons and have been used for that purpose by several investigators.

Wilson and Waldron (1968) used Lewis' neuromimes to simulate the generation of rhythmicity observed in patterned flight control in locusts. These authors note that in invertebrates it often seems that individual component cells are considerably different from one another, so that models with homogeneous systems of identical cells seem inappropriate. Thus, they feel that the use of Lewis' models, wherein one has considerable flexibility in adjusting parameter values from one neuron to another, is quite appropriate. In the control of locust flight, two major groups of motoneurons are involved which alternate in firing. One system is a wing elevator system, whereas the other is the wing depressor system. These major groups of motoneurons alternate in firing bursts of spikes, as shown

Figure 8-8. Block diagrams for neuromime built by E. R. Lewis. (a) Circuit equivalent of the combined Eccles, Hodgkin–Huxley descriptions. The elements on the left of the dashed line represent the synaptically induced conductances described by Eccles. Those on the right of the dashed line represent the elements of excitable membrane, as described by Hodgkin and Huxley. (b) Block diagram of the electronic system used to simulate the Hodgkin–Huxley model. The potassium and sodium conductance time-course networks are active electronic filters designed to fit the Hodgkin and Huxley data. The pulse-period modulators and conductance networks transform the filter outputs into equivalent conductances.

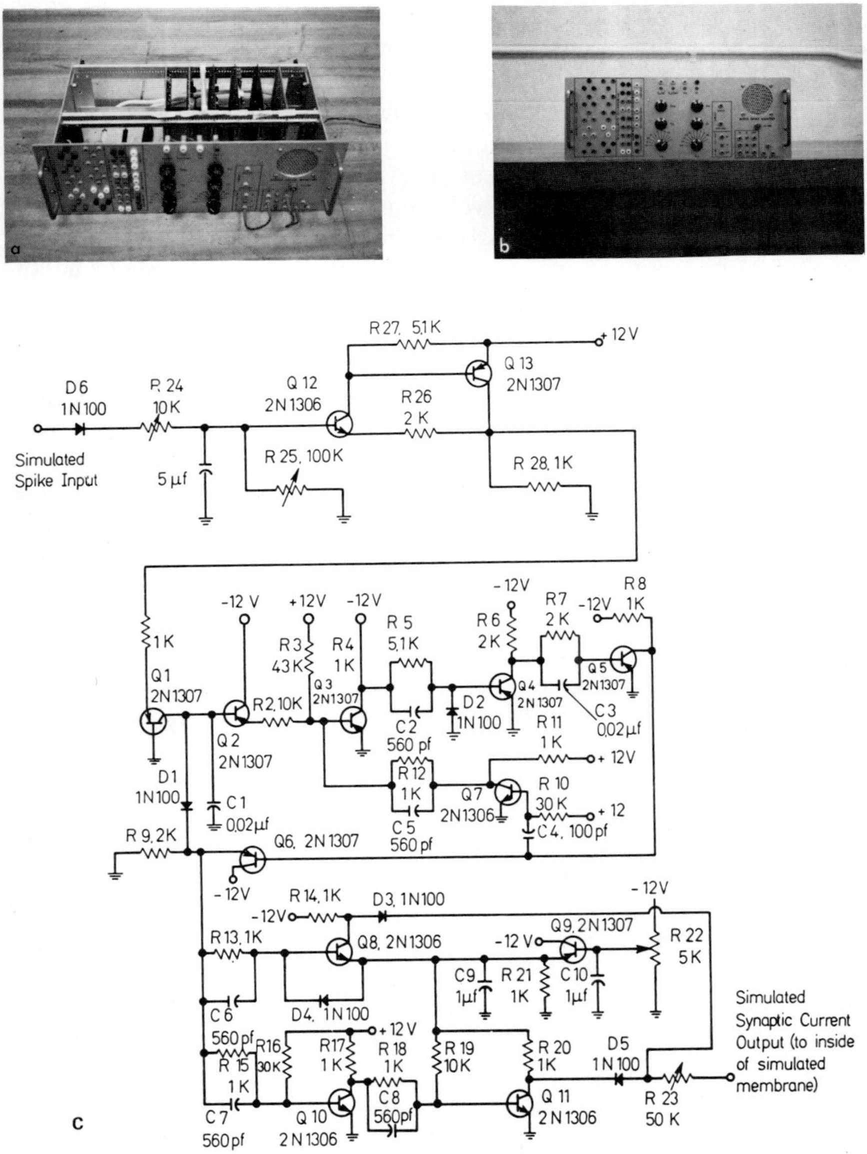

Figure 8-9. (a and b) Electronic neuromime designed by E. R. Lewis. (c) An electronic circuit simulating synaptic chloride-ion conductance.

in Figure 8-11. Within each burst there exists acceleration of spikes in that successive intervals are progressively smaller. As the level of excitation increases, the frequency of burst occurrence increases and, moreover, the

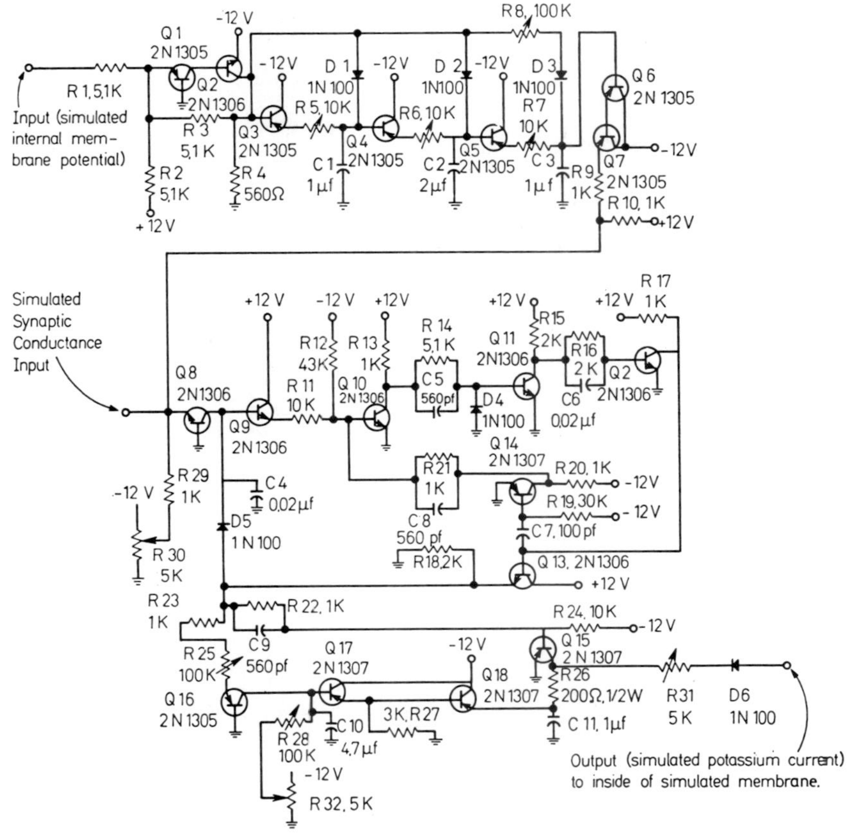

Figure 8-9. (d) An electronic circuit simulating the voltage-dependent potassium-ion conductance of the Hodgkin-Huxley description.

number of spikes per burst also increases. The basic flight pattern seems to be generally independent of the input and generated rather by interaction within and between these groups of motoneurons. Wilson and Waldron then tried to simulate this behavior, using first a pair of mutually inhibiting cells driven a common source, as indicated in Figure 8-12. They found that with this arrangement they could indeed get the alternations of bursts as Harmon and Reiss had described earlier, but that the arrangement was not satisfactory. In particular, there were no long pauses between the oscillations; i.e., one cell would start firing immediately when the other stopped. Moreover, as one increased the level of excitation, longer bursts occurred, but the burst rate decreased rather than increased as it did in the real data.

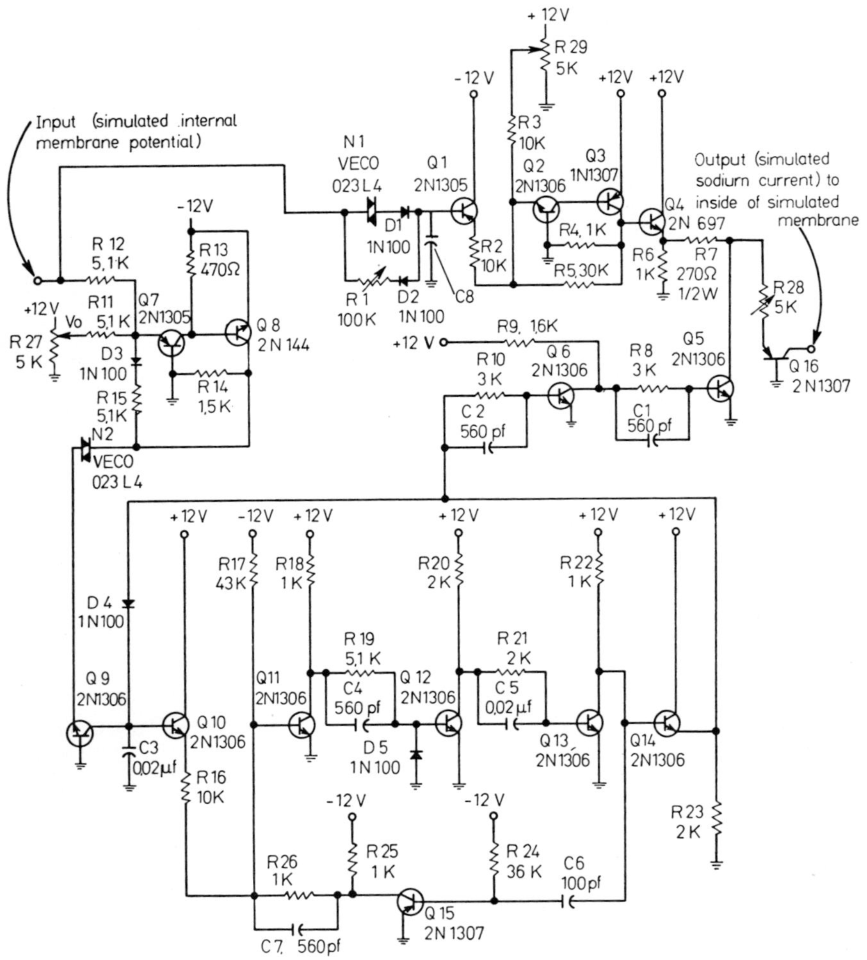

Figure 8-9. (e) An electronic circuit simulating the voltage-dependent sodium-ion current of the Hodgkin–Huxley description.

Moreover, there was a deceleration within the individual bursts.

As an alternate model, they considered an arrangement with positive feedback, as shown in Figure 8-13. This arrangement also led to the generation of bursts. Here, when the input activates the cell sufficiently to elicit action potentials, this in turn facilitates the production of more action potentials. This continues until the cell becomes sufficiently fatigued and refractory to stop the burst. A pause in firing is then associated with recovery from this fatigue. Within the bursts there is acceleration. By arranging mutual inhibition between two such positive feedback systems, they were able to match the observed behavior rather well. In this system cross-inhibition was not needed to stop the bursts and thereby generate the rhythm but rather to ensure antagonistic phase setting. Finally, they noted that this mechanism of pattern generation seemed quite similar to that proposed for the vertebrate respiratory system.

Fernald (1971), with help from Gerstein, also used Lewis' neuromimes to construct models to match output spike trains obtained from neurons in the cochlear nucleus of the cat in response to a tonal stimulus whose pitch was systemically varied. The data were in the form of poststimulus histograms (PST—see Chapter 10). The units were arranged to represent neurons having various patterns of dendritic trees. Figure 8-14 shows a block diagram of the system Fernald used to control the input stimulation of the model and a representative model experiment. With these simulations Fernald was able to match various PST patterns, including mirror symmetry, translation symmetry, and asymmetry, and to show PST shape constancy for various frequency-modulation rates.

Another electronic neuromime was constructed by French and Stein (1970). A flow diagram for this neuromime and its electronic circuit are shown in Figure 8-15. This model includes accommodation according to Hill's model wherein the threshold is increased according to the integral of membrane potential. The model includes absolute refractoriness by resetting the membrane potential at resting level and holding it there for the duration of the output pulse. Relative refractoriness is included by elevating the threshold upon occurrence of an output spike and letting the threshold decay exponentially to resting level. French and Stein used this neuromime to investigate basic unit input–output behavior with particular emphasis on noisy inputs. A representative example is shown in Figure 8-16. The main finding is that low-output firing rates are obtainable from low-level inputs if the inputs are noisy rather than constant.

Recently, MacGregor (1974) has developed a neuromime in an attempt to represent more accurately repetitive firing properties of neurons. The model is a four-state-variable representation defined here as

$$\begin{aligned} \frac{dE}{dt} &= -E - g_K(E - E_K) + [\mathrm{SC} - \mathrm{GE}(E - E_{ex}) - \mathrm{GI}(E - E_{in})] \\ \frac{d\theta}{dt} &= -\frac{(\theta - 1)}{\lambda} + \frac{cE}{\lambda} \\ \frac{dg_K}{dt} &= -\frac{g_K}{\tau} + bS \\ S &= \begin{cases} 1 & \text{if } E \geqq \theta \\ 0 & \text{otherwise} \end{cases} \end{aligned} \tag{8-5}$$

The four state variables are E, θ, g_K, and S, which represent, respectively,

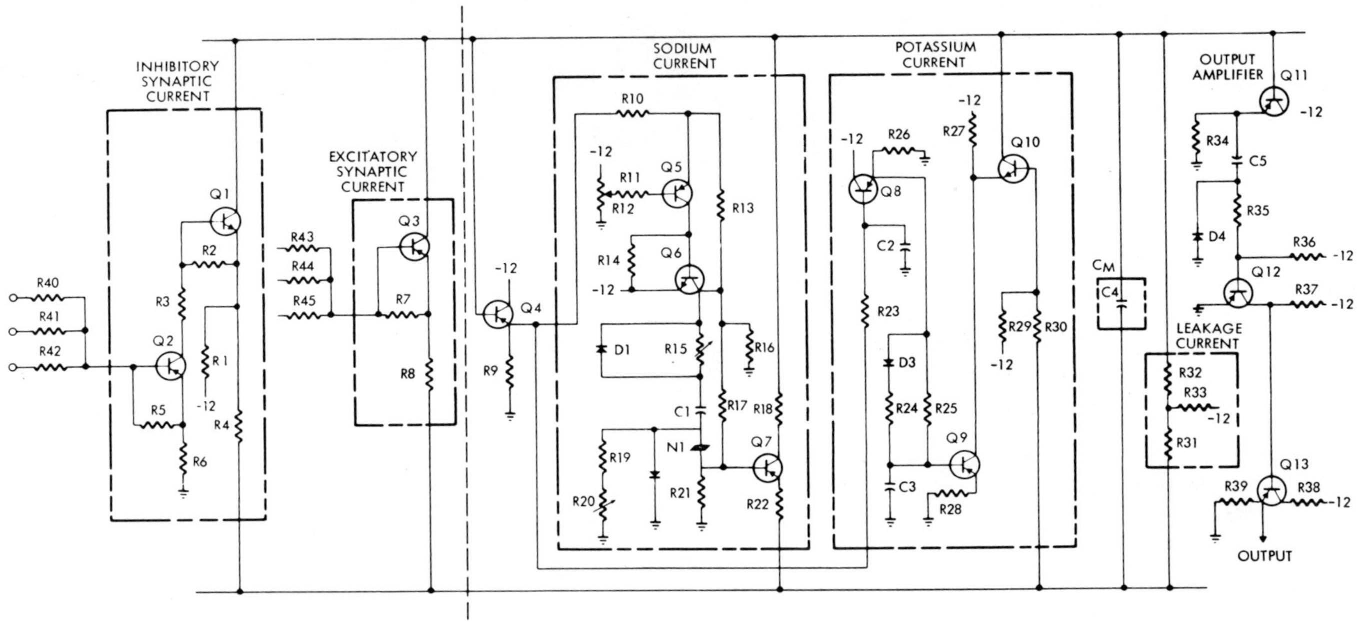

Figure 8-10. Simplified neuromime design by E. R. Lewis. A circuit designed to simulate a single patch of electrically excitable membrane (on the right of the vertical dashed line) with a contiguous patch of subsynaptic membrane (on the left of the vertical dashed line). Each subcircuit enclosed in a dashed box represents a shunt element across the neural membrane. The top of the circuit represents the inside of the membrane; the bottom represents the outside. Component values are given in Tables 8-1 and 8-2.

TABLE 8-1. Component Values for the Circuit of Figure 8-10 as a Trigger Region[a]

$R1$	1,000 Ω	$R24$	390 Ω	$Q1$	2N1306
$R2$	3,000 Ω	$R25$	10,000 Ω	$Q2$	2N1307
$R3$	1,000 Ω	$R26$	1,000 Ω	$Q3$	2N1307
$R4$	3,000 Ω	$R27$	560 Ω	$Q4$	2N1307
$R5$	10,000 Ω	$R28$	1,000 Ω	$Q5$	2N1307
$R6$	30,000 Ω	$R29$	1,000 Ω	$Q6$	2N1306
$R7$	1,000 Ω	$R30$	2,000 Ω	$Q7$	2N1307
$R8$	1,000 Ω	$R31$	1,000 Ω	$Q8$	2N1307
$R9$	3,000 Ω	$R32$	20,000 Ω	$Q9$	2N1307
$R10$	5,100 Ω	$R33$	1,000 Ω	$Q10$	2N1306
$R11$	1,000 Ω	$R34$	7,500 Ω	$Q11$	2N1307
$R12$	10,000Ω[b]	$R35$	7,500 Ω	$Q12$	2N1307
$R13$	20,000 Ω	$R36$	47,000 Ω	$Q13$	2N1307
$R14$	5,100 Ω	$R37$	2,000 Ω	$D1$	1N100
$R15$	100,000Ω[c]	$R38$	1,000 Ω	$D2$	1N100
$R16$	1,000 Ω	$R39$	10,000 Ω	$D3$	1N100
$R17$	100,000 Ω	$R40$	1,000,000 Ω	$D4$	1N100
$R18$	100 Ω	$R41$	1,000,000 Ω	$N1$	02134[d]
$R19$	100 Ω	$R42$	1,000,000 Ω	$C1$	0.5 μF
$R20$	10,000Ω[c]	$R43$	100,000 Ω	$C2$	0.22 μF
$R21$	2,000 Ω	$R44$	100,000 Ω	$C3$	1.0 μF
$R22$	510 Ω	$R45$	100,000 Ω	$C4$	0.2 μF
$R23$	3,900 Ω			$C5$	0.5 μF

[a] From Lewis (1968*a*).
[b] Potentiometer $R12$ controls the threshold for spike generation. The potential at the center tap of $R12$ should be between −6 and −7.
[c] Variable resistor $R20$ normally should be set at 1 kΩ or less; $R15$ should be set at 20 kΩ. Increasing these resistances will introduce pronounced cumulative refractoriness in the model.
[d] Varistor.

TABLE 8-2. Component Values for the Circuit of Figure 8-10 as an Integrative Region[a]

$R9$	1,000 Ω
$R13$	10,000 Ω
$R23$	20,000 Ω
$R24$	2,000 Ω
$R27$	2,400 Ω
$R28$	3,900 Ω
$C1$	5 μ F
$C4$	2 μ F

[a] From Lewis (1968*a*).

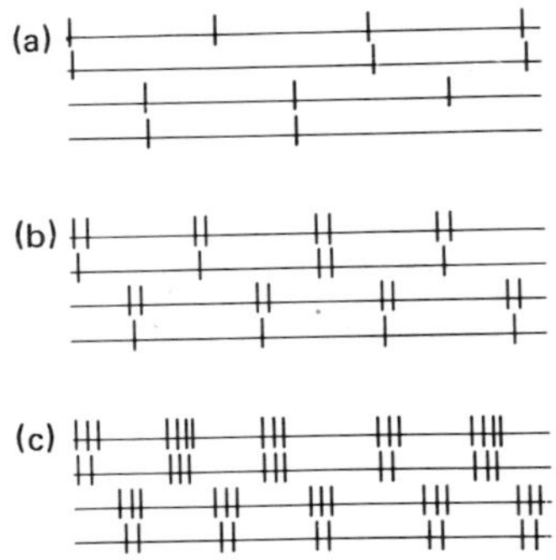

Figure 8-11. Schematic presentation of the motor output pattern of the locust flight-control system. Each set [(a), (b), or (c)] depicts four motoneurons divided into two synergistic groups that alternate with each other. (In the locust there are actually tens of motoneurons in each synergistic group.) In weak flight (a) not all motoneurons are excited, and most fire only once per wingbeat cycle. The phase of less excited motoneurons lags the more excited ones. During more energetic flight [(b) and (c)] the wingbeat frequency is higher, and the individual motoneurons each discharge short bursts of impulses in each wingbeat cycle. When there are three or more impulses per burst, the successive interspike intervals tend to decrease. Again the phase of the less active units lags the more excited ones.

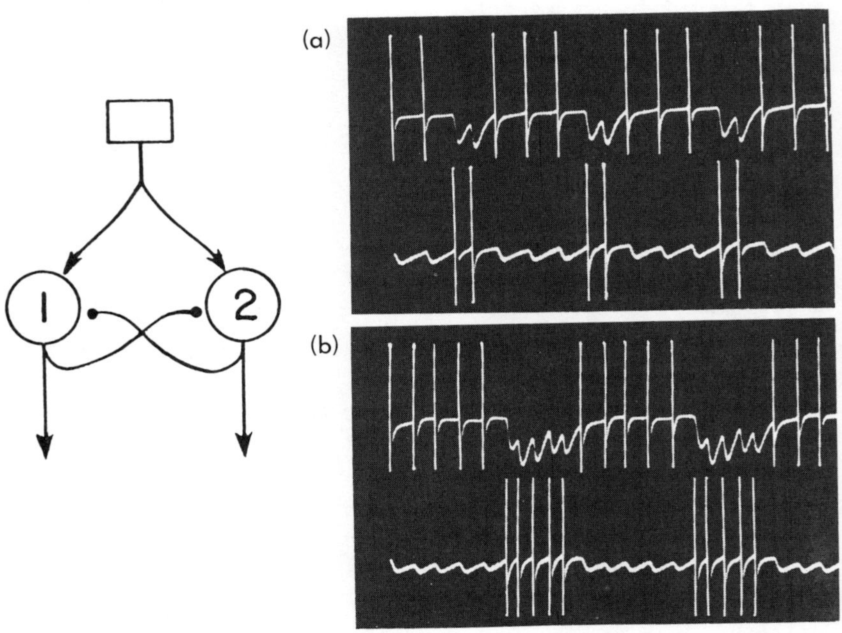

Figure 8-12. A two-membered reciprocal inhibition network may produce alternating bursts of impulses. Increasing input [(a) to (b)] gives rise to longer bursts, but at a lower rate of alternation, unlike the locust motor output. The pause between the end of one impulse burst and the first impulse in the antagonistic burst is not usually significantly greater than interspike intervals within the bursts. This feature is also unlike the locust pattern.

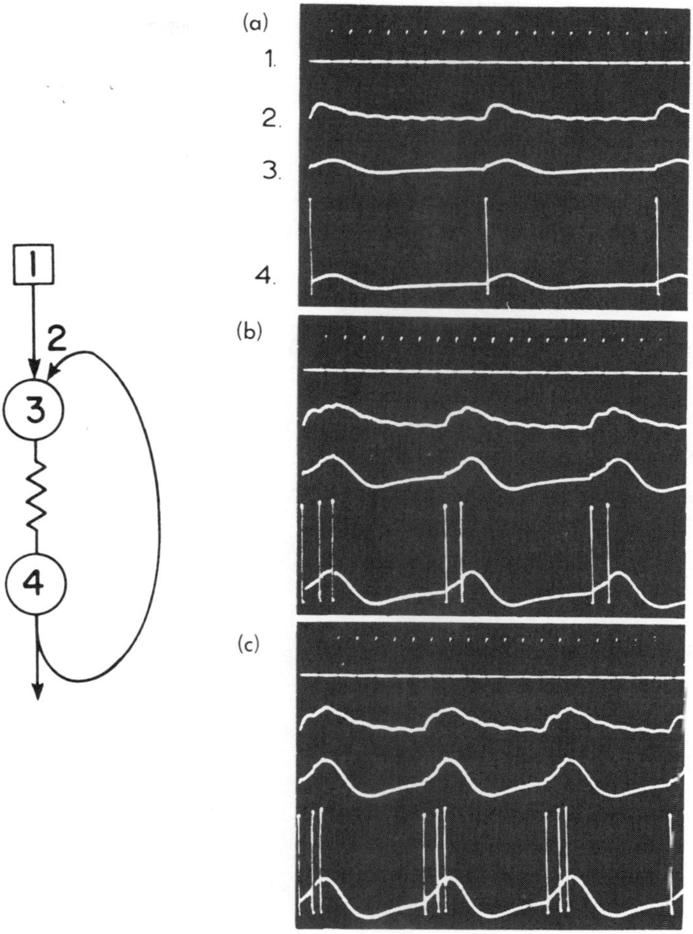

Figure 8-13. Some properties of mutually excitatory networks can be simulated by a single unit with positive feedback on itself. In the schematic network, *1* is an input source, *2* is an excitatory feedback synapse on *3*, *3* is a non-spike-producing, but active, membrane patch connected electrotonically to *4*, and *4* is a spike-initiating patch. In (a), (b), and (c), these four sites are monitored on successive oscillograph traces. The top traces show increasing strengths of high-frequency input pulses. The next lower traces show the nearly smoothly filtered resulting synaptic potential with the superimposed larger effect of each feedback impulse. The next-to-bottom traces show the potential of the active postsynaptic membrane. This potential is communicated effectively to the spike-initiating patch through a nonrectifying low-pass filter. The high-frequency spikes are barely seen in *3*. When the network is sufficiently excited, it produces impulse bursts. The bursts themselves show acceleration of the interspike intervals. Both these features are easily achieved in positive-feedback networks built of our neuromimes. This illustration shows that the locustlike relationship of the number of impulses per burst versus burst repetition rate can be achieved too. However, this relationship is relatively hard to produce, and it requires the presence of the active non-spike-producing patch *3* which introduces a slower regenerative process that is not reset by each spike.

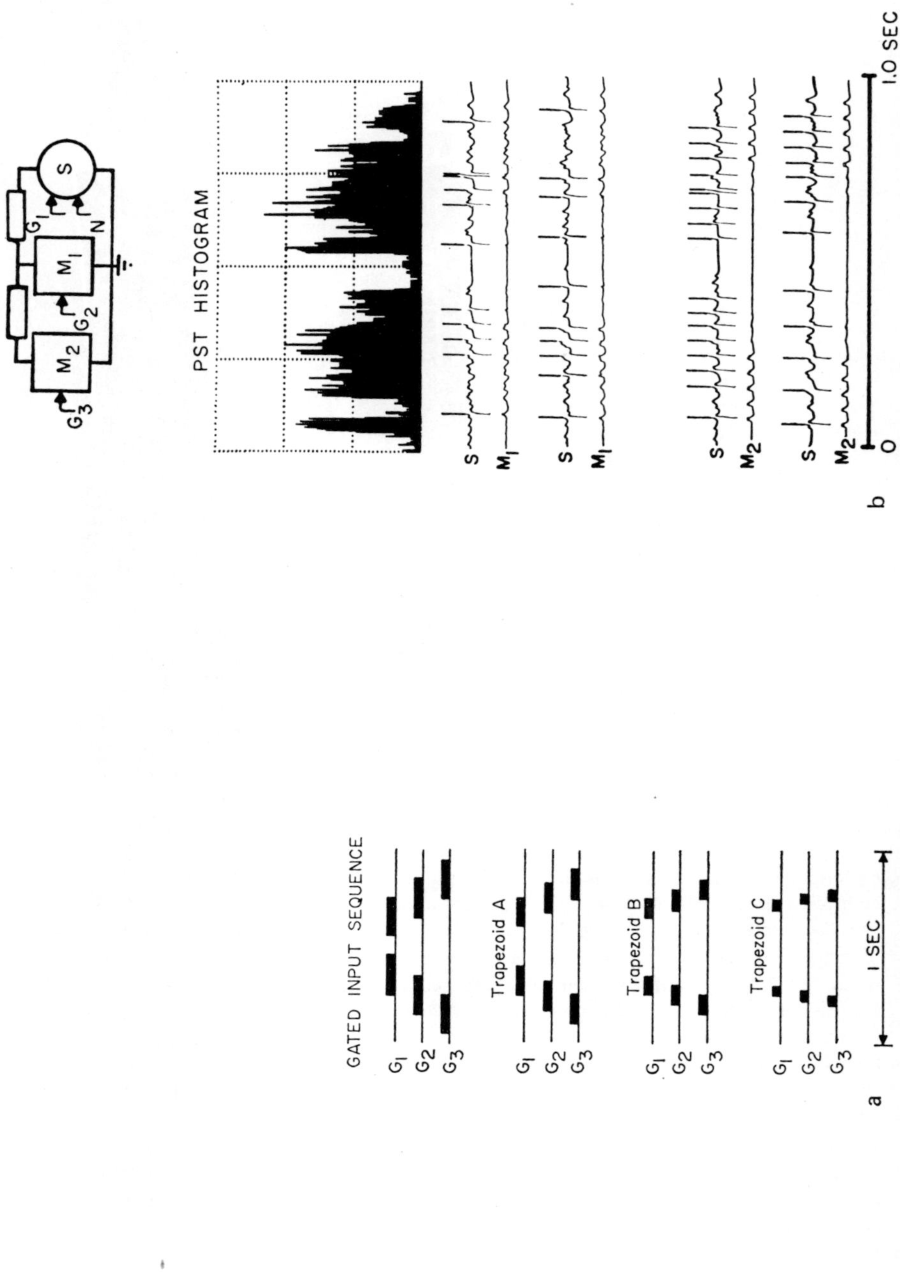
PST HISTOGRAM
S
M1
M2
G1
G2
G3
N
0
1.0 SEC
b
GATED INPUT SEQUENCE
Trapezoid A
Trapezoid B
Trapezoid C
I SEC
a

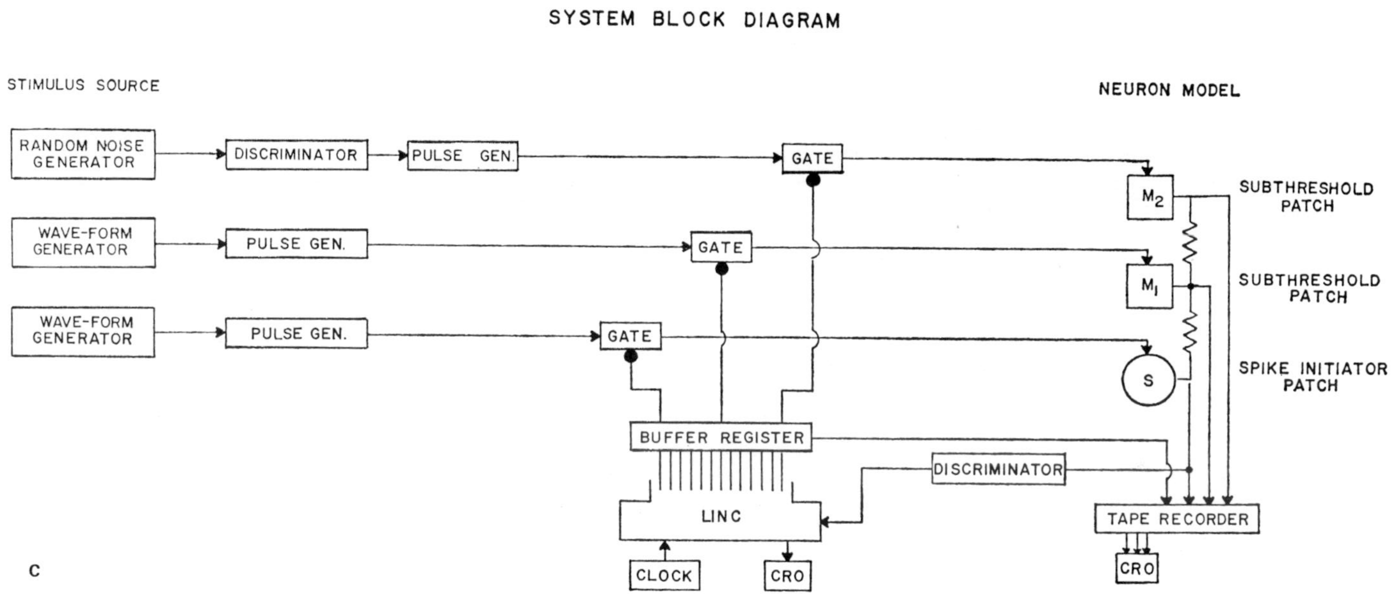

c

Figure 8-14. Neuromime as studied by Fernald. (a) Temporal conditions for the model stimuli. Each signal line, indicated by G, is turned on and off according to a programmed pattern. While the signal line is on, a pulse train of fixed frequency is delivered to the synaptic input. (b) Typical model configuration and the resulting output. The records indicate membrane potential measured at the spike-initiation patch and each subthreshold patch. The signal-line (gate) timing is shown in (a), top. The input labeled N is a random (Poisson) noise input. (c) Block diagram of system used to control the neuron model.

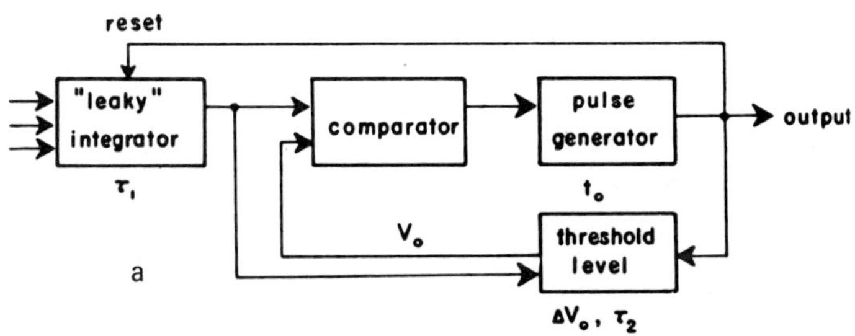

a

Figure 8-15. Neuromime presented by French and Stein. (a) Block diagram of the neural analog. (b) Detailed circuitry of the neural analog. Compensation components as specified by the manufacturers have been omitted from the diagram. Transistor T_1 is a 2N1305, and T_2 and T_3 are 2N3906s. The gates are all in one Texas Instruments 7400 integrated-circuit chip. All times indicated on switch positions are in milliseconds. The values of condensers were calculated and then trimmed to within 5% of the desired values. Similarly, resistors in the gain controls were calculated and checked to see that they were producing gains within 5% of values specified.

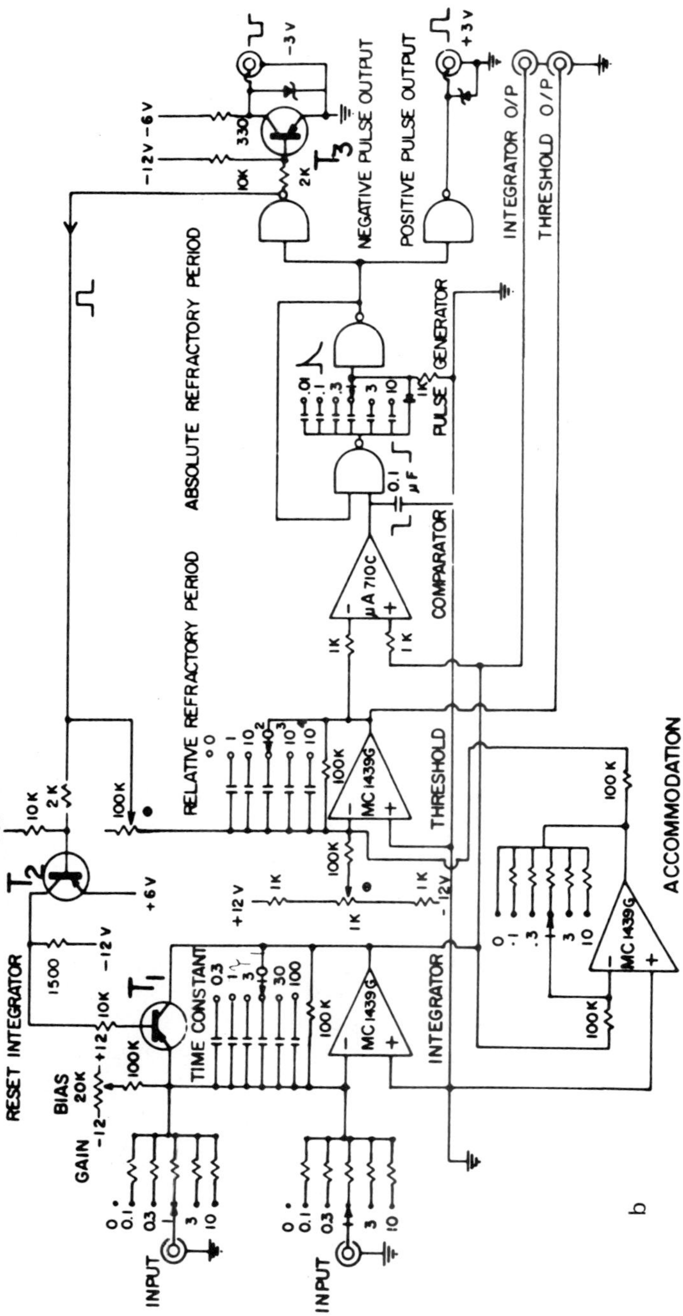
RESET INTEGRATOR
BIAS
GAIN
INPUT
INPUT
TIME CONSTANT
INTEGRATOR
MC1439G
THRESHOLD
RELATIVE REFRACTORY PERIOD
ABSOLUTE REFRACTORY PERIOD
μA 710C
COMPARATOR
PULSE GENERATOR
NEGATIVE PULSE OUTPUT
POSITIVE PULSE OUTPUT
INTEGRATOR O/P
THRESHOLD O/P
ACCOMMODATION

b

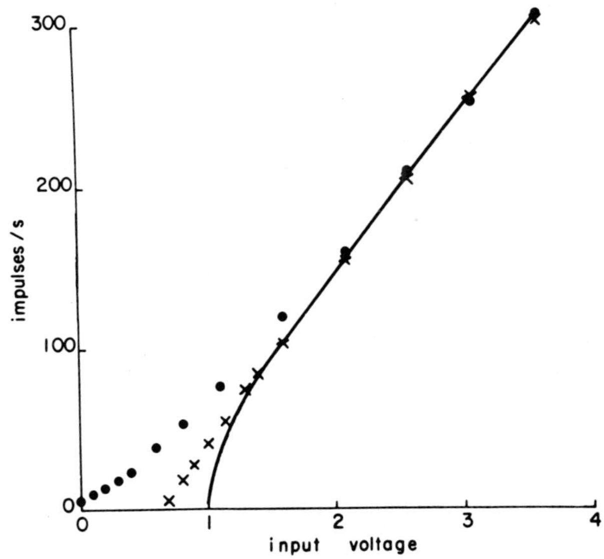

Figure 8-16. Firing rate versus current strength for neuromime presented by French and Stein. Output frequency of the neural analog as a function of the input voltage in the absence (solid line) and presence (data points) of noise. The white noise added has a root-mean-square amplitude of 0.4 V (filled circles) or 0.1 V (crosses) with a bandwidth effectively limited by the analog's time constant of 10 msec.

membrane potential, threshold, potassium conductance above resting level, and output spike train. In the last equation it is understood that S is set equal to unity and held there for one time unit when E is greater than θ. This neuromime has been realized by numerical integration on the digital computer and by the electronic circuitry shown in Figure 8-17. The model is activated by input current SC, excitatory conductance change GE, and inhibitory conductance change GI. Four parameters, λ, τ, b, and c, allow one to create specific behavior patterns of various types. The model includes nonlinear interaction of PSPs. Accommodation is included according to an extended version of Hill's model wherein the threshold rises at a rate proportional to the instantaneous value of the membrane potential. The parameter c determines the sensitivity of threshold rise; the parameter λ is the time constant of this rise. The strength-duration curve corresponding to this accommodative mechanism for a ramp current of slope m is given here as

$$I^* = \frac{t^*}{(1-c)t^* - \left\{(1-e^{-t^*}) + c\left[-(\lambda+1) + \frac{\lambda^2}{\lambda-1}e^{-t^*/\lambda} - \frac{1}{\lambda-1}e^{-t^*}\right]\right\}} \tag{8-6}$$

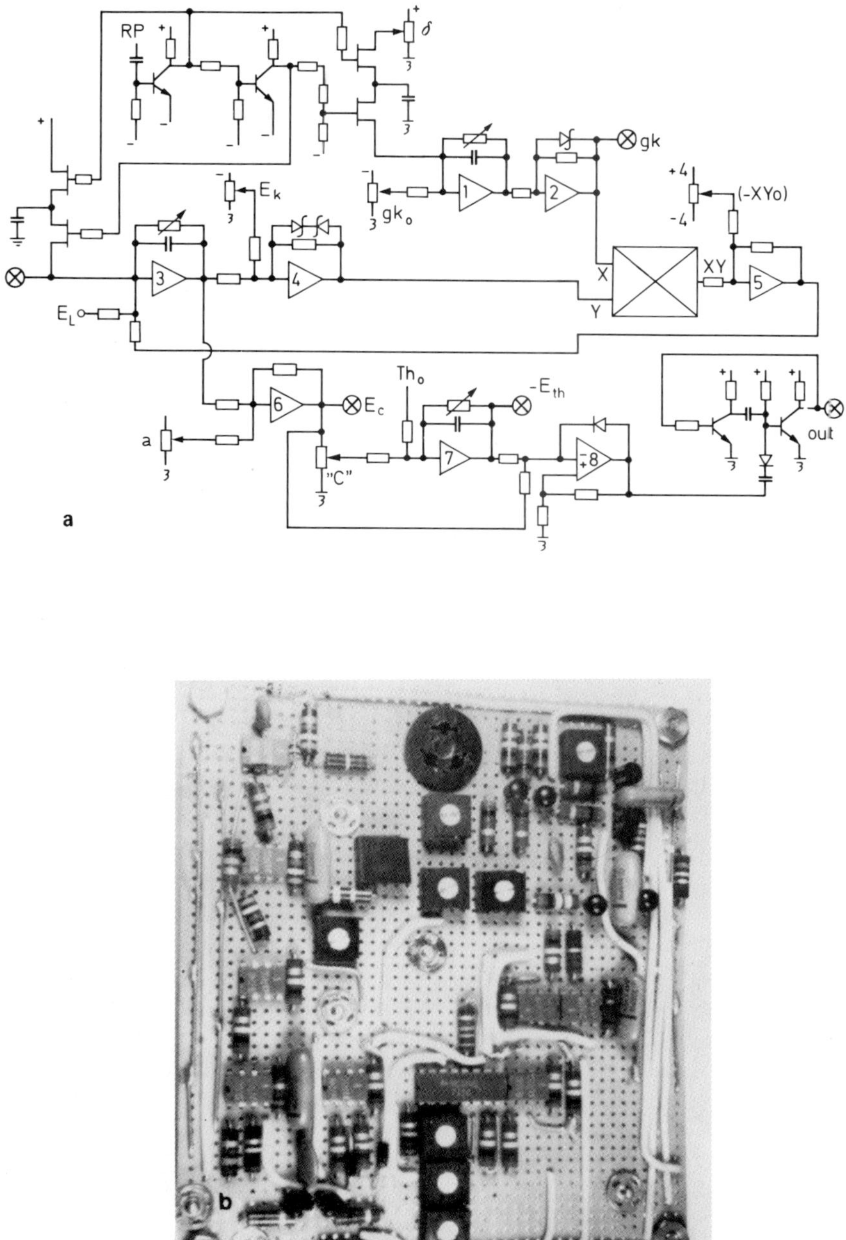

Figure 8-17. (a) Electronic neuromime design by MacGregor and Oliver. (b) "Cell" circuit. All Fet 3819. All Trs 5133. All Op Amps μA 741. Multiplier Mot. MC1594L.

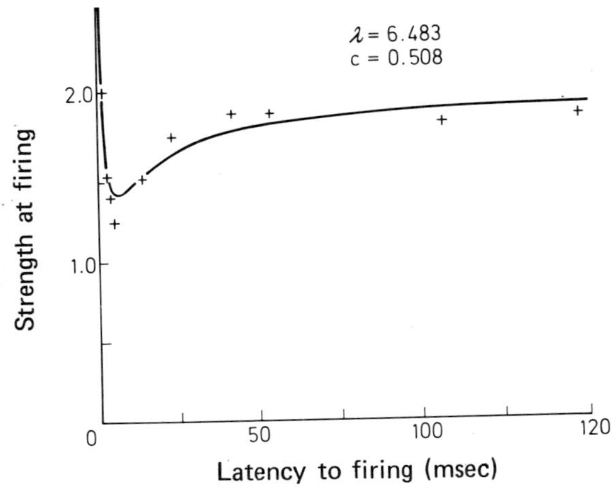

Figure 8-18. Strength-duration curve from neuromime matched to data from a "high-ceiling" unit presented by Bradley and Somjen. Data points are obtained from ramp-current stimuli with differing slopes.

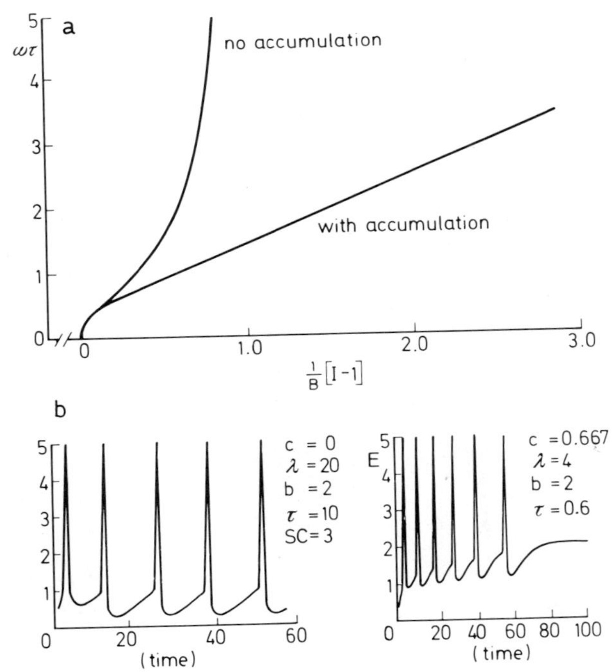

Figure 8-19. Adaptive behavior of model. (a) Representative after hyperpolarization curves. (b) Steady firing rate with and without accumulation of after-hypopolarization conductance.

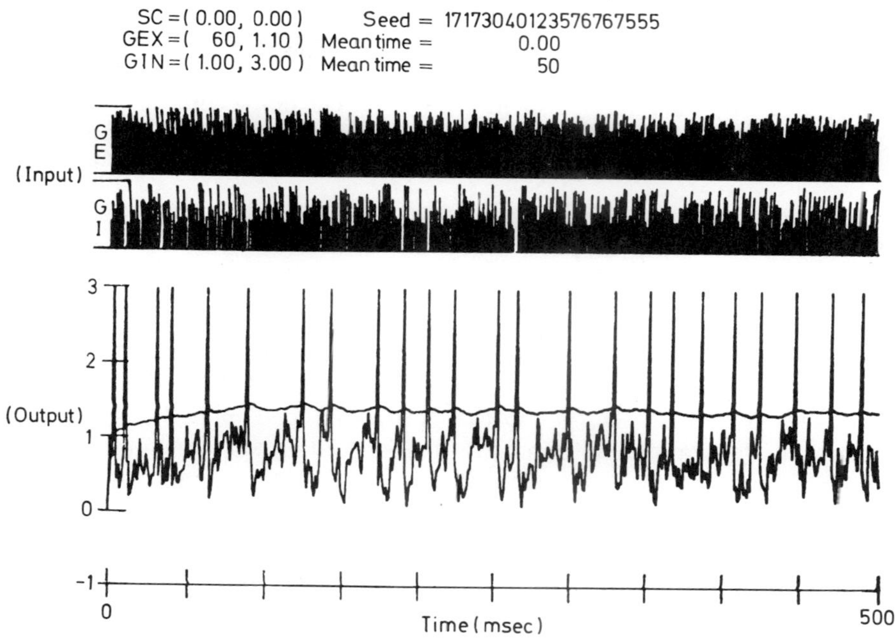

Figure 8-20. One-half-second sample of activity of neuromime when subjected to high-intensity excitation and inhibition.

This curve has been used to experimentally measure accommodative properties of real neurons by plotting the strength of current at excitation, I^*, versus the latency of activation, t^*, for ramp currents. By adjusting the parameters c and λ, one can match a wide variety of observed strength-duration curves including, for example, that shown in Figure 8-18 for a so-called high-ceiling cell presented by Bradley and Somjen (1961).

An absolute refractory period of one time unit is included in the digital representation since in that model firing can occur at most every time unit. In the electronic model the membrane potential must cross the threshold from below to determine firing. Before another firing can occur the membrane potential must first cross the threshold downward, at which time firing does not occur. By this mechanism the electronic model can exhibit "inactivation" for a sustained high level of input wherein the membrane potential crosses threshold from below and remains above threshold for perhaps a considerable time without additional firing. Relative refractoriness is determined in both models by an elevated level of potassium conductance corresponding to the output history of the model. At each output pulse firing, g_K is elevated a certain amount and subsequently decays exponentially with the time constant τ. The increments in g_K from separate output spikes add, so that the value of g_K at any particular time represents the history of firing activity in the model for several time constants τ. In this model, the output firing rate ω is related to the amplitude I of a maintained steady activating current according to

$$\frac{e^{-1/\omega\tau}}{1 - e^{-1/\omega\tau}} = \frac{I/\theta - 1}{b[1 - V_K/\theta]} \tag{8-7}$$

This relationship is shown in Figure 8-19. This model for adaptation is an extension of a model for adaptation suggested by Kernel (1968). A representative sample of activity from the neuromime for combined high-intensity excitatory and inhibitory inputs is shown in Figure 8-20.

This particular neuromime has been developed in both digital and electronic realizations as a module for simulation of large numbers of interconnected units. It is described further in Chapter 11 on neuron pools.

9

Stochastic Models of Neuron Activity

Several mathematically oriented modelers have tried to use simple stochastic models to represent spontaneous firing in neurons. These modelers have typically tried to predict the relative occurrence of time intervals of various durations between the adjacent spikes in an output spike train which is produced by applying the threshold rule to some assumed random distribution of input events. This problem has been found to be virtually unsolvable mathematically, despite its apparent physical simplicity and its extremely modest goal.

9.1. Gerstein's Model

The earliest and most illustrative of such models was presented by Gerstein in 1964 who applied techniques from random-walk theory to describe interspike interval histograms in neurons. Gerstein and Mandelbrot (1964) defined the problem as follows:

> The random walk problem can be restated in the language of neurophysiology in order to define a model for the spike activity of a single neuron:
>
> 1. Let the electrical state of polarization of the somatic and dendritic membrane of the neuron be specified by a single number. As time passes and the electrical state of the membrane varies, the state point will move back and forth along a straight line.
>
> 2. Choose a particular point on the line as the resting potential.
>
> 3. Choose another point on the line, z_0 units away, as the threshold. (This corresponds to the absorbing barrier in the random-walk problem.) If at any time the state point reaches the threshold, the neuron will produce an action potential.
>
> 4. Assume that each incoming elemental EPSP (*E*xcitatory *P*ost *S*ynaptic *P*otential) moves the state point one unit toward the threshold, and that each incoming

elemental IPSP (*I*nhibitory *P*ost *S*ynaptic *P*otential) moves the state point one unit away from the threshold.

5. Let the average rate of incoming elemental EPSP and elemental IPSP be the same; i.e., at any time there is an equal probability that the state point move either a unit toward or a unit away from the threshold.

6. Immediately after the state point has attained the threshold and caused the production an action potential, it returns to the resting potential, only to begin again on its random walk.

This problem is mathematically equivalent to the classical "ruin" problem of probability theory in the case where one is playing an infinitely rich adversary. The threshold corresponds to the player's total amount of money, an EPSP corresponds to a loss by the player, and an IPSP corresponds to a win by the player. The player is "ruined" when he has lost all his money.

Gerstein notes that if one supposes that each elemental movement of the state point in the model is very small compared with the distance Z_0 between the resting potential and the threshold, and if one does not need a detailed description of the motion of the state point, then the problem can be treated as a diffusion process with an absorbing barrier. For this problem the density function for the interspike intervals is given by*

$$I(Z_0,t) = \begin{cases} \dfrac{Z_0 e^{-Z_0^2/4t}}{(4\pi)^{1/2} t^{3/2}} & \text{if } t \geqslant 0 \\ 0 & \text{if } t < 0 \end{cases} \tag{9-1}$$

In this equation, t is the duration of the interspike interval and Z_0 is the threshold level. Unfortunately, this expression is not accurate in describing interspike interval histograms. In particular, long intervals are much too likely to occur, and even the mean value is infinite. This situation reflects in part the fact that this simple model is just as likely to move away from the threshold as it is to move toward it. In neurophysiological terms, it was assumed that the average rates of incoming elemental EPSPs and IPSPs are identical.

Gerstein then made a slight adjustment in this model and was able to come up with an expression for the interspike interval density function which approximated observed first-order interval histograms very well. He noted that it was much more reasonable in a physiological model to assume that there is generally a higher rate of EPSPs than of IPSPs. In terms of a diffusion process, this can be considered as a drift velocity toward threshold and can be framed mathematically by the boundary-value problem

*The reader interested in the mathematical derivation of this equation is referred to Feller's book (1971) or to Griffith's book (1971).

$$\begin{aligned}
\frac{\partial\omega}{\partial t} &= \frac{\partial^2\omega}{\partial Z^2} - c\frac{\partial\omega}{\partial Z} \\
\omega(Z,0) &= \sigma(Z - Z_0) \qquad \text{if } t = 0 \\
\omega(0,t) &= 0 \qquad \text{if } t \geqslant 0
\end{aligned} \tag{9-2}$$

In this equation, c represents the drift velocity and ω represents the spatial density of diffusing particles; i.e., each such particle represents the state point of the neuron model during the time between two action potentials. The normal diffusion current at the absorber corresponds to the probability density of first passage time in the model. The diffusion coefficient has been taken to be unity. The solution to this problem is given by

$$I(Z_0, c, t) = \begin{cases} \dfrac{Z_0 e^{-(Z_0+ct)^2/4t}}{(4\pi)^{1/2} t^{3/2}} & \text{if } t \geqslant 0 \\ 0 & \text{if } t < 0 \end{cases} \tag{9-3}$$

$$I(a, b, t) = Z\frac{e^{-(a/t+bt)}}{t^{3/2}}$$

Equation (9-3) is a convenient analytical expression with which one can match observed first-order interspike interval histograms. It is a two-parameter curve wherein each of the parameters has a specific physical interpretation with respect to the model. The parameter a is associated with the square of the resting potential; the paramater b is associated with the difference between the rates of EPSPs and IPSPs which are incident upon the neuron (the drift velocity). By matching this curve to existing histograms, one can find those values of a and b which are required to match the particular curve, and then make inferences concerning the threshold of the neuron and the input which it is receiving. Figure 9-1 shows several sets of experimental data which Gerstein matched with equation (9-3).

Gerstein's model results in a very simple analytical expression and, moreover, allows the inference of fundamental characteristics of the neuron (threshold, input level) from analysis of the observed results. It is thereby extremely illustrative. Unfortunately, in this particular case, the model is too simplified; one cannot take the inferences too seriously. For example, there is no decay of the postsynaptic potentials in this model. Thus, a given excitatory event moves the state point one step toward or away from the threshold, at which point it remains dependent only upon subsequent input events. Moreover, refractory properies are supposed merely to reset the potential at zero level.

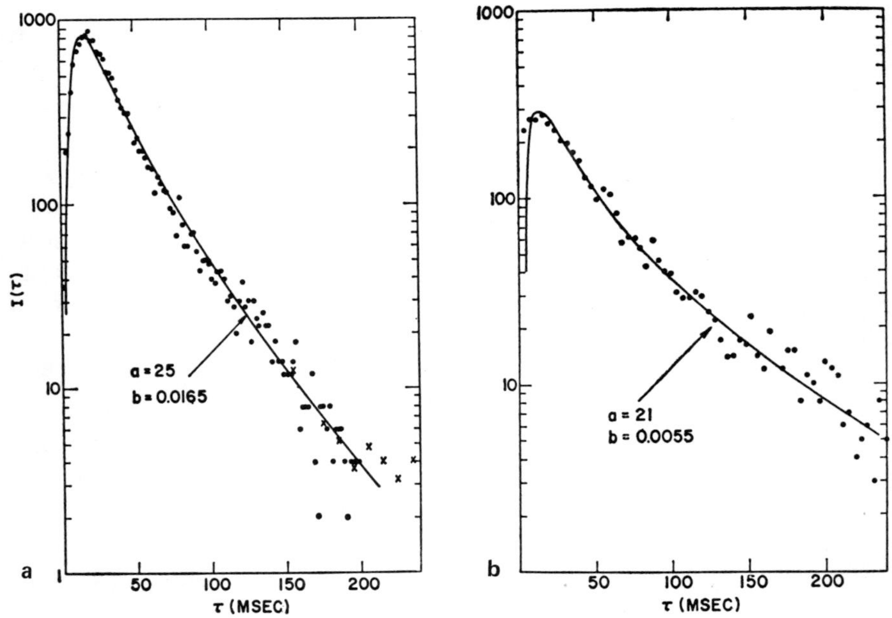

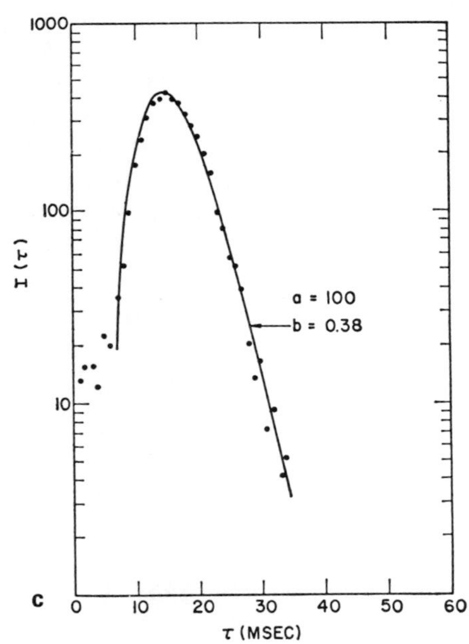

Figure 9-1. Fit of equation (9-3) to experimental data. (a) Unit 240-1, (B) Unit 6-2, (c) Unit R-4-100.

9.2. More General Models

The more general problem where these and other deficiencies are accounted for, however, has remained intractable. For example, consider now a diffusion approximation for single-neuron firing patterns presented by Fienberg. Fienberg (1970) is also considering a two-state variable model membrane potential and threshold which he defines as follows:

> The membrane potential of the neuron $X(t)$ will be governed by the following stochastic model:
>
> (a) At time zero, the membrane potential is set (or reset) to $X(0) = x_0$.
>
> (b) The neuron receives excitatory pulses of magnitude ϵ and according to a Poisson process with parameter α, and inhibitory pulses of magnitude $-\epsilon$ according to a Poisson process with parameter β.
>
> (c) The membrane potential is subject to exponential decay with time constant τ, to a resting level x_r.
>
> (d) When $X(t)$ reaches a fixed threshold level S an output pulse or firing occurs and the potential is reset to x_0.

Fienberg then shows that, under these assumptions, the probability density function for the membrane to be at the potential level x at time t, given that $X(0) = x_0$, is $f(t, x)$, where f is given by

$$\frac{\partial f}{\partial t} = \tfrac{1}{2}(\alpha\epsilon^2 + \beta i^2)\frac{\partial^2 f}{\partial x^2} - \frac{\partial}{\partial x}\left[\left(\alpha\epsilon - \beta i - \frac{x - x_r}{\tau}\right)f\right] \qquad (9\text{-}4)$$

In deriving this expression, Fienberg supposed that terms of the order of ϵ^3 and i^3 could be ignored. This expression is a well-known Fokker-Planck equation. Unfortunately, as Fienberg points out, Cowan (1972) has shown that closed solutions for the first passage-time problem for this expression do not exist.

However, Gluss (1967) has shown that when the threshold (or absorbing barrier) is neglected, the density function given by

$$f_w(x, t) = [2\pi D\zeta^{-1}(1 - e^{-2\zeta t})]^{-\frac{1}{2}} \exp\left[\frac{-\{(x - c/\zeta)e^{\zeta t} - (w - c/\zeta)\}^2}{2D\zeta^{-1}(e^2\zeta t - 1)}\right] \qquad (9\text{-}5)$$

can be obtained for a particle starting at position $x = w$ at $t = 0$. Here D, C, ζ are constants related to α, β, ϵ, and i. This expression does not have the difficulty with long intervals which plagues the results presented by Gerstein.

Many other workers, including Ricciardi and Esposito (1966), Capocelli and Ricciardi (1971, 1972), Gluss (1967), Cowan (1972), and various other colleagues, working primarily at the University of Chicago with the Committee on Mathematical Biology or the Laboratorio di Cibernetics del C.N.R. in Naples, have also struggled with this problem. Ricciardi, in particular, has tried to generalize the problem presented by Fienberg to the case where input terms could be of any magnitude (recall that Fienberg assumed that inputs of the order of ϵ^3 and higher could be ignored) and where at least some sort of refractory properties might be included. These more general approaches also result in integral or differential equations for which no analytical closed-form solution exists.

Ricciardi, in a somewhat earlier paper, presented a slightly different approach to the problem in an attempt to include the idea of "dead time" or absolute refractoriness. This model, rather than using the idea of no decay of elemental PSPs as in Gerstein's model or the more accurate exponential decay used in the more general problem, uses instead the idea of instantaneous decay of PSPs. That is, his model requires the simultaneous occurrence of a sufficiently large number of pulses in order to trigger an action potential. (As we shall see in Chapter 13 this is a neuronic model presented by Caianiello in 1961, with whom Ricciardi worked.)

Ricciardi then attempted to get an expression for the density function for output interspike intervals for this artificial neuron in the case of a Poisson distribution of input pulses, i.e., the probability that n pulses arrive at a time interval dt, between t and $t + dt$, is given by

$$P_n(t)dt = \frac{(\alpha t)^n}{n!} e^{-\alpha t} dt \tag{9-6}$$

Riccardi shows that for this case the output distribution is given by

$$p_n(T,\tau) = l[T-(n-1)\tau]\left\{1 - e^{-\alpha[T-(n-1)\tau]} \sum_{k=0}^{n-1} \frac{\alpha^k[T-(n-1)\tau]^k}{k!}\right.$$

$$\left. - \, l[T-n\tau] \times [1 - e^{-\alpha(T-n\tau)}] \sum_{k=0}^{n} \frac{\alpha^k (T-n)^k}{k!}\right\} \tag{9-7}$$

$$l(X) = \begin{cases} 1 & \text{if } X > 0 \\ 0 & \text{if } X \leqslant 0 \end{cases}$$

He also shows that the mean firing rate is given by

$$\langle n(t,\tau)\rangle = \sum_{n=0}^{\gamma} \sum_{k=n+1}^{\infty} \frac{\alpha^k (t-n\tau)^k}{k!} e^{-\alpha(t-n\tau)}$$

$$\gamma + 1 = \{\text{greatest integer} < \frac{t}{\tau} + 1\} \tag{9-8}$$

In these equations, n is the number of output pulses, τ is the length of the refractory period (width of the output pulses), and γ is defined in equation (9-8).

We can see that there has been a considerable amount of effort expended in the attempt to obtain a mathematical model for the stochastic firing properties of artificial neurons under the conditions of random input distribution. This work is highly mathematical and has been limited generally to the attempt to get a density function for the distribution of times required for a membrane potential to go from a resting level to a fixed threshold level. The main conclusion is that even this most basic problem appears mathematically cumbersome. We refer the student who wishes to explore this topic in more detail to an excellent recent review by Fienberg (1972). In the meantime, it is expeditious to study the problem with computer-simulation techniques.

10

Statistical Analysis of Neuronal Spike Trains

Certain concepts from the information sciences have been particularly useful in helping describe nervous activity. One can picture the dynamical activity of a nervous network as consisting of a large number of ongoing simultaneous time series of events (spike trains) emerging from all the constituent cells. The output train of any particular cell reflects the ongoing input trains at each of some several hundred or even thousands of synapses. Various synapses are weighted in significance according to the strength of the metabolic transmitter apparatus and proximity to the soma. Whether or not one assumes that under some conditions certain components of the input or output patterns might be considered as "background noise" with respect to some other component which might comprise a "signal," it is certainly clear that the methods and concepts of stochastic processes applied on an operational level can help to reveal many characteristics of neuroelectric events not readily apparent in other ways.

These tools have been applied to neuronal activity primarily to help elucidate patterns in neuronal spike trains. Investigators have been concerned primarily with how these patterns might relate to information, how they might relate to stimulation patterns being experienced by a cell at a particular time, and how they might relate to functional organization and interconnections within particular networks. If one considers a time series of neuronal action potentials as illustrated in Figures 1-5 and 1-6, one can specify characteristics at each of three levels: rate, interspike intervals, and the order of intervals. Such analyses can be used to characterize the output firing patterns of cells and thereupon lead to consequent classification of cells; to provide information concerning underlying neuronal mechanisms; and, perhaps most importantly, to spell out functional interrelationships between cells. The development of the techniques discussed below has been carried out primarily by Perkel, Gerstein, and Moore, (1967) Segundo (1973), and various colleagues over the last decade.

10.1. Statistical Measures for Single Trains

Consider first spontaneous firing activity in a single unit. One can characterize such activity by the rate of firing, the number of spikes occurring per unit time. A somewhat more penetrating measure considers the set of interspike intervals, i.e., the set of time intervals in, say, milliseconds, between adjacent spikes in the train. This measure includes rate as a special case and also specifies the degree of regularity in the firing pattern. A convenient way to display this information is the first-order interspike-interval histogram (IH). Suppose t_i is the set of numbers which correspond to the time intervals between adjacent spikes in the train. The number of t_i which fall in various size ranges are plotted against the size ranges themselves. This histogram approaches the probability density for interspike intervals as the bin size for the time ranges approaches zero and can be considered as an approximation to it; i.e., the histogram measures the actual distribution of intervals in the train as they occurred, whereas the density function is an abstraction corresponding to the mathematical model one might adopt to describe the underlying process.

Examples of certain canonical interspike-interval histograms are shown in Figure 10-1. If a particular firing train is precisely regular, all the interspike intervals will have the same value. The histogram for this case will consist of a single very large peak at that particular time value. This is illustrated in Figure 10-1a. This particular pattern is called a pacemaker. If one introduces a certain amount of irregularity into such a train by supposing that the intervals occur with a given mean value, but that the actual individual intervals may be slightly less or greater than this value with, say, a Gaussian distribution, one gets an interspike-interval histogram which approaches a Gaussian distribution as illustrated in Figure 10-1b. Such a process may be considered a noisy pacemaker. A firing pattern more relevant to the majority of actual neuronal spike trains is the so-called Poisson process. This is a process wherein the interspike intervals are distributed according to a decaying exponential curve as given in equation (11-1) and shown in Figure 10-1c:

$$p(\tau) = \mu e^{-\mu\tau} \tag{10-1}$$

In this expression, μ is the mean rate of activity. The Poisson process is

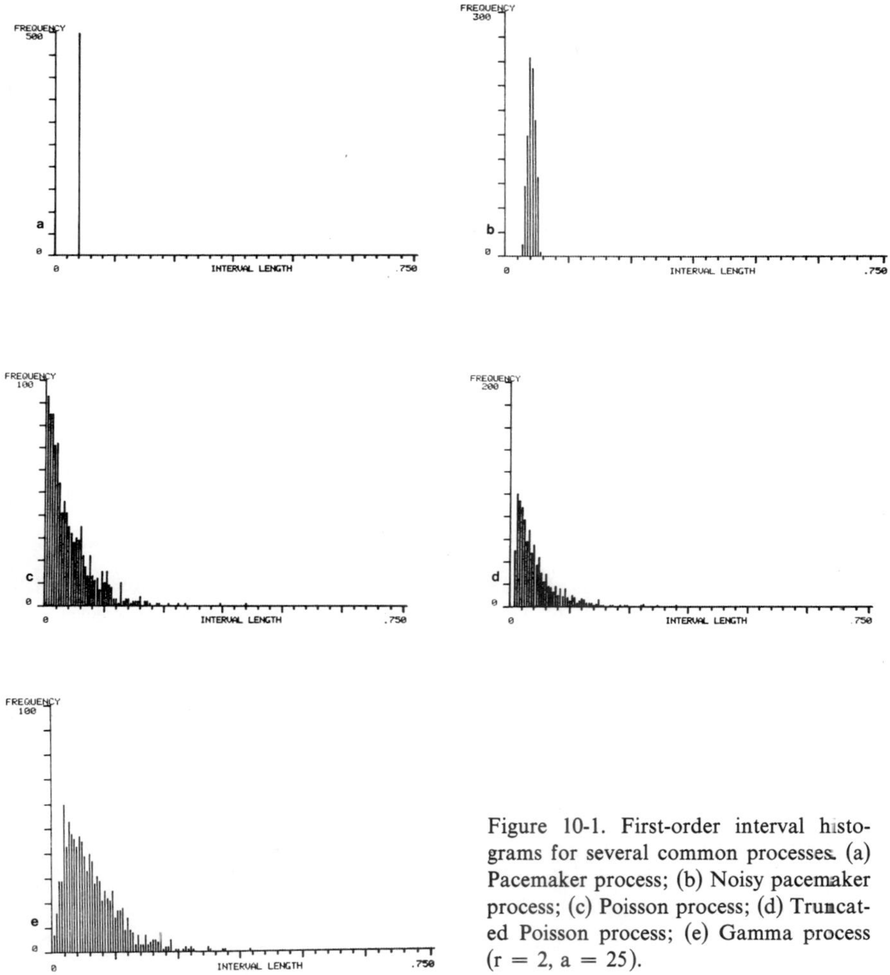

Figure 10-1. First-order interval histograms for several common processes. (a) Pacemaker process; (b) Noisy pacemaker process; (c) Poisson process; (d) Truncated Poisson process; (e) Gamma process (r = 2, a = 25).

highly irregular. Most neuronal firing patterns reflect the refractory properties of neurons by exhibiting a paucity of very small intervals. A typical such record is shown in Figure 10-1d. Such a record might be conceptualized as reflecting a Poisson process whose smaller intervals are then curtailed because of neuronal refractory properties. A similar and related distribution is the gamma distribution which is described in equation (10-2) and illustrated in Figure 10-1e:

$$p(\tau) = \frac{a^r \tau^{(r-1)} e^{-a\tau}}{(r-1)!} \tag{10-2}$$

Neuronal firing patterns are often characterized and classified according to how their interval histograms compare to these basic canonical forms. The first-order interval histogram is particularly useful for measuring the overall degree of regularity within the train.

The distribution function $F(\tau)$, the survivor function $\Gamma(\tau)$, and the age-specific failure rate $\Phi(\tau)$ are other functions which operate at the level of interval analysis. These functions are defined by

$$\begin{aligned} F(\tau) &= \int_0^\tau p(t)\,dt \\ \Gamma(\tau) &= 1 - F(\tau) \\ \Phi(\tau) &= p(\tau)/\Gamma(\tau) \end{aligned} \tag{10-3}$$

The probability density $p(\tau)$ times a differential time interval measures the probability that any particular interval will fall in the range between τ and $\tau + d\tau$. The distribution function $F(\tau)$ measures the probability that any given interval will fall in the range between 0 and τ. The survivor function $\Gamma(\tau)$ measures the probability that any particular interval will have a value larger than τ. The age-specific failure rate $\Phi(\tau)$ times a differential time element measures the probability that a spike will occur in the interval between τ and $\tau + d\tau$, given that one has not occurred in the interval between 0 and τ.

A somewhat more detailed consideration of neuronal firing patterns takes into account not only the mean rate and the degree of regularity of intervals, but also the temporal ordering of intervals within the train. A convenient display in this category is the autocorrelation histogram. This histogram is constructed by considering the set of times t_i which correspond to the intervals between any given spike paired with all other spikes following it within a certain time interval. These intervals can be measured directly or can be obtained from higher-order-intervals defined as indicated in Figure 10-2. Moreover, one considers such sets of times for every spike within the train and subsequently plots a histogram for the combined set of

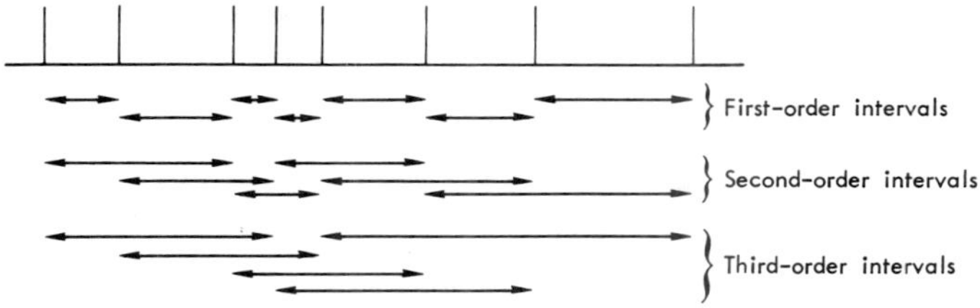

Figure 10-2. Calculation of interspike intervals for autocorrelation histogram.

all such intervals. It is easy to see that such a measure approximates the following probability distribution: Given that the cell fired at time $t = 0$, what is the probability that it fired again at any time t after that time, irrespective of how many intervening spikes may or may not have occurred? This measure is very useful in determining the occurrence or nonoccurrence of rhythmicities in firing patterns and for other recurrent temporal patterns which reflect on temporal relationships within the train. Such measures can be used, for example, to make inferences concerning possibilities of reverberatory loops in which a given cell might be participating, or sources of endogenous rhythmicities and the like.

Examples of autocorrelation histograms are shown in Figure 10-3. The autocorrelation histogram for a pacemaker neuron shown in Figure 10-3a exhibits very pronounced peaks at integral multiples of the period of the pacemaker. A noisy pacemaker also exhibits peaks at integral multiples of the mean period of the pacemaker, but the peaks progressively diminish and the distributions progressively flatten out corresponding to the increase in standard deviation for higher-order occurrences. In particular, for a Gaussian distribution of intervals in an order-independent process, the standard deviation increases as the square root of n, where n is the order of the recurrence. The autocorrelation histogram of a Poisson process is entirely flat. This reflects the interpretation of a Poisson process as follows: Given that the cell fired at time $t = 0$, one gains no information as to the likelihood of its firing at any other point in time. Many neurons exhibit essentially Poisson-like autocorrelation histograms with the exception that very small values of intervals rarely occur corresponding to refractoriness in individual neurons. The autocorrelation histogram for the gamma distribution is very similar to that for the truncated Poisson distribution.

The number of events per time bin in the autocorrelation histogram approaches the mean firing rate as time gets very large. This is true for all stationary processes. The autocorrelation histogram may show deflections from the constant level corresponding to the Poisson process, which reflect either the particular shape of the interspike-interval histogram or which

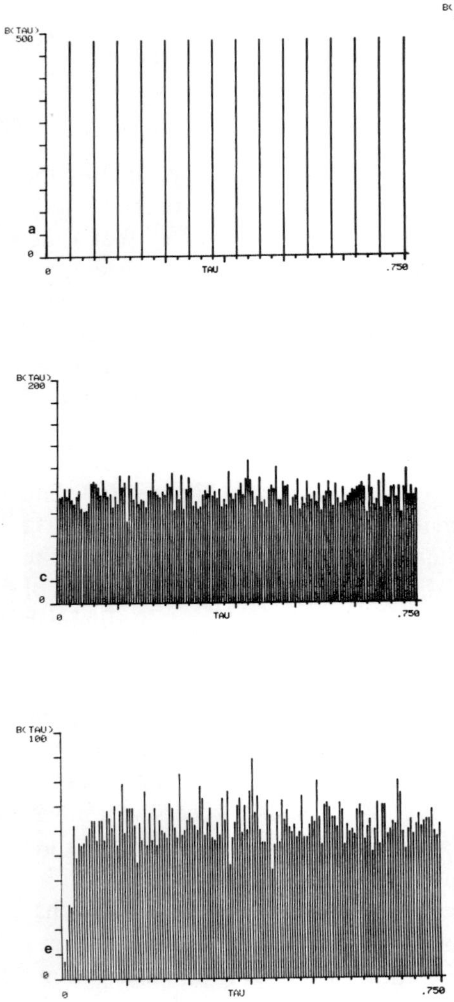

Figure 10-3. Autocorrelation histograms for the processes whose first-order interval histograms are shown in Figure 10-1. (a) Pacemaker process; (b) Noisy pacemaker process; (c) Poisson process; (d) Truncated Poisson process; (e) Gamma process (r = 2, a = 25).

reflect the sequencing of intervals within the train. It is important to be able to distinguish between these possible sources in particular autocorrelation histograms. There are two ways to determine which properties of a given autocorrelation histogram are order dependent. The first and very commonly used technique is to simply shuffle the intervals in a given train and recompute the autocorrelation histogram on the shuffled train. This autocorrelation histogram serves as a control against which the autocorrelation histogram of the unshuffled train can be interpreted. Moreover, one can then easily subtract the autocorrelation histogram for the shuffled train from the original autocorrelation histogram to obtain a so-called difference autocorrelation histogram which contains only the order-dependent characteristics. This technique is simple, straightforward, and generally reliable. However, there is always the possibility that the shuffled cases may provide a rare artificiality. Therefore, when using this technique, particularly with regard to important conclusions, one should apply the method repeatedly using several different controlled shuffled cases.

A second and more rigorous technique to establish the order dependence or independence of particular characteristics observed in the autocorrelation histogram is to construct the autocorrelation histogram from the interspike-interval histogram under the assumption that the intervals are independently ordered. The difference between this constructed autocorrelation histogram for the order-independent case, on the one hand, and the actual observed autocorrelation histogram, on the other, serves to measure those characteristics which are order dependent. Such a constructed autocorrelation histogram can be constructed according to

$$
\begin{aligned}
f_i(t) &= \int_0^t f_{i-1}(t) f(t - \tau)\, d\tau \\
\text{Autocor}\,(t) &= \sum_{i=1}^{\infty} f_i(t)
\end{aligned}
\tag{10-4}
$$

Here the autocorrelation histogram is recognized as the sum of all the ith-order histograms, where i refers to the intervals between two spikes which are separated by $i - 1$ intervening spikes. The assumption of order independence then allows one to construct the second-order interval histogram as the autoconvolution of the first-order interval histogram and, more generally, to construct the jth-order interval histogram from the convolution of the $j - 1$ interval histogram with the first-order interval histogram. This technique then is a more rigorous way to obtain an order-dependent histogram for a given observed train than by mere shuffling. It is, however, somewhat more difficult to effect.

Certain other statistical measures are sometimes used to characterize neuronal firing patterns with particular reference to the order dependence of their constituent intervals. Rodieck (1967) and Gerstein and Kiang (1960) introduced the scatter diagram for displaying the so-called joint interval distribution. In this diagram the length of a given interspike interval t_i is plotted on the abscissa, and the length of the next interval t_{i+1} is plotted on the ordinate. A train in which short intervals tend to follow long intervals, and vice versa, would exhibit a joint interval scatter diagram with a "downward trend." A train in which short intervals tend to follow short intervals and long intervals to follow long intervals would show an "upward trend" in the joint interval distribution. In a train whose intervals are independently distributed, the mean number of events in all rows and columns have a common constant expected value.

Serial correlation coefficients are sometimes used to describe temporal sequences in neuronal spikes. The serial correlation coefficient ρ_j of order j is defined by

$$\begin{aligned} \mu &= 1/E(t_i) \\ \sigma &= (E[(t_i - \mu)^2])^{1/2} \\ C_j &= E[(t_i - \mu)(t_{i+j} - \mu)], \qquad j = \ldots, -1, 0, 1, \ldots \\ \rho_j &= C_j/\sigma^2 \end{aligned} \tag{10-5}$$

Here μ is the mean rate of the train, σ is its standard deviation, and C_j is the covariance of interval lengths of lag j. The set of serial correlation coefficients is called the serial correlogram and may be plotted as in Figure 10-4. When intervals are independently distributed, the expected values of the serial correlation coefficients of all orders are zero. Deviations from zero measure temporal dependence in the train. The serial correlation coefficient of lag 1 summarizes an entire scatter diagram of the joint interval distribution of lag 1.

The joint interval histogram and the serial correlogram are sometimes misleading and should be used only in conjunction with other statistical measures. Moreover, it should be noted that these measures relate to numbers assigned to intervals rather than to time explicitly. Thus, they are not as directly interpretable and are somewhat more sensitive to errors in data processing.

Nakahama, Ishii, and Yamamoto (1972) have recently developed

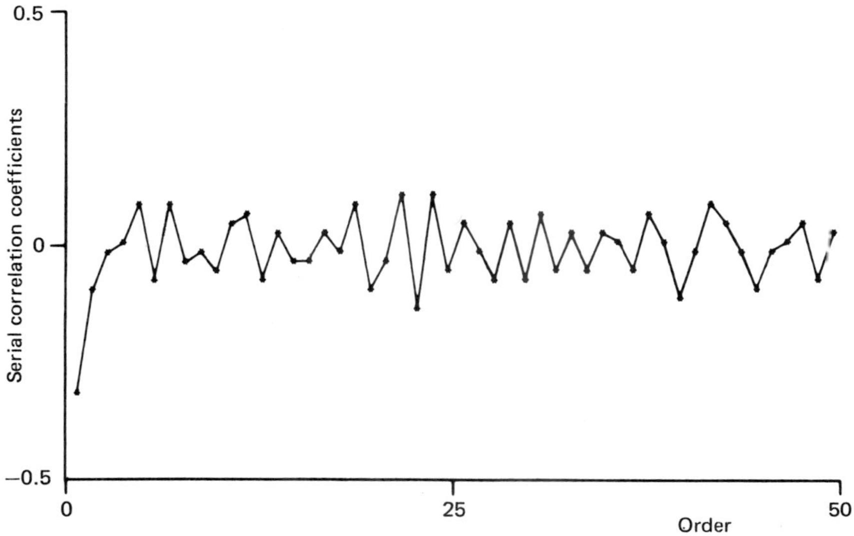

Figure 10-4. Serial correlogram for simulated neuron producing irregular bursts.

related measures of dependences in neuronal spike trains based on Shannon's concept of entropy. The first-order "dependence" statistic $D(t)$ is defined in terms of the entropy H_0 and the first-order entropy H_1 of the spike distribution. These are defined in terms of the interpike-interval histogram $p(i;\tau)$ and the first-order interval transition matrix $p_i(j;\tau)$ according to

$$H_0(\tau) = -\sum_i p(i;\tau)\log_2 p(i;\tau)$$

$$H_i(\tau) = -\sum_i \sum_j p(i;\tau)\log_2 p_i(j;\tau) \qquad (10\text{-}6)$$

$$D(\tau) = \frac{H_0 - H_i}{H_0}$$

Here $p_i(j;\tau)$ approximates the conditional probability of the jth event under the condition that the ith event occurred just before the jth event; τ is the bin size upon which the measures are defined. The measure $D(\tau)$ ranges between 0 and 1. When the adjacent intervals are independent, then $H_i = H_0$ and $D = 0$. These authors then generalize these first-order entropy and dependence measures to provide scalar measures for higher-order Markov dependence in an analogous fashion.

10.2. Statistical Measures for Simultaneously Recorded Trains

Consider now the case of two spike trains recorded simultaneously from each of two neurons. The cross-correlation histogram is a very convenient device for measuring the temporal relationship in firing patterns between the two neurons. Suppose two spike trains are labeled A and B, respectively, as indicated in Figure 10-5. The cross-correlation histogram is obtained by considering the set of numbers t_i which represent the intervals from a given single spike in train A to the times of occurrences of neighboring spikes in train B within some time range. Such a set of time intervals is obtained for every spike in train A. The histogram constructed on the basis of the combined sets of times is the cross-correlation histogram. It is easy to see that the cross-correlation histogram can be interpreted as follows: Given that train A fired at time $t = 0$, what is the probability that train B fired as a function of time either before or after that event?

It can be shown that the cross-correlation histogram from A to B is a mirror image of the cross-correlation histogram from B to A. More

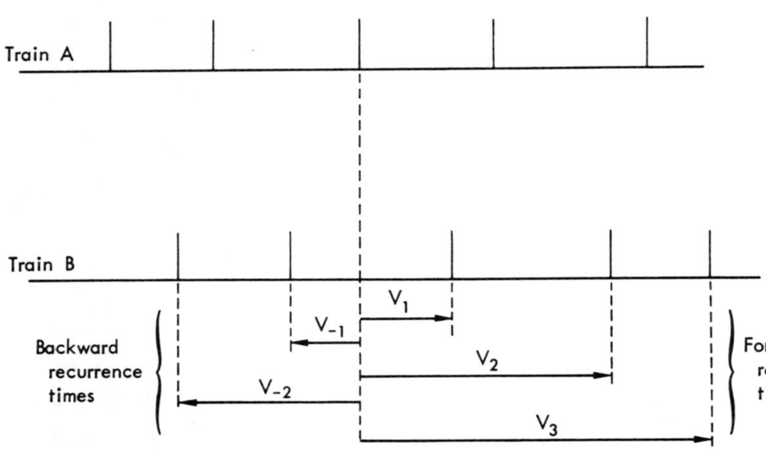

Figure 10-5. Calculation of interspike intervals for cross-correlation histogram.

precisely, these histograms are related as

$$\frac{XC_{AB}(\tau)}{\mu_A} = \frac{XC_{BA}(-\tau)}{\mu_B} \tag{10-7}$$

Here μ_A and μ_B represent the firing rates in trains A and B, respectively; $XC_{AB}(\tau)$ is the cross-correlation function from train A to train B. Thus, in interpreting the left side of the cross-correlation histogram, one can say either: Given that A fired at zero, how many times did B fire before that as a function of time, or given that B fired at zero, how many times did A fire after that as a function of time? One can interpret the autocorrelation histogram discussed above as being a special case of the cross-correlation histogram where the trains A and B are one and the same train. Under this assumption equation (10-7) tells us that the autocorrelation histogram is symmetric with respect to the origin $t = 0$. Therefore, in constructing the autocorrelation histogram one typically considers only positive times. However, the cross-correlation histogram is not necessarily symmetric with respect to the origin. Therefore, when constructing the cross-correlation histograms, it is useful to include both negative and positive time ranges from the referent spike.

The cross-correlation histogram indicates the degree of interdependence between two units according to peaks or troughs. One can show that the cross-correlation histogram for independent units is flat. If, on the other hand, cell A excites cell B with the time lag of, say, τ, then there will be a peak in the cross-correlation histogram from A to B at the time τ. Similarly, an inhibitory influence on B from A with time lag τ shows up as a trough in the cross-correlation histogram from A to B at the time τ.

The interpretation of deviations from flatness in observed cross-correlation histograms, however, is sometimes ambiguous, as discussed in the papers by Perkel *et al.* (1967). Deviations from flatness can reflect either direct or diffuse interconnections between the two cells, the experiencing of shared activation of various varieties, or artificial interdependences. Some examples of cross-correlation histograms for certain canonical interrelations between pairs of units are shown in Figure 10-6. The most common source of error in observed cross-correlation histograms occurs when one is recording from two independent cells one or both of which may have rhythmic firing characteristics. Under these conditions, it is quite common to observe false interdependences, as indicated in Figure 10-6. More cases are discussed in the papers by Perkel *et al.* (1967).

Consider now the relation between firing in a single unit and a repeatedly presented stimulus. This relationship can be displayed with the so-called poststimulus time histogram (PSTH). This histogram may be

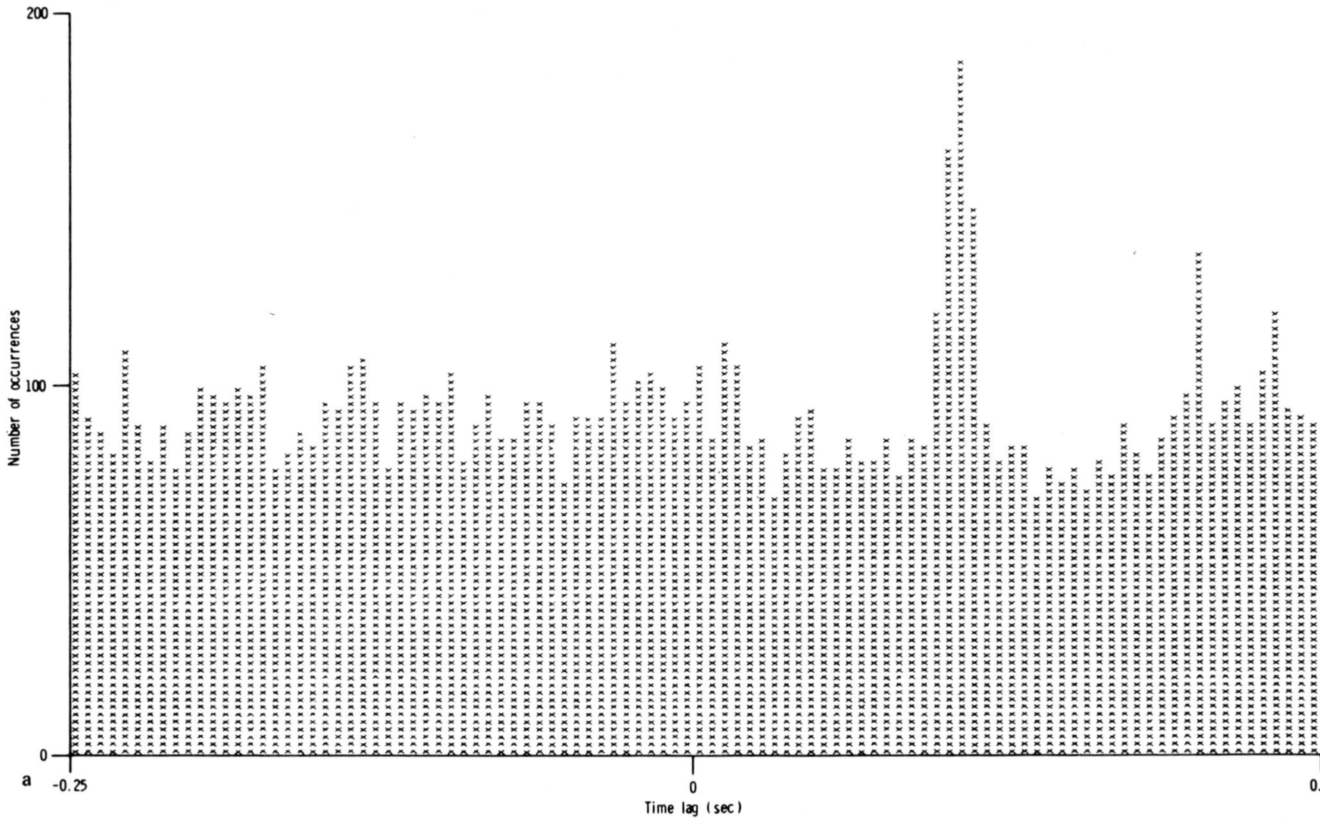

Figure 10-6. Cross-correlation histograms for interacting neurons; data from computer simulations. (a) Direct excitation. Each neuron excited by an independent Poisson source of EPSPs, amplitude normally distributed with mean 7.5 mV, standard deviation 1.0 mV; mean arrival rate 100/sec. Cell A makes excitatory synapse on cell B with latency of 100 msec; EPSP mean amplitude 10 mV, standard deviation 1 mV. In 250 sec, 2053 spikes are produced by cell A, and 2286 spikes by cell B.

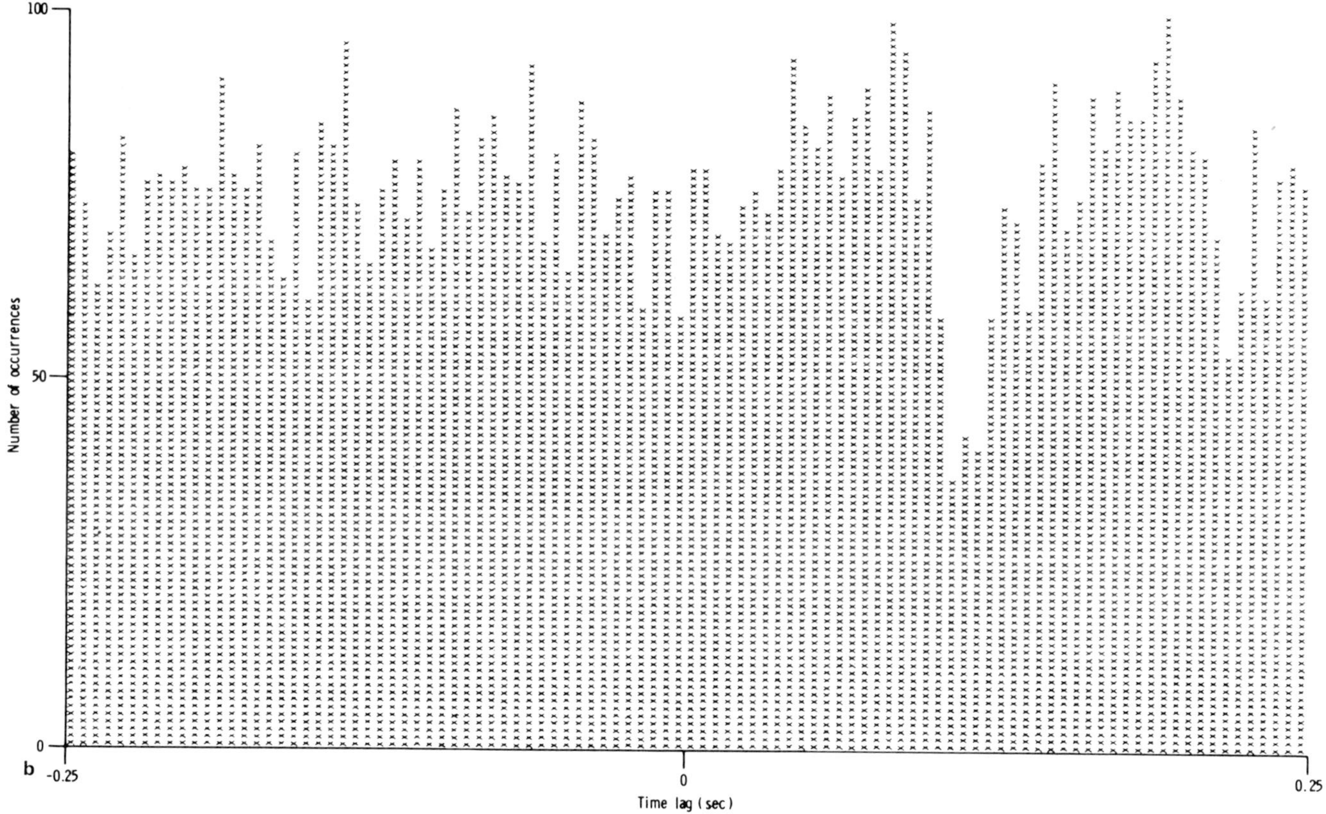

Figure 10-6. (b) Direct inhibition. Same as case (a) except A–B synapse is inhibitory (mean amplitude −10 mV, etc.). In 250 sec, 2068 spikes are produced by cell A, and 1858 spikes by cell B.

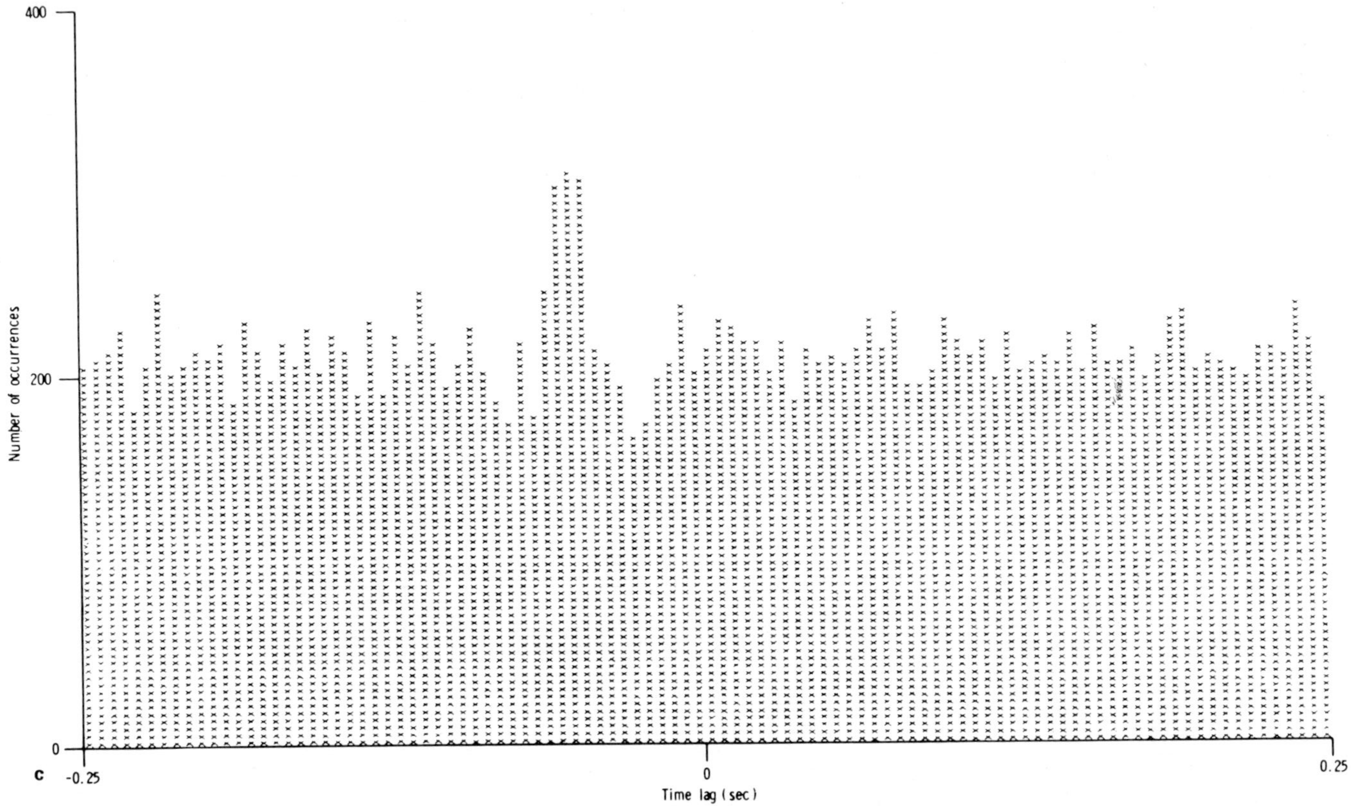

Figure 10-6. (c) Shared excitation. Simulated neurons similar to those above, but with no synaptic connections between them. Each independent Poisson source now produces 10-mV EPSPs, and is reduced in rate to 25/sec. An additional Poisson source with a mean rate of 150/sec delivers 10-mV EPSPs to both cells, each arriving at cell A 50 msec earlier than at cell B; i.e., each cell receives an average of 175 EPSPs per second, of which 150 (86%) are shared with the other cell. Each cell produced about 2500 spikes in 150 sec.

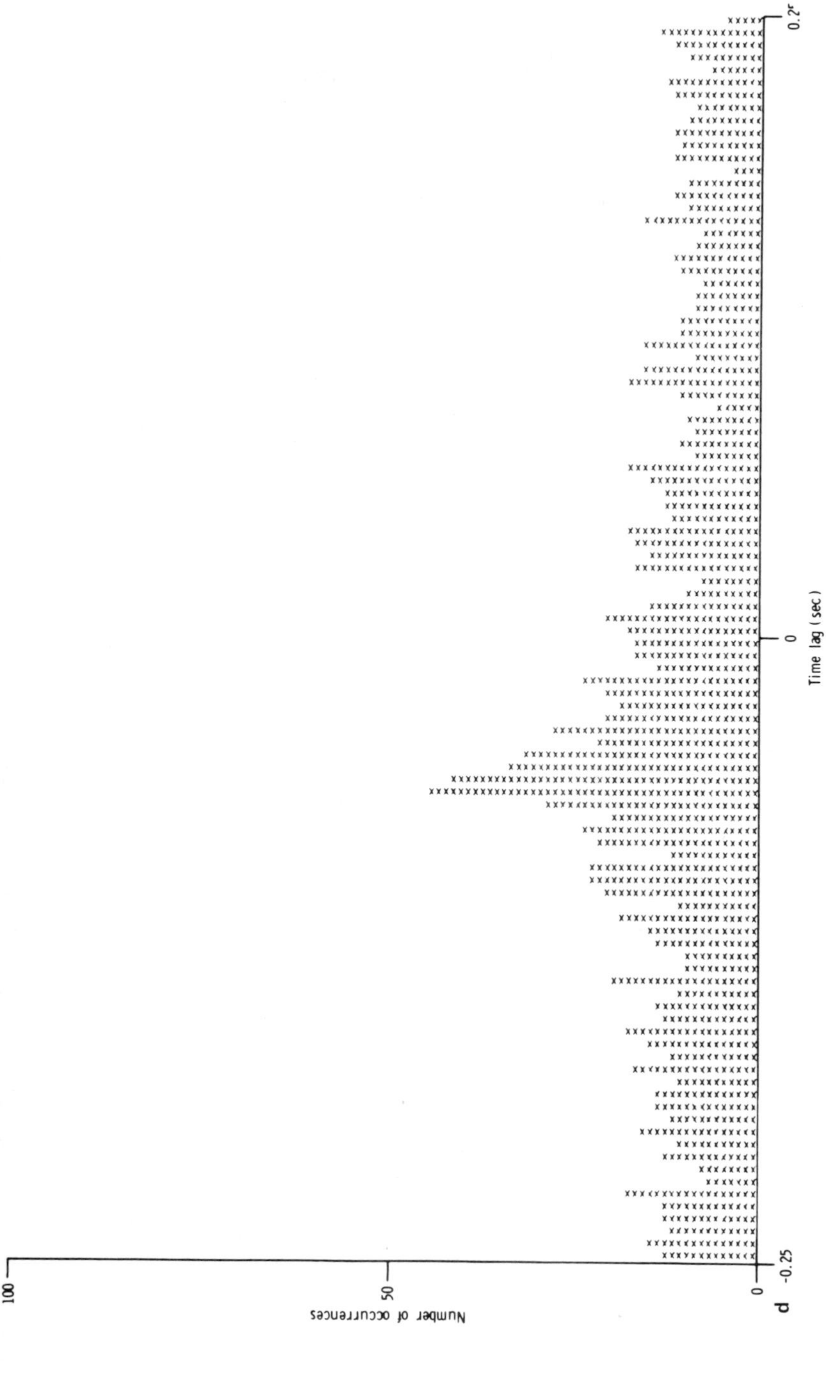

Figure 10-6. (d) Shared inhibition. Similar to previous example. Each independent excitatory input produces 10-mV EPSPs at a mean rate of 175/sec, and shared channel produces 30-mV IPSPs at a mean rate of 100/sec. Both cells depolarized by dc input to produce spikes. About 550 spikes are produced by each cell in 150 sec.

Figure 10-6. (e) Shared inhibition and excitation. Similar to previous cases; cell A is depolarized and receives IPSPs from shared 100/sec channel as in case (d); cell B is excited as in case (c). Cell A produced 732 spikes, and cell B, 2600 spikes in 200 sec.

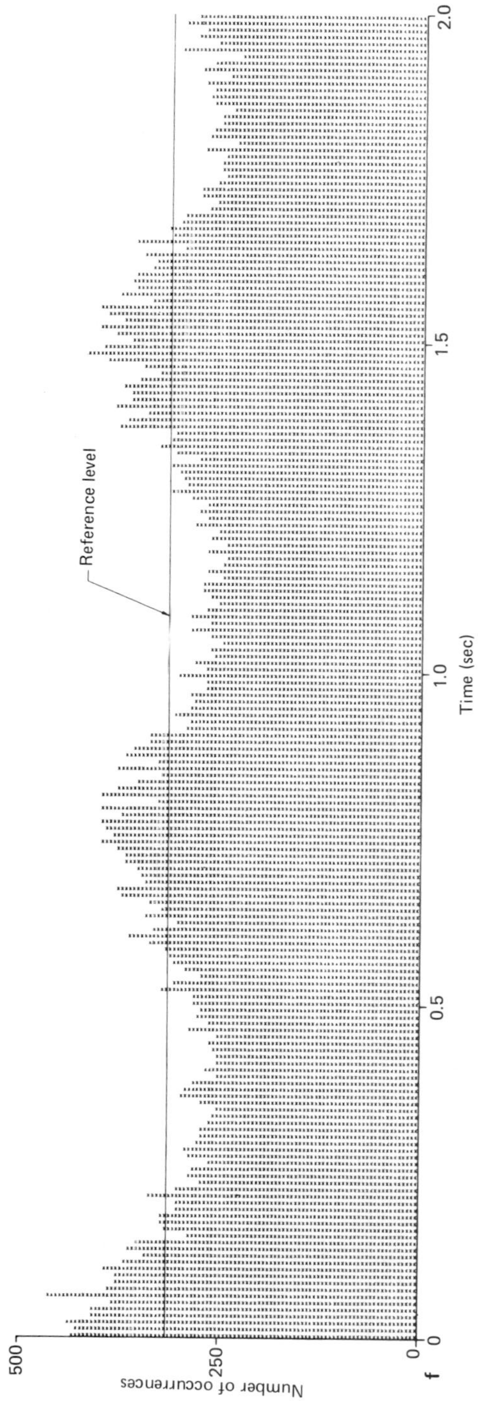

Figure 10-6. (f) False correlations between independent trains with independent time variation in mean rate. (Adapted from D. H. Perkel *et al.*, 1967.)

considered as a special case of the cross-correlation histogram discussed above wherein train A is identified as the stimulus. The set of times considered in train A is the set of times of stimulus onset. Thus, the poststimulus time histogram gives the following measure: Given that the stimulus was initiated at time $t = 0$, what was the probability of firing in the unit as a function of time? This measure is a much more sensitive measure of a unit's tendency to fire in response to stimulation than is mere visual observation of several repeated trials. As Gerstein and Kiang pointed out in 1960, as more single-unit activity is measured by the poststimulus time histograms, the number of units falling into the "no response" category diminishes considerably.

The usefulness of statistical techniques is that they can be applied on an operational basis to help reveal properties of nervous activity not readily apparent by other means. To use the words of Moore, Perkel, and Segundo (1966):

> Our principal working assumptions are ... (a) that there is an enormous wealth of information about the structure and function of the nervous system which can be derived from careful study of the deailed timings of spike events; (b) that analysis of these signals can shed light on mechanisms of spike production within the observed cell, on the presynaptic input to the cell, and on the mechanisms by which the latter is transformed into a postsynaptic output; and (c) that observation of multiple units can reveal details of interconnections and functional interactions. ... and the appreciation that neuronal processes at all levels involve a probabilistic element which must be adequately incorporated if quantitative hypotheses or models of neuronal functions are to be valid. Finally, it is held that only the more detailed analyses of spike timings are appropriate to any quantitative theory of information processing by the nervous system.

These techniques are being used by an increasing number of electrophysiologists and their more quantitatively oriented colleagues. We will now survey the extent to which they have been used thus far to spell out firing patterns and functional interactions in specific neuronal networks.

10.3. Applications of Neuronal Spike Train Analysis

The only definitive analysis of auto- and cross-correlation statistics for various known canonical arrangements of interconnection patterns has been done on computer-simulation models by Perkel *et al.* (1967). Applications to specific real networks of neurons have bcen considerably less studied. However, there have been numerous studies describing the statistical properties of neurons in various networks of the mammalian nervous system.

One of the first groups to use statistical methods to analyze neuronal spike activity was Werner and Mountcastle (1965), and later Poggio and

Vernstein (1964). These investigators used Macaque monkeys which were unanesthetized and paralyzed with gallamine triethiodide. They recorded from neurons of the ventrobasal thalamus which responded to rotation of limbs. On the basis of interspike-interval histograms from these cells, Werner and Mountcastle reported that the standard deviation for spontaneous activity in these cells is an irregular proportion of the mean interval. They suggest that receiving cells might use this relation for sensory discrimination rather than dealing only with differences in absolute level of activity. In autocorrelation histograms constructed over a range of about 3 to 4 sec and using 200-msec bin sizes, Werner and Mountcastle discovered marked periodicities in about 50% of these units. A typical result is given in Figure 10-7a. For this unit the rhythmic period is about 1.7 sec. In different periods of activity analyzed for the same single neuron, the frequencies of significant peaks sometimes changed. Poggio and Vernstein, reporting autocorrelation histograms from the same preparation, describe spontaneous activity as being strikingly rhythmic with periods ranging from 30 to 50 msec. A typical example is shown in Figure 10-7b. They note that, although the mean rate of firing generally increased with increased intensity of stimulation, the period of rhythmicity was sometimes independent of the intensity of stimulation. They conclude that since the mean period and the rhythmic period seem to be independent, there must be independent mechanisms generating them.

Another early study of statistical properties of spontaneous activity was presented by Rodieck (1967) who studied the maintained activity of ganglion cells in the retina of the cat. The cats were either decerebrated or anesthetized. Rodieck compared autocorrelation histograms from these cells with the so-called autoconvolutions which were computed from the interval histogram according to our equation (10-4) in order to determine the order dependence of the trains. In all of these cases there was generally excellent agreement between the two, indicating there was no statistically significant order-dependence in these trains. Rodieck also constructed cross-correlation histograms for five simultaneously recorded pairs of neurons. In each case the pair was obtained from a single microelectrode. Rodieck found no significant interdependencies except in one pair which showed a postreferent trough lasting some 16 msec. Rodieck's main results are shown in Figure 10-8.

An interesting study of firing properties of cerebellar Purkinje cells has been reported by Bell and Grimm (1969) with the help of Perkel and Segundo. They studied records obtained in anesthetized cats with single and separate electrodes. Single Purkinje cells generally exhibit two types of discharge: a simple spike which is similar to typical extracellularly recorded spikes and which is thought to reflect activation by mossy fibers, and complex spikes in which the initial spike is immediately followed by a series

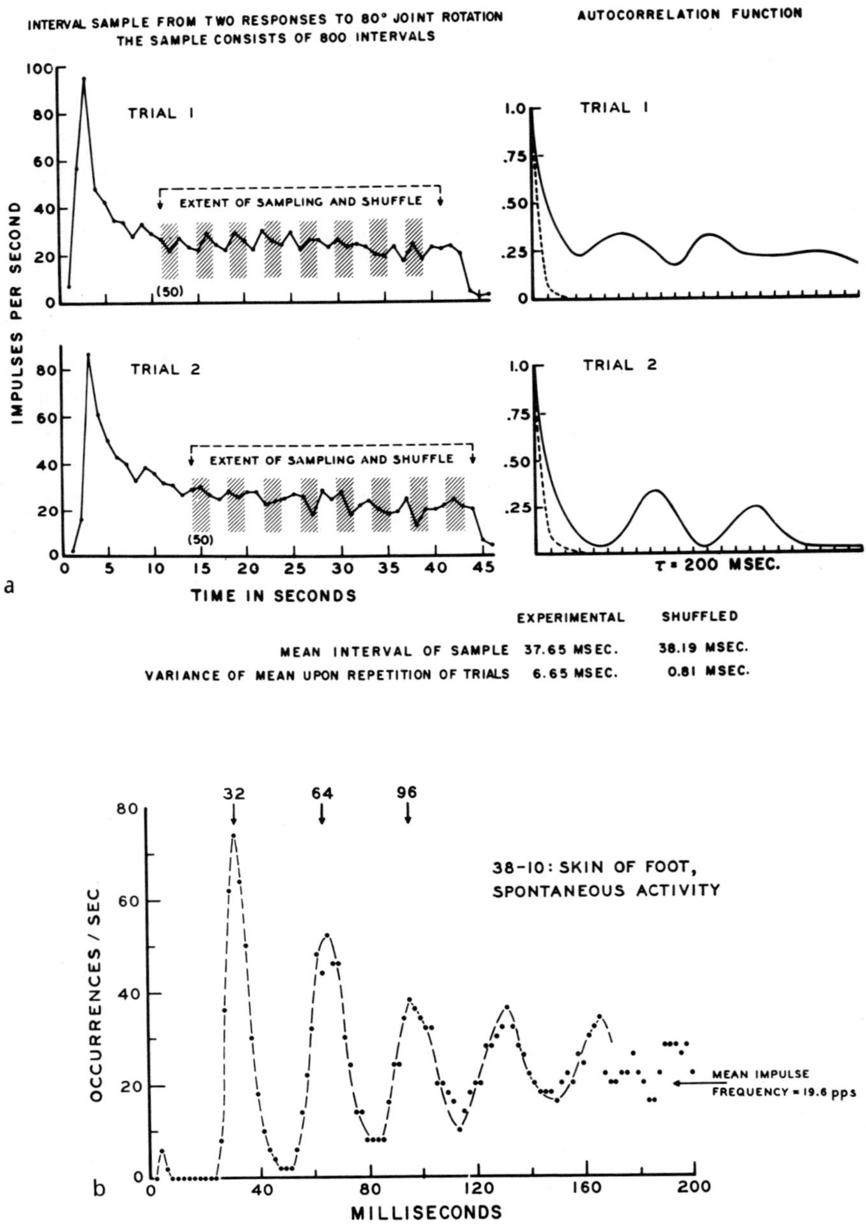
INTERVAL SAMPLE FROM TWO RESPONSES TO 80° JOINT ROTATION
THE SAMPLE CONSISTS OF 800 INTERVALS
AUTOCORRELATION FUNCTION
TRIAL 1
TRIAL 2
EXTENT OF SAMPLING AND SHUFFLE
(50)
IMPULSES PER SECOND
TIME IN SECONDS
τ = 200 MSEC.
a
EXPERIMENTAL
SHUFFLED
MEAN INTERVAL OF SAMPLE 37.65 MSEC. 38.19 MSEC.
VARIANCE OF MEAN UPON REPETITION OF TRIALS 6.65 MSEC. 0.81 MSEC.
32
64
96
38-10: SKIN OF FOOT,
SPONTANEOUS ACTIVITY
OCCURRENCES / SEC
MEAN IMPULSE
FREQUENCY = 19.6 pps
MILLISECONDS
b

of small spike oscillations lasting about 10 msec and which is thought to reflect activation by climbing fibers. On the basis of interval histograms and autocorrelation histograms, they describe simple responses as occurring rather regularly and complex spikes as occurring more irregularly and at a slower rate. Examples are shown in Figure 10-9. They also constructed cross-correlation histograms on five pairs of cells recorded simultaneously from separate electrodes. These cross-correlation histograms show high correlations between the simple spikes of the Purkinje cells, very suggestive of cells experiencing shared excitatory input. On the other hand, cross-correlation histograms between the complex spikes of neighboring Purkinje cells showed no relationship, suggesting a more specific input from the climbing fibers. Examples are also shown in Figure 10-9. They also note that within single cells there is a tendency for a pause in simple spike firing to occur following the occurrence of complex spikes. However, the climbing-fiber response of the cell did not seem to influence the simple spike train of a nearby Purkinje cell. A sketch of the main lines of influence suggested by these studies is shown in Figure 10-9. Finally, Bell and Grimm performed statistical analysis on cells in the region of the olivary nuclei, which are known to be an origin of climbing fibers. These cells showed slow and irregular firing characteristics similar to those of the complex spikes in the cerebellar cortex.

Several groups of investigators have studied the firing properties of neurons in the cerebral cortex. Holmes and Houchin (1966) presented autocorrelation histograms and cross-correlation histograms taken from pairs of cells and recorded with separate electrodes in the cortex of anesthetized rats. Their main finding is the occurrence of a slow rhythm

Figure 10-7. Autocorrelation histograms from thalamic neurons. (a) Estimation of the variability of responses to identical stimuli. In the left part of the figure, two responses of a ventrobasal thalamic joint neuron (33-6) to 80 ° flexion of the knee are plotted in impulses per second. The discharge intervals were sampled in groups of 50 each over the bracketed period, as indicated by the shaded blocks. The computation of the variance of the mean upon repetition of trials is based on this array of blocks. The corresponding sampling and computation after random shuffle of the discharge intervals in the bracketed period resulted in values listed under "shuffled." On the right, the solid lines plot the autocorrelation functions of the responses in the bracketed periods, with time lags of 200 msec. The corresponding spectral-density functions (not shown) had significant peaks at frequencies of 0.68 cycle/sec (trial 1), and 0.78 cycle/sec (trial 2). In 28 similar trials, to various angles, the frequencies varied between 0.23 and 1.0 cycle/sec. The respiratory rate was 81/min. The autocorrelation functions after random shuffle within the bracketed response periods are plotted as dashed lines. (b) Plot of the expectation density function typical of that found for the majority of impulse sequences of thalamic lemniscal neurons. Points indicate the average likelihood for an impulse to occur (occurrences/sec) at successive time segments (2 msec width) after the occurrence of any one impulse in the sequence. Broken line was fitted by hand to the computed points.

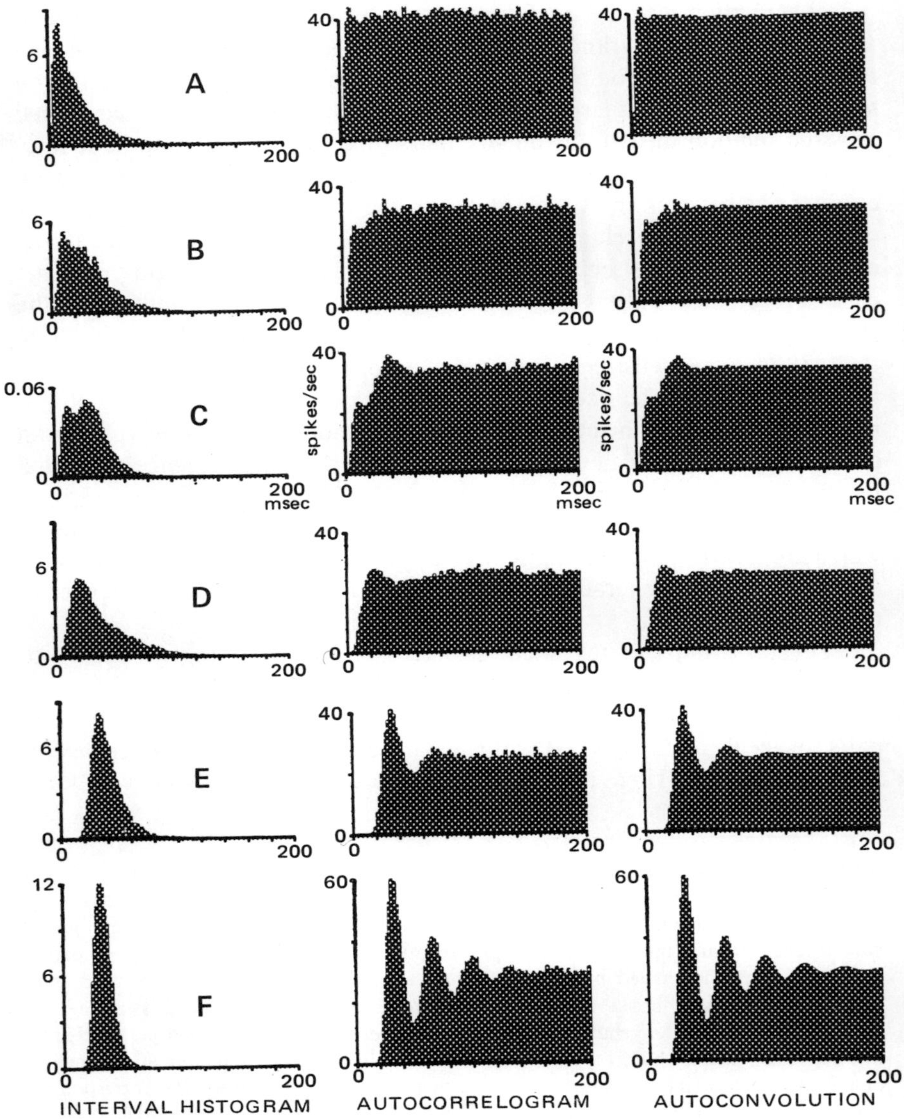

Figure 10-8. Histograms from retinal ganglion cells. (Above) Normalized interval histograms, autocorrelograms, and autoconvolutions compiled from six sample records of maintained activity. (Right) Cross-correlograms of three pairs of units recorded simultaneously. A: two off-center units recorded in the light. B: off-center unit (time moves to the right) and on-center unit (time moves to the left). C: off-center unit (time moves to the right) and on-center unit (time moves to the left).

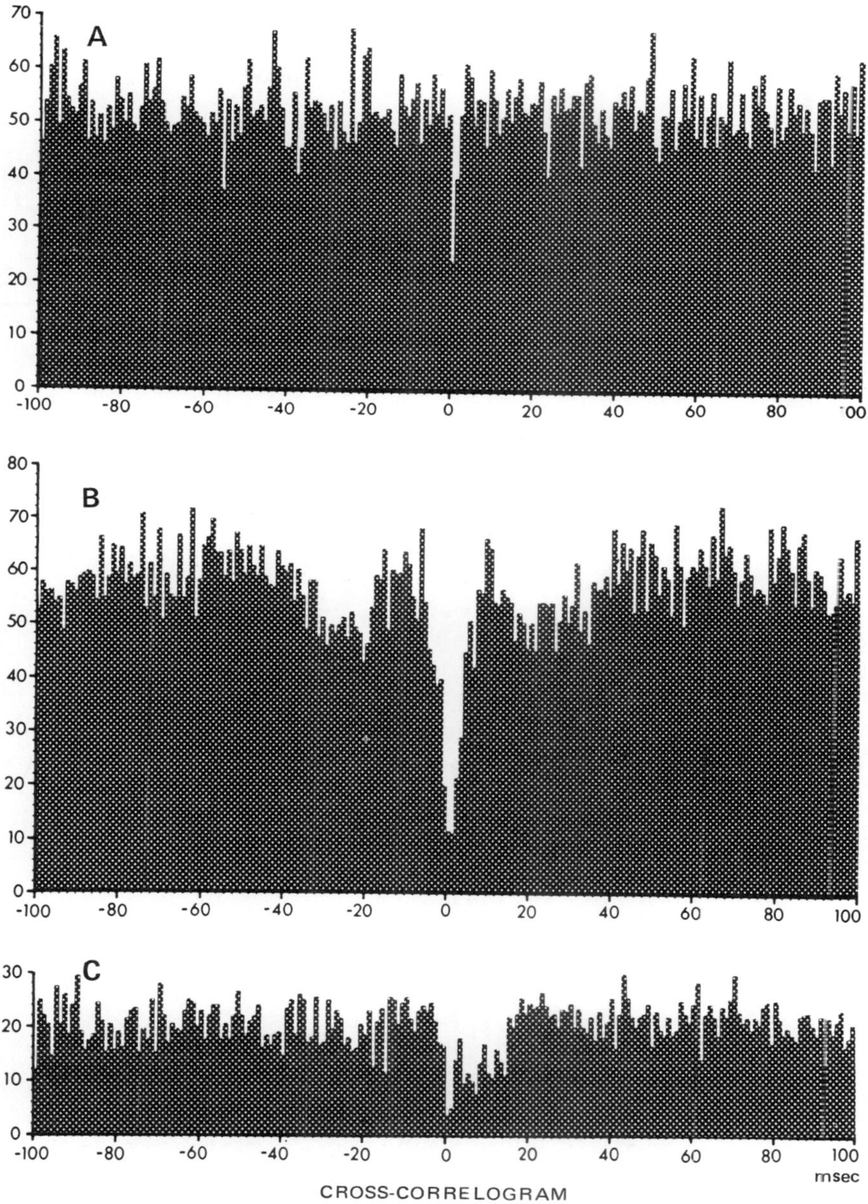

Figure 10-8 (continued)

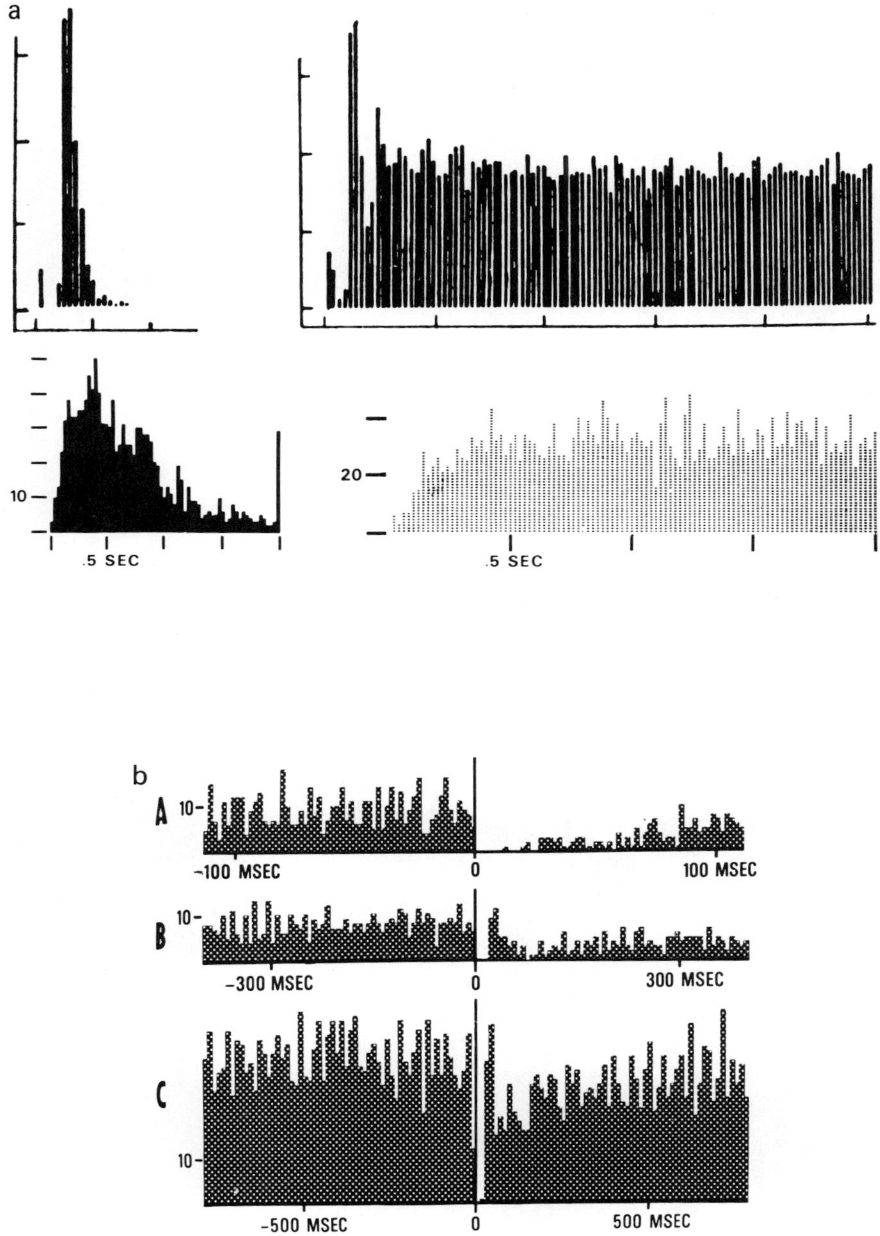

Figure 10-9 a,b

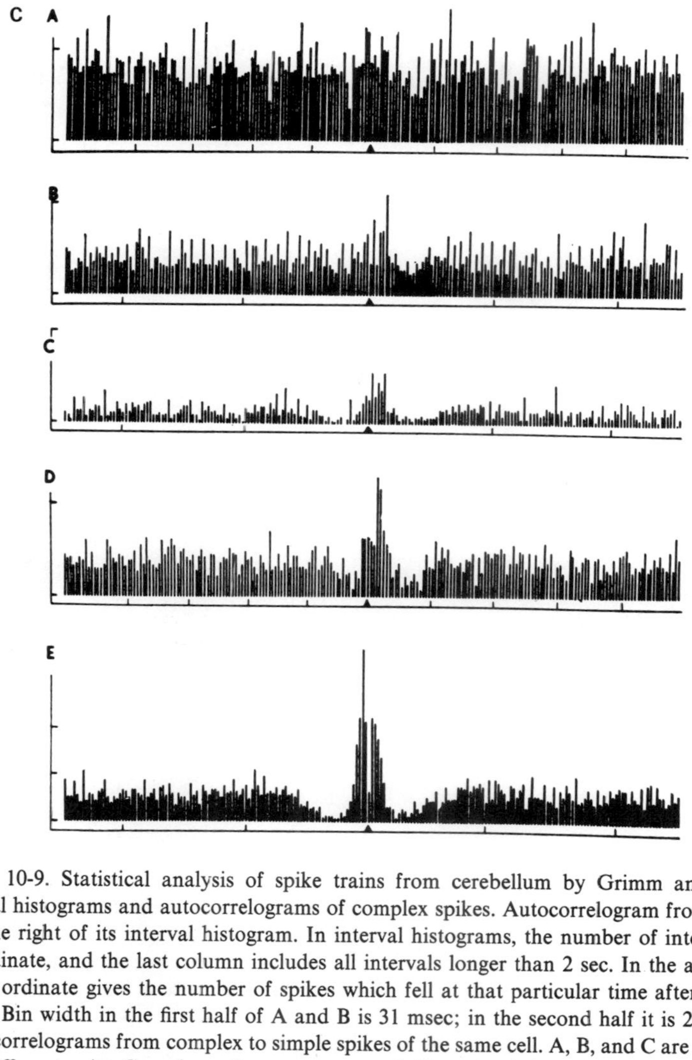

Figure 10-9. Statistical analysis of spike trains from cerebellum by Grimm and Bell. (a) Interval histograms and autocorrelograms of complex spikes. Autocorrelogram from each cell is to the right of its interval histogram. In interval histograms, the number of intervals is on the ordinate, and the last column includes all intervals longer than 2 sec. In the autocorrelograms, ordinate gives the number of spikes which fell at that particular time after a spike at $\tau = 0$. Bin width in the first half of A and B is 31 msec; in the second half it is 20 msec. (b) Cross-correlograms from complex to simple spikes of the same cell. A, B, and C are taken from three different units. Complex spike occurs at $\tau = 0$. Note that in B and C the long pause does not begin immediately but is preceded by a brief pause and a period in which the discharge rate is the same as that preceding the CFR. The latter is referred to in the text as activation. (c) Cross-correlograms between neighboring Purkinje cells. Each graph shows the relationship between firing times of the two Purkinje cells. Members of each pair were recorded on separate microelectrodes. Electrodes were 50–100, 50, 65, 40, and 70 μ apart in A, B, C, D, and E, respectively. Solid triangle in the center of each graph indicates times $\tau = 0$. Marks on abscissa are 10 msec apart; on ordinate, 50 counts apart.

with period of about 1.9 sec which seems to permeate the entire structure during deep anesthesia. In this preparation, groups of relatively frequent discharges are separated by periods of quiescence. This rhythmicity shows up in the auto- and cross-correlation histograms, as indicated in Figure 10-10. The rhythmic correlations were in phase if the neuron pairs were in the same coronal section, and out of phase otherwise. They suggested that their data are consistent with a traveling wave propagating at 5 to 10 cm/

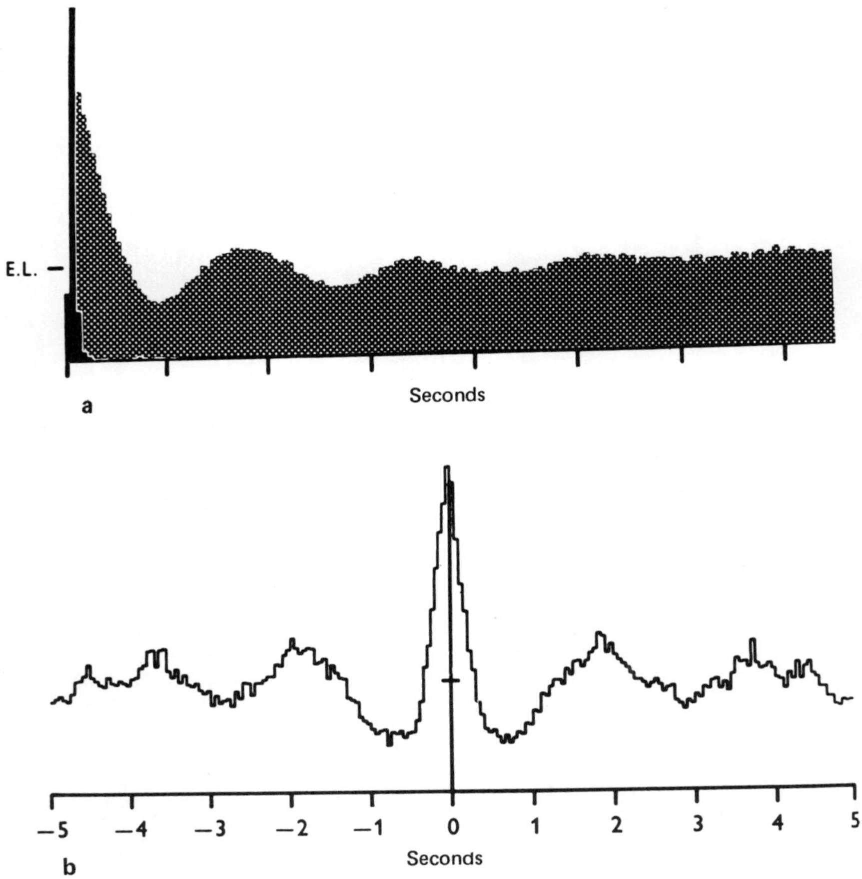

Figure 10-10. Auto- and cross-correlation histograms for rat cortical units by Holmes and Houchin. Cross-intensity functions between two units recorded simultaneously from a single microelectrode. (a) Animal lightly anesthetized; (b) 4–7 min after supplementary urethane administered intraperitoneally. In both cases the functions were computed on 1500 intervals. Expected levels were 302 and 141, respectively.

sec anterior to posterior. The degrees of both the rhythmicities and cross-correlations increased with depth of anesthesia. The amplitude of the periodic fluctuation diminishes as the depth of anesthesia diminishes, such that it is generally absent in the unanesthetized preparation. The period, however, does not change with depth of anesthesia. Moreover, increasing the sensory bombardment of the cortex reduces or abolishes the correlation between the activities of pairs of units. Holmes and Houchin suggest that these results probably reflect a single control process which causes synchronous excitability changes in all the cortical units. It is not clear from their work whether the rhythmic process reflects interconnections of the cortical units themsees or is imposed by some ascending influence from subcortical structures.

The correlation in spike activity between adjacent cortical neurons was studied by Efremova *et al.* (1950). They recorded two or three neurons simultaneously from one microelectrode in unanesthetized and nonimmobilized rabbits. A total of 213 pairs of neurons were analyzed: 73 in the visual cortex, 15 in the auditory cortex, and 125 in the sensorimotor cortex. These authors then evaluated a single cross-correlation coefficient as defined by

$$R = \frac{\sum_{i=0}^{n} (a_i - \bar{a})(b_i - \bar{b})}{\left[\sum_{i=0}^{n} (a_i - \bar{a})^2 \sum_{i=0}^{n} (b_i - \bar{b})^2\right]^{1/2}} \tag{10-8}$$

Here a_i and b_i represent, respectively, the events in train a and train b; $\bar{a}$ and $\bar{b}$ represent their respective means. Using this measure, they find a significant degree of correlation in about 10% of pairs in the sensorimotor cortex and 15% of pairs in the visual cortex. These relationships are generally maintained unchanged in time whether during background activity or during presentation of weak stimulation. However, when the animals are subject to rhythmic photic stimulation or injected with sedatives such as bromide and morphine, an increase in the number of pairs which fire synchronously occurs.

Noda and Adey (1970*a–c*, 1973) have recently presented studies of the firing properties of single neurons and pairs of neurons in the cerebral cortex of the cat in various stages of sleep and wakefulness. In deep and intermediate sleep, the single-cell firing patterns tend to consist of clusters or bursts of activity separated by relatively long, quiet intervals. Moreover, the long, quiet intervals tend to occur at the same time in pairs of units. Thus, autocorrelation histograms obtained in deep or intermediate sleep often show a preponderance of relatively small intervals corresponding to the burst activity. Cross-correlation histograms in these states generally

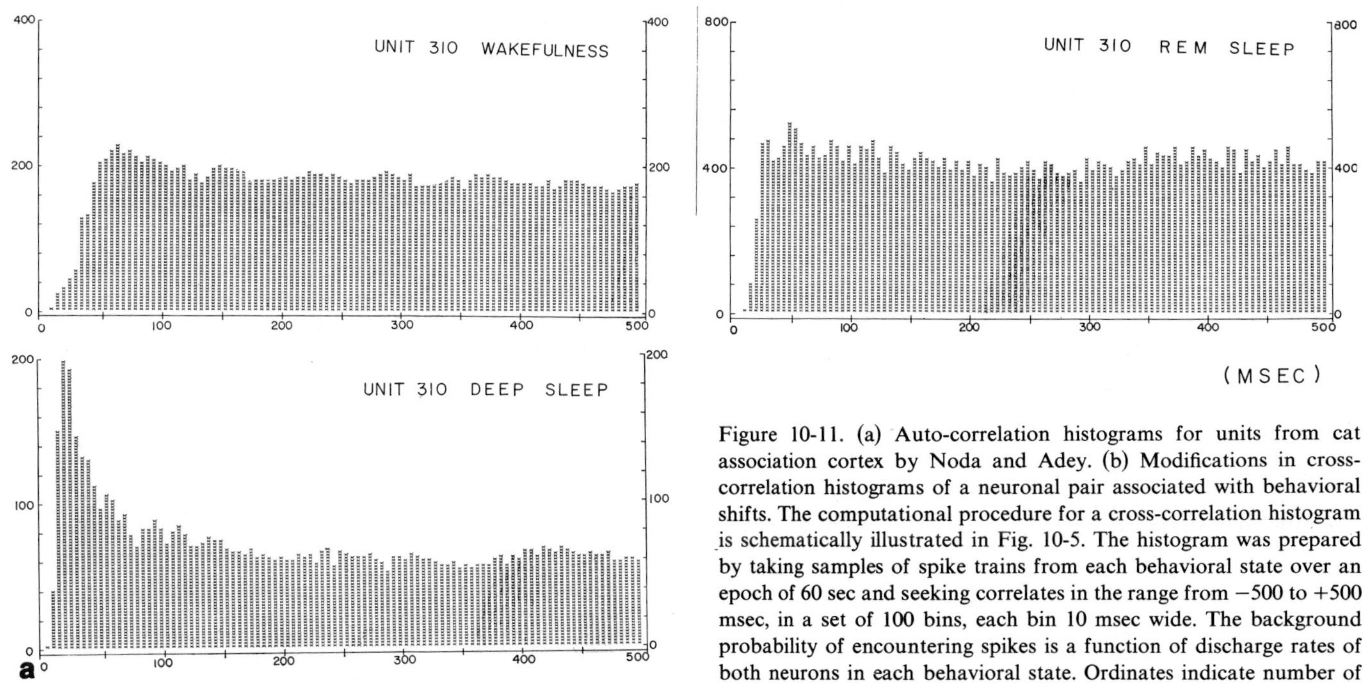

Figure 10-11. (a) Auto-correlation histograms for units from cat association cortex by Noda and Adey. (b) Modifications in cross-correlation histograms of a neuronal pair associated with behavioral shifts. The computational procedure for a cross-correlation histogram is schematically illustrated in Fig. 10-5. The histogram was prepared by taking samples of spike trains from each behavioral state over an epoch of 60 sec and seeking correlates in the range from −500 to +500 msec, in a set of 100 bins, each bin 10 msec wide. The background probability of encountering spikes is a function of discharge rates of both neurons in each behavioral state. Ordinates indicate number of spikes found in each bin. (c) A: cross-correlation histogram showing a broad correlation peak and marked fluctuations on both sides. B, C: autocorrelation histograms for individual neurons in the same period of deep sleep seen in the cross-correlation histogram A. D: cross-correlation histogram of the same neuronal pair during REM sleep. Ordinate units indicate number of spikes found in each bin.

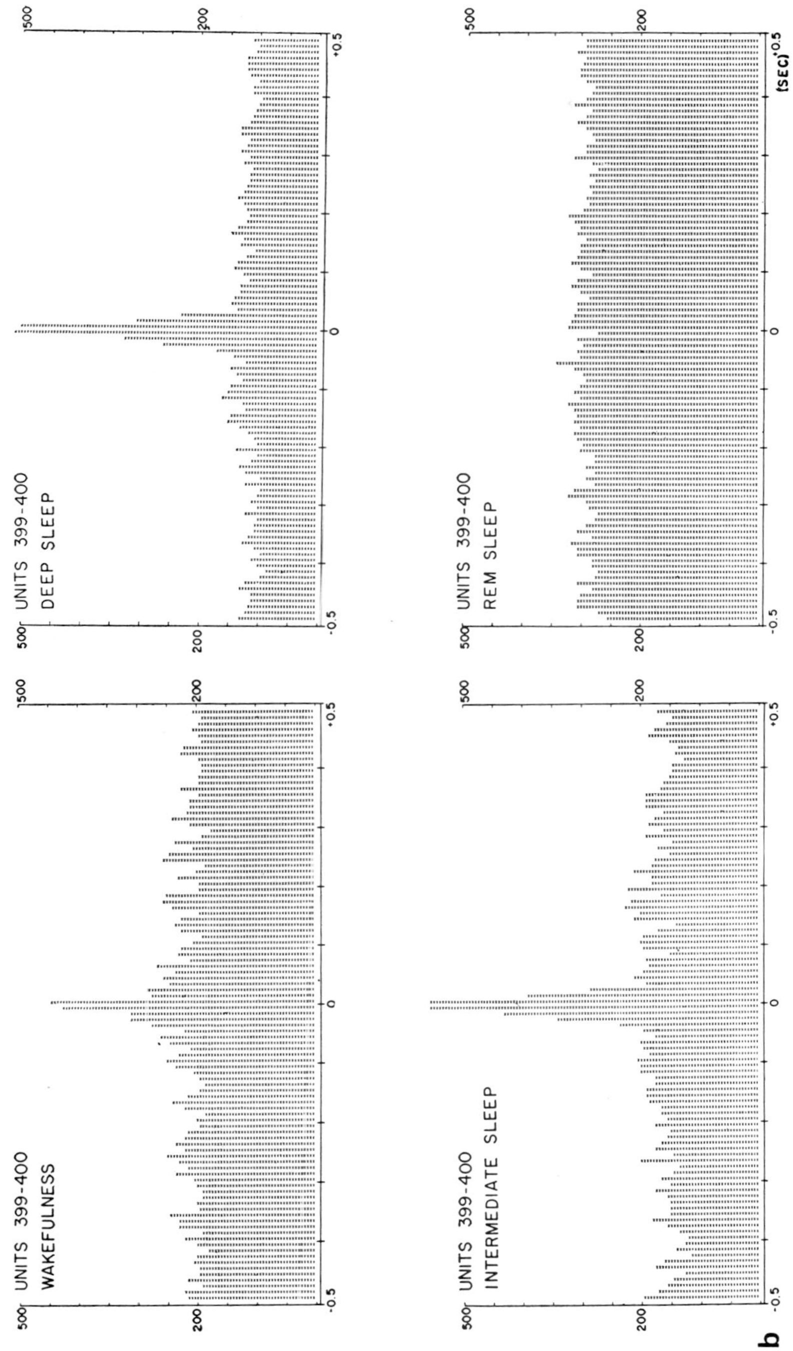

Figure 10-11b.

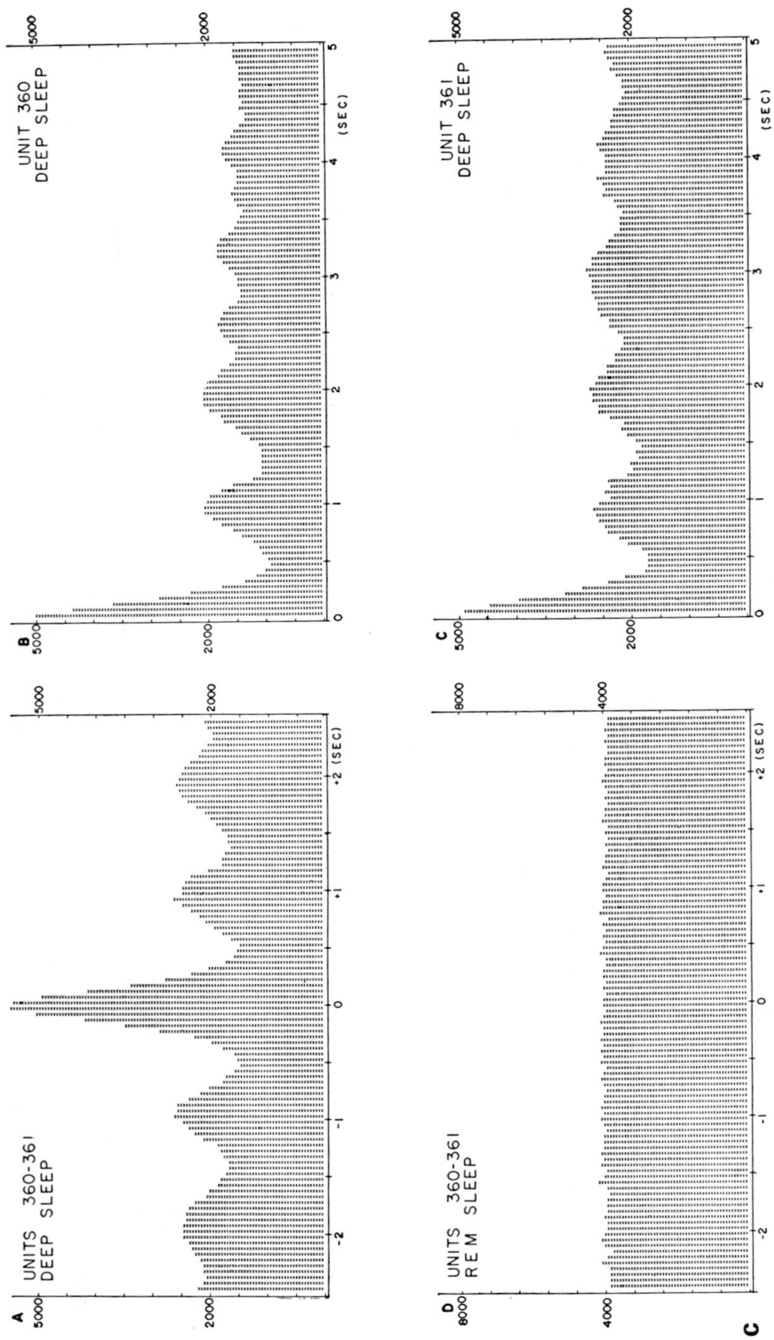

Figure 10-11c (See caption on page 261.)

exhibit a high correlation with generally zero lag. Examples are shown in Figure 10-11. On the other hand, in rapid-eye-movement (REM) sleep and in the aroused state the single-cell patterns are typically more regular and tend not to manifest the bursting activity shown in deep and intermediate sleep. Moreover, in arousal and REM sleep the coupling as shown by the cross-correlation histograms is generally absent. Examples are shown in Figure 10-11. These authors suggest the existence of some common control mechanism active during deep and intermediate sleep which generates the slow rhythmicity and correlated firing properties of these units. The cross-correlation histograms were taken from a sample of 81 pairs of simultaneously recorded units. Each pair was taken from a single electrode. Performing similar analyses on records from hippocampal units, Noda, Monahar, and Adey (1969) found very similar results. Moreover, many of these units exhibit slow rhythmicity in the autocorrelation histograms in deep sleep.

The firing patterns of neurons in the brain stem reticular formation have also been investigated by several groups of investigators. In 1958, Amassian and Waller reported on correlations between firing patterns in reticular neurons of the unanesthetized, paralyzed cat. These data were presented in terms of the number of spikes on one channel that occur within a fixed number of milliseconds after a spike had occurred on another channel. This measure was given as a function of time and called an "interval for coincidence." Amassian and Waller (1958) presented data for several pairs, each obtained from separate electrodes. In all cases, these curves were essentially straight lines and deviated insignificantly from what one would expect on the basis of mean firing rates. They concluded that these neurons were independent in their spontaneous activity.

Noda, Monahar, and Adey (1969) have extended their study of spontaneous behaviour of neuron population in sleep–wakefulness cycles to the mesencephalic reticular formation. Their study describes activity from single and simultaneously recorded pairs of cells in chronically prepared cats. In this preparation, the firing patterns are considerably different from what they found in the cerebral cortex and hippocampus. In this network they find many cells which change their rate and pattern according to the state of wakefulness or sleep period. Loosely, they describe activity in arousal as being more regular than in REM sleep, which is characterized as more random, and that occurring in slow sleep as being in between these patterns. They also find many units whose firing patterns and rates do not change with state of sleep or wakefulness. Some 10% of their units fire rather regularly and might be considered as noisy pacemakers. These have mean intervals between 15 and 80 msec. Some examples are shown in Figure 10-12. These authors present cross-correlation histograms from seven pairs of neurons simultaneously recorded in each case with a single microelectrode. On the basis of these histograms they are able to come to

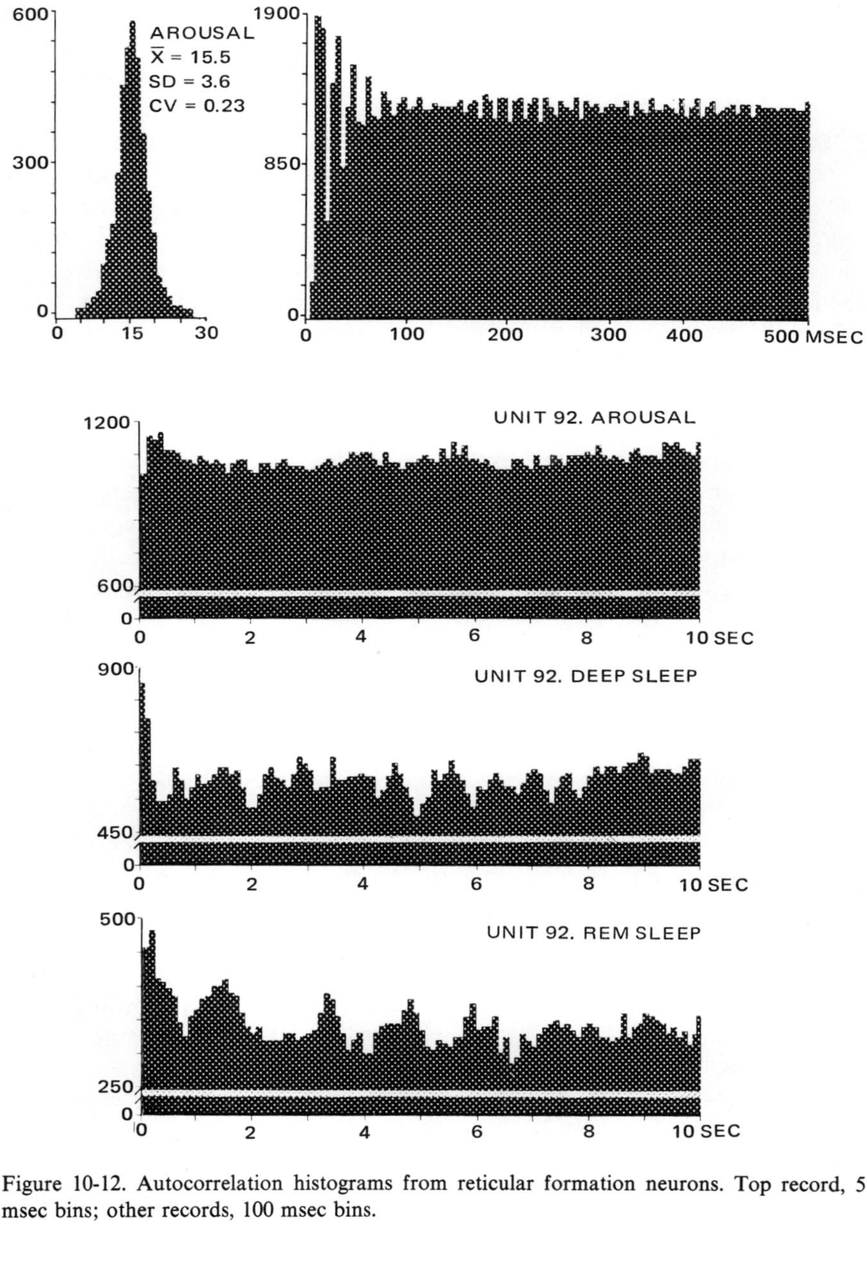

Figure 10-12. Autocorrelation histograms from reticular formation neurons. Top record, 5 msec bins; other records, 100 msec bins.

no conclusions concerning cross-correlations of reticular neurons except that the kind of correlation they observed in the cortex and hippocampus is not readily apparent in the reticular formation.

Recently we have presented (MacGregor *et al.*, 1973, 1975*a*) statistical analyses of spontaneous activity in midbrain reticular neurons of the unanesthetized, immobilized rat, obtained in the laboratories of our colleague P. M. Groves. These unit results were obtained from a total of 57 single cells and 32 pairs of cells. Nine of the pairs were from single electrodes, and 23 pairs were from separate electrodes. The first impression one has is that reticular firing patterns are, for the most part, random. Interspike interval histograms are approximately exponential, and visual inspection of records reveals an unremarkable randomness consistent with a Poisson process. Firing rates range from 1/sec to more than 100/sec, with many firing between 1 and 10/sec and most firing at about 15 to 25/sec. A few scattered cells exhibit noticeable slow rhythmicities in firing. In a typical group of about 13 cells, for example, 11 would exhibit unremarkable random patterns, while one would exhibit pacemaker activity firing a single spike with clocklike regularity with periods of about 0.64 sec, and another single unit would exhibit marked bursts of activity recurring rhythmically every 1.3 sec. The pacemaker units have firing rates of from 1 to 6/sec, while the bursting cells have firing rates of 4 to 48/sec with an average of about 20/sec.

On closer inspection, however, it becomes clear that many of the randomly firing cells contain an underlying slow rhythmicity at about the same interval as that of the pacemaker and bursting cells. These rhythms are not easily (and often not at all) discernible on visual or auditory screening of the records. However, autocorrelation histograms show clear, although low-level, rhythmicities in about two-thirds of the randomly firing cells. The slow rhythmic character of these 57 reticular cells can be categorized into four main groups: pacemaker (P), marked rhythmicity (M), moderate or marginal rhythmicity (m), and no rhythmicity (n). Four of 57, or 7%, are pacemakers (regular intervals of 0.64, 0.64, 0.585, and 0.625 sec); five, or 9%, exhibit marked rhythmicities (these had periods of 0.64, 0.64, 01.28, 01.76, and 1.88 sec); 30, or 53%, show moderate or marginal rhythms; and 18, or 32%, have no slow rhythms. A representative example of each of these four classes of units is shown in Figure 10-13. A notable feature of these slow rhythms is that the most prominent intervals seem to be integer multiples of a common base interval of 0.64 sec. One of our two most marked pacemakers has a firing interval of this value, while the other pacemaker, although also clearly firing at this interval, tends to quite often skip two firings. This results in a very prominent peak at 1.92 sec in the autocorrelation histogram for this unit. Many of the slower rhythms in marked or moderate/marginal cells occur with intervals of about 1.3, 1.9, and 2.5 sec, with 1.3 sec by far the most common. Autocorrelation histograms also showed that most of these cells display an initial period of

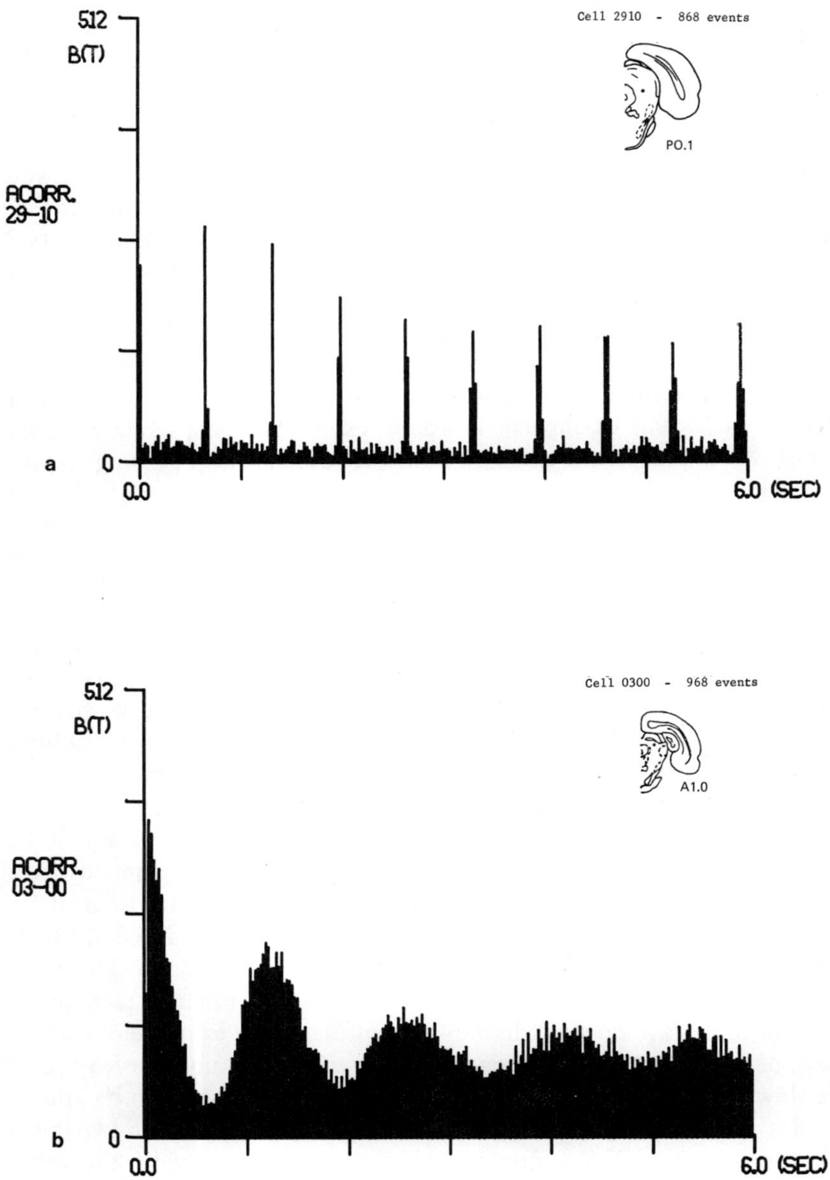

Figure 10-13a,b

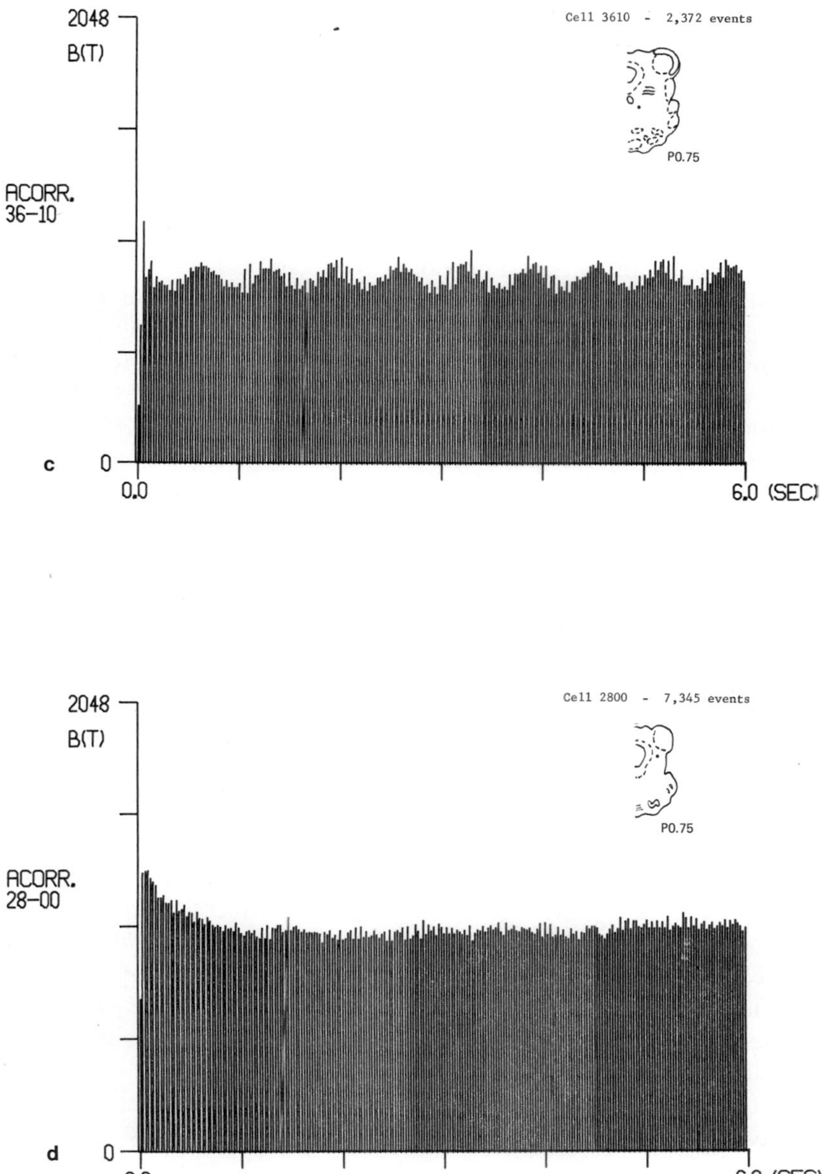

Figure 10-13. Autocorrelation histogram from (a) representative pacemaker cell, (b) representative markedly rhythmic cell, (c) representative moderately rhythmic cell, and (d) representative arrhythmic cell.

subdued firing and a more or less pronounced peak followed by a gradual decay upon which the rhythmicity, when it occurred, was superimposed. By comparing autocorrelation histograms to those obtained on the shuffled data, it was ascertained that the so-called peak and decay phenomenon and the observed rhythms were order dependent. Moreover, in subsequent modeling studies (MacGregor and Radcliffe, 1973), it was shown that the peak and decay phenomenon might be mediated by recurring reverberations within reticular nets.

The fact that previous investigators of the reticular formation have not remarked on the pervasive rhythmicity we have described may reflect a species difference in that most previous investigations have been performed on cats. Also, we should note that pilot studies on a small number of anesthetized rats in our laboratory provided a very markedly reduced occurrence of slow rhythms although these did occur in this preparation in the absence of artificial respiration. The rhythm may relate to the animals' respiration, but if this is so it is a subtle relationship and not so simple as one might suppose. For example, the rhythmic firing intervals do not correspond to the period of the respirator.

Cross-correlation histograms using 1-msec bin size and 150-msec range reveal a general paucity of correlated firing. The general impression is that particular synaptic pathways between midbrain neurons and from afferent fibers to neurons which are prepotent and repeatedly successful in eliciting postsynaptic action potentials are at a minimum. However, there do indeed exist some correlations in firing in pairs of reticular cells. In particular, at least three and at most six of nine neuron pairs recorded with single electrodes show short latency correlations in firing. At least two and at most five of 23 pairs recorded with separate electrodes show short latency correlations in firing.

A notable feature of these data is that the tendency for a midbrain neuron to exhibit correlations in firing with other midbrain neurons seems to increase with the degree of rhythmicity apparent in the cell. Moreover, broader-based cross-correlation histograms show that neighboring cells (e.g., recorded with a single electrode) tend to exhibit in-phase broad rhythmic correlations whereas cells separated by hundreds of microns (recorded with separate electrodes) tend to exhibit out-of-phase broad rhythmic correlations. Examples are shown in Figure 10-14.

A tentative picture consistent with these data is that local regions in midbrain tissue sometimes exhibit synchronized firing activity upon which any other afferent activation is superimposed. Slightly distal regions, however, would be out of phase with each other. This situation could reflect either the propagation of a slow wave through the midbrain or the existence of separate subsystems of cells each tending to fire synchronously within itself but remaining not necessarily synchronized with other systems. If this picture is generally valid, there are some exceptions in that some local

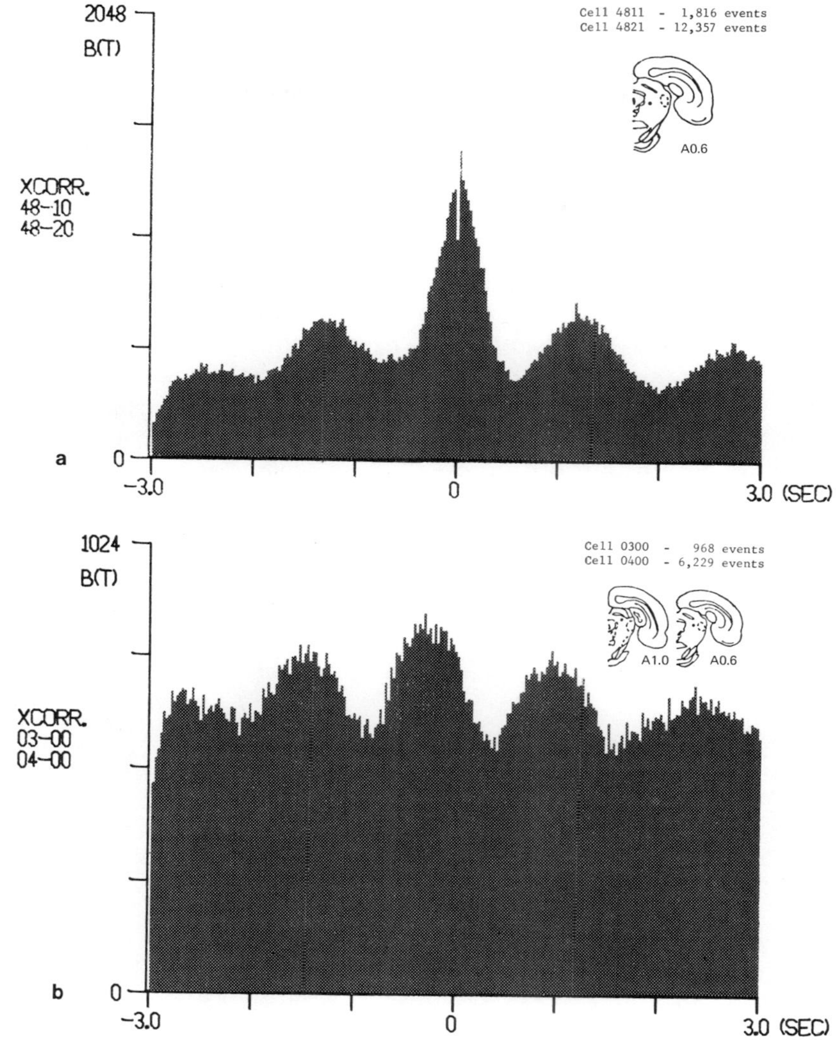

Figure 10-14. Slow rhythmic correlation between two cells recorded with (a) one electrode, and (b) separate electrodes.

neuron pairs are out of phase and some distal cell pairs are in phase. For example, one pair of cells including a markedly rhythmic cell and a moderate or marginally rhythmic cell obtained from separate electrodes exhibited broad out-of-phase relationships consistent with the generalization but also exhibited weak but clear short latency correlations in firing.

Another unit was out of phase with a local neighboring cell but in phase with two distal cells recorded with the same electrode. This cell did not exhibit short latency correlations with any of the other three. One might speculate that the hypothetical wave advanced in a not-totally homogeneous fashion extending isolated branches of synchronizing influence into as yet uninvaded regions or, alternatively, that separate functionally synchronized subsystems might interpenetrate anatomically.

The tools of statistical analysis have only begun to be applied to the main networks of the mammalian central nervous system. However, many investigators have come to appreciate the value and necessity of studying pools of neurons simultaneously and the network context of electrical records generally. Segundo *et al.* (1973) have recently published a study of auto- and cross-correlation histograms for cells of the abdominal ganglion in aplysia. This system is particularly valuable as a model system since a great deal of its anatomy, electrophysiology, and neuropharmacology are known. We can confidently expect these tools to greatly enhance our understanding of the functional organization of entire neuronal networks or assemblies as they operate in the ongoing functioning of the brain. In the development of this work, help from continued studies on simpler systems with known interconnection patterns and from studies on artificial canonical systems will be very valuable.

11

Models of Neuron Pools

In this chapter we consider efforts directed toward understanding the dynamic properties of neuron pools of intermediate size, containing some hundreds to thousands of neurons. The motivation behind this work is the possibility of understanding the overall dynamics of large masses of neural tissue in terms of such pools, on the one hand, and to understand the properties of such pools in terms of their individual constituent neurons, on the other. Thus, such pools might serve as an intermediate conceptual device in linking the dynamic properties of single neurons to the behavior of extensive masses of neural tissue—a gap that has not been successfully bridged by other means at least thus far.

An insightful and thorough critique of this point of view and its justification in terms of nervous structure and physiology is given by Freeman (1972). Freeman points out several limitations involved in considering the nervous system as consisting of a vast number of distinct individual neurons which interact solely via discrete action potentials. For example, the numbers of neurons and of possible interactions in vertebrate central nervous systems are astronomically large; the significant decision making in nervous tissue seems to occur in the graded generator potentials of neurons, which are essentially continuous analog signals, rather than in the discrete trains of action potentials; groups of neurons tend to fire coordinated activity patterns such that in considering the output or input to any given region the natural measure seems to be the simultaneous activity patterns in several or large numbers of fibers rather than the spike train of a single fiber. Freeman also points out that the idea of considering nervous activity as continuous and graded is not new. For example, Sherrington (1906) used such a concept in his so-called "central excitatory state" as the main variable for describing the activity of pools of motoneurons; Lashley's (1960) concept of "mass action" is also suggestive of continuous variables in masses of tissue rather than discrete pulses in discrete neurons. It is rather the pulse logic hypothesis which is new. This point of view became widespread only after the ubiquitous use of micro-electrodes after 1940 and the subsequent intensive development of single-unit electrophysiology. After his cogent review and arguments, Freeman concludes that the "search must undertake the description of neural masses in terms peculiar to their own properties, arising out of but distinct from the properties of single neurons."

Freeman has also developed extensively a viable approach to this problem. He defines a class of measures denoted as "activity density functions" as representing more natural measures of nervous activity than individual spike trains. These functions are conceived to be continuous and graded functions of position and time. The value associated with a particular action density function at a particular point in space may be thought to represent the average of single-cell activity appropriate to that density function in all appropriate single cells in a small volume centered at the point. Freeman suggests that such activity density functions are the "state variable" of neural pools and defines three main functions of this type: first, the so-called "pulse density function" which is defined over axon terminals and represents the density of input pulses incident on any dendrite at the point in question; second, a so-called "wave density function" which is defined over the peripheral regions of dendritic trees and represents the response in the dendritic tree to the "pulse density function"; third, a second wave density function which represents activity over triggering sections in neurons and which results from a transformation of the first wave density function as it is propagated from the dendrites to the soma of neurons. In Freeman's structuring of this problem, there would be, corresponding to these three density functions, three transformations representing the transfers from the pulse density function to the first wave density function, from the first wave density function to the second wave density function, and from the second wave density function to the pulse density function. There would also be three classes of parameters of a given neuron pool: first, those parameters representative to the pulse-to-wave and wave-to-pulse transformations which would determine the overall output to input ratio or "forward gain"; second, the time delays associated with dendritic and axonal conduction and synaptic delay; and, third, information concerning the spatial connectivity of the network.

Freeman also takes initial steps in categorizing various types of neuron masses. The most primitive in his view is the so-called "aggregate," which is a neural mass with common input and zero functional connection density even if anatomical connections exist within it. A neuron "population," on the other hand, is a set of densely interconnected neurons with common input and the same sign of output. Populations can be one of two types: excitatory with positive excitatory feedback, or inhibitory with positive inhibitory feedback. Here, Freeman is relying on the simplified inference from Dale's principle [see Nachmansohn (1976)] that all cells are either excitatory or inhibitory. Finally, the so-called "assembly" of neurons results from the dense interconnection of an excitatory with an inhibitory population.

Freeman has also taken some preliminary steps in attempting to obtain a quantitative description of the dynamics of such pools. Thus, he suggests that the pulse-to-wave conversion by dendrites might be approximately quantified by the equations

$$\begin{aligned} \frac{dV}{dP} &= -\zeta(v - v_i) \qquad \text{for } v < 0 \\ \frac{dV}{dP} &= \frac{\zeta}{r_d}(v - v_e) \qquad \text{for } v > 0 \end{aligned} \tag{11-1}$$

Here P is a continuous graded variable representing the magnitude of the pulse input as a function of time, and V is a corresponding graded continuous function corresponding to the wave output in dendrites. Note that in these equations time is contained only implicitly. The constants ζ and r_d represent empirically determined rate constants. The constants v_e and v_i represent the level of extracellular wave potential corresponding to the excitatory and inhibitory equilibrium potentials for the dendritic membranes, respectively. The solutions to these equations as given by Freeman are

$$\begin{aligned} v &= v_i(1 - e^{-\zeta(P-P_0)}) \qquad \text{for } P > P_0, v < 0 \\ v &= r_d v_e(1 - e^{-(\zeta/r_d)(P-P_0)}) \qquad \text{for } P < P_0, v > 0 \end{aligned} \tag{11-2}$$

Freeman suggests that the wave-to-pulse conversion by axons can be approximately quantified as

$$\begin{aligned} \frac{dP}{dv} &= r_a \gamma P \qquad \text{if } v < 0 \\ \frac{dP}{dv} &= \gamma P \qquad \text{if } v > 0 \end{aligned} \tag{11-3}$$

Here the interpretations of P and V are the same as above, and r_a and γ are empirically determined excitatory inhibitory rate constants. The solutions to these equations as given by Freeman are

$$P = \overline{P}_0 e^{r_a \gamma V} \qquad V < 0$$
$$P = P_{max} - (P_{max} - \overline{P}_0)e^{-\gamma V} \qquad V > 0 \qquad (11\text{-}4)$$
$$P = \overline{P}_0 \qquad V = 0$$

Here, $P_{max} = \overline{P}_0(r_a + 1)$. Freeman goes on to calculate various properties associated with these equations, including forward gain, open-loop response, characteristics of positive and negative feedback, and related topics for certain simplified populations. He reports, for example, that the impulse response of neuron populations is monotonic, with decay rate approaching zero as feedback gain approaches unity, and that the impulse response of the assembly is oscillatory. He remarks that, in general, the positive inhibitory feedback loop contributes an internal inhibitory bias in an assembly and, at unity gain, determines its stability characteristics. For normal operation, a source of external excitatory bias is required.

A particularly valuable component of Freeman's approach to coding in neural pools is that his theoretical work is coupled with extensive experimentation on the vertebrate olfactory system. His theoretical results are related to these experimental data. The interested reader is referred to Freeman's publications for descriptions of this work.

We would like to suggest here a systematic approach to considering the ways excitatory and inhibitory neurons can be arranged in pools. We will consider a highly simplified model case, based on the not-always-warranted simplification of Dale's chemical principle, namely, that there exist only two classes of neurons—excitatory and inhibitory—and correspondingly, that there exist only two kinds of input fibers—excitatory and inhibitory. We will define a pool of neurons as a set of neurons containing one population of excitatory neurons and one population of inhibitory neurons. It is then easy to recognize that there are 16 different kinds of pools differentiated according to their interconnections. The population of excitatory cells can send collateral connections to no other cells in the pool, to some fraction of (SFO) the excitatory population only, to (SFO) the inhibitory population only, or to (SFO) both the excitatory and inhibitory populations. The total number of permutations available by matching these four possibilities for the inhibitory population with those of the excitatory population is 16. These arrangements are indicated in Figure 11-1.

Consider now the various ways such pools might be activated by small numbers of input fiber systems. For the purpose of classification, assume that spontaneous activity which may be endogenously generated within the cells can be included within the framework of the excitatory input fiber system. (For example, excitatory membrane currents associated with leaky membranes might be included in an activation current term associated with the excitatory input fibers.)

To systematically include all the ways of activating such neuron pools, we can proceed sequentially through the cases involving one input fiber system, two input fiber systems, and so on up to N input fiber systems. For the case of a single excitatory input fiber system, there exist three possible arrangements: The fiber system may activate SFO only the excitatory cells, SFO only the inhibitory cells, or SFO both the excitatory and inhibitory cells. These possible arrangements are shown in Figure 11-2. The case of a single inhibitory fiber system gives no output activity (since we have assumed the case of endogenous spontaneous activity to be included within excitatory input systems in our classification scheme). Although such situations may have some functional significance, the situation gives no output activity, and we will not consider it here. The various cases involving two separate fiber systems provide nine candidate possibilities corresponding to whether each constituent fiber system activates SFO the excitatory

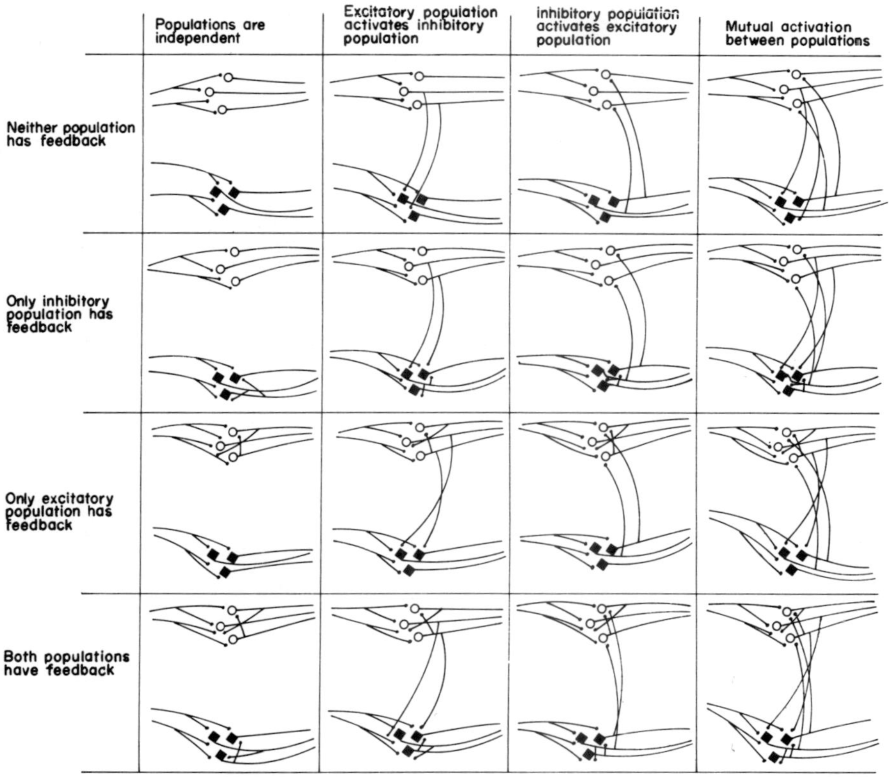

Figure 11-1. Sixteen types of neuron pools with excitatory and inhibitory populations.

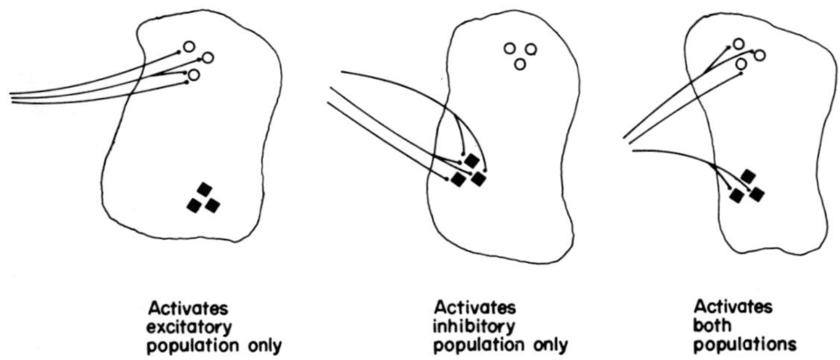

Figure 11-2. Three ways of activating neuron pools with one afferent fiber system.

population only, SFO the inhibitory population only, or SFO both the excitatory and inhibitory populations. For the case where both the two input systems are excitatory, some of these cases are redundant, and six possibilities exist. These are shown in Figure 11-3. For the case of one

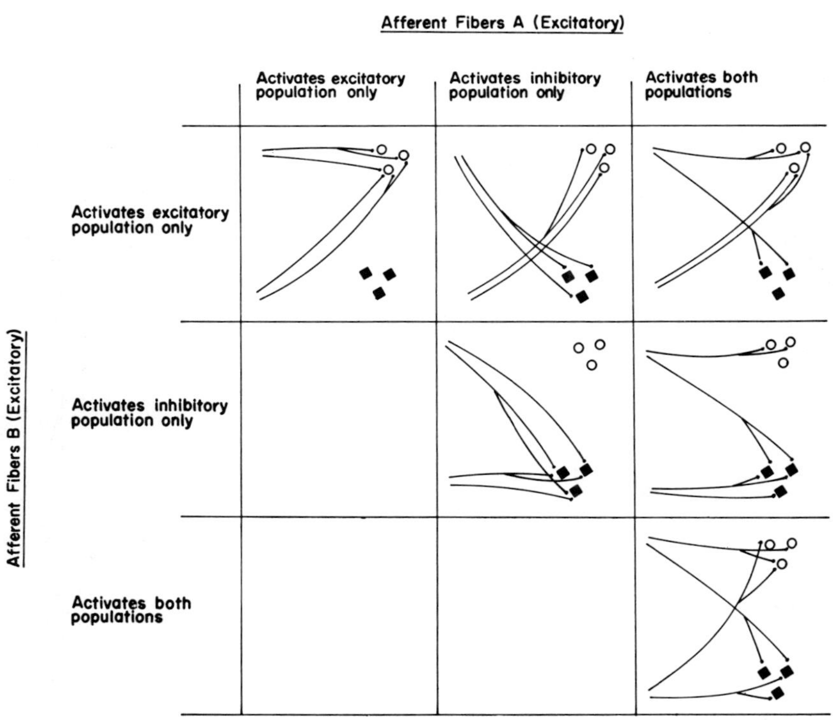

Figure 11-3. Six ways of activating neuron pools with two excitatory afferent systems.

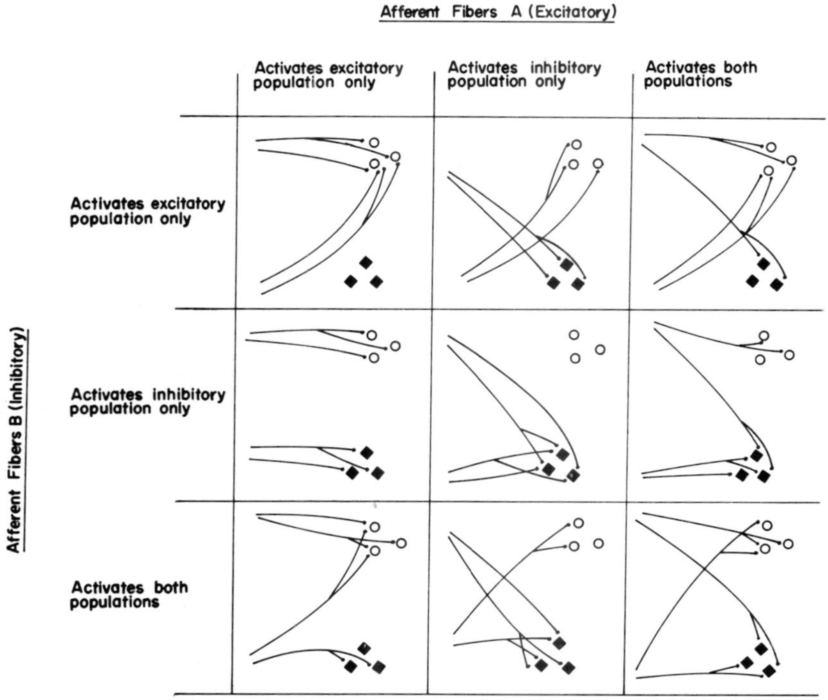

Figure 11-4. Nine ways of activating neuron pools with one excitatory and one inhibitory afferent system.

excitatory input system and one inhibitory input system, all nine possible cases are distinct. These are indicated in Figure 11-4. The case of two inhibitory fiber systems does not produce output spikes and will not be considered here.

These considerations show that even for such a simplified interpretation of neuron pools, there exist only slightly less than 288 different arrangements involving up to two different afferent fiber systems: 48 cases from three ways of activating 16 pools with one excitatory system only; 96 from six ways of activating 16 pools with two excitatory systems; and 144 ways from nine ways of activating 16 pools with one excitatory and one inhibitory system. Some of the possible cases are redundant, so that the total number of possibilities is somewhat less than 288.

There also exist other degrees of freedom for such neuron pools according to how their output fibers are distributed. This feature represents the manner in which such pools might be used in the nervous system. One may conceive that the axons from the excitatory population constitute the only output of the system. Alternatively, one might suppose that axons from the inhibitory population constitute the system output. Finally, it is possible that axons from both populations can contribute to the output. A further distinction can be made for this combined case as to whether or not the axons from excitatory populations and those from inhibitory populations activate the same receiving population or separate receiving populations. In this context, it is easy to see how a wave response measured within a pool by a macroelectrode might be quite misleading. For example, the electrode might often record the combined responses of both excitatory and inhibitory constituent populations, whereas the actual output fibers of the pool may emanate from only one of the populations, or the separate populations may send their axons to distinct receiving regions.

Only a very small number of researchers have addressed this problem, and only the most rudimentary results have appeared. Moreover, no systematic approach to this problem had been suggested before Freeman's publications. In the 1950s and 1960s several investigators undertook digital-computer simulation of relatively large numbers of interconnected neuromimes. One of the first of these studies was presented by Rochester in 1956. These studies were motivated largely by the desire to test and spell out Hebb's (1949) hypothetical "cell assemblies." Hebb had suggested that certain subsets of cells within cortical networks develop a tendency to behave as functional units with coordinated activity patterns according to the differential development of synaptic strength throughout the net. These subsets were called assemblies. Hebb supposed that the strength of a synapse between cells A and B increased slightly wherever firing in A was followed by firing in B with very small time delay.

Rochester's model included artificial neurons with a conventional two-state variable representation. Adaptation and refractoriness were incorporated by increasing the threshold upon the occurrence of an output spike and allowing a relatively slow decay to resting level. Excitatory and inhibitory synapses were used. The synapses were supposed to change in efficiency according to use, and the neurons were interconnected in a many-to-many fashion. In the first experiments performed by Rochester, 64 randomly connected excitatory neurons were subjected to random background input. In this arrangement, Rochester found no development of cell assemblies and noted that the model was extremely sensitive to parameter values. For example, slight changes in the value for the threshold in just one unit could markedly change the activity pattern of the entire net. With the model, he described what he called "diffuse reverberations," i.e., a periodic diffuse pattern circulating throughout the network which may or may not suddenly stop. In a second simulation study described by Rochester, 512 neurons were used. In this case, the probability of a connection between two cells separated by a distance x decayed as a negative exponential of x. This model included inhibition as well. Input was provided through four blocks, each of which consisted of four constituent neurons of the net. In this model, Rochester encountered what one could reasonably describe as cell assemblies. Intraassembly connections were largely excitatory, and interassembly connections were largely inhibitory. Moreover, he found that with this model there existed interassembly coupling.

Similar simulations were performed by Farley and Clarke (1961). These investigators used a two-dimensional array including 1296 elements. These, again, were two-state-variable neuromimes including a cell membrane potential and a threshold; absolute and relative refractory properties were included in the units through adjustments on the threshold. Spatial and temporal summation of inputs occurred in their model. Connections were excitatory only. The simulation used a discrete time interval, but the time intervals were very small to approximate an analog representation. Output was displayed on a cathode ray tube. The probability of a connection between two cells separated by x fell off as the reciprocal of x. Farley and Clarke distinguished two types of nets. The so-called "tightly connected nets" had connection parameters, and delay and refractory parameters, such that dense wave fronts occurred throughout the net. In these nets, they found that either a pattern moves throughout the whole net or it died out, depending on the threshold, and, moreover, with low threshold and repetitive stimulation, sustained oscillatory activity prevailed. Moreover, in "loosely connected" networks they found that wave fronts became fuzzy, that scattered activity throughout the whole network is commonly seen, and that under some conditions they could get diffuse oscillations either synchronously over the entire net or transferring from one part of a net to the other. Most of these behaviors are attributed to refractoriness in the units. The results are discussed with respect to slow wave rhythms in the EEG, and to the recruiting response observed in the EEG upon stimulation of the thalamus. Similar simulations were provided by Smith and Davidson in 1962 who added inhibitory synapses and explored their influences on oscillatory activity.

Other simulation studies in the later 1960s considered the spread of excitation through the so-called diffuse conducting systems representative of the nervous systems of coelenterates. These systems are peculiar in that all synapses are axoaxonal or axosomatic. There are no dendrites. Moreover, these synapses conduct in either direction. Also, and indeed therefore, the individual cells conduct in both directions. The entire network then consists of sets of pathways along which individual pulses may conduct in either direction.

Simulation programs to study this system were first presented by Josephson, Reiss, and Worthy in 1961 and later by T. H. Bullock, Fehmi, and others. The main question under investigation concerned how the spatial spread of activity in the networks might depend on the temporal patterns of activation. The model used by these investigators consisted of a

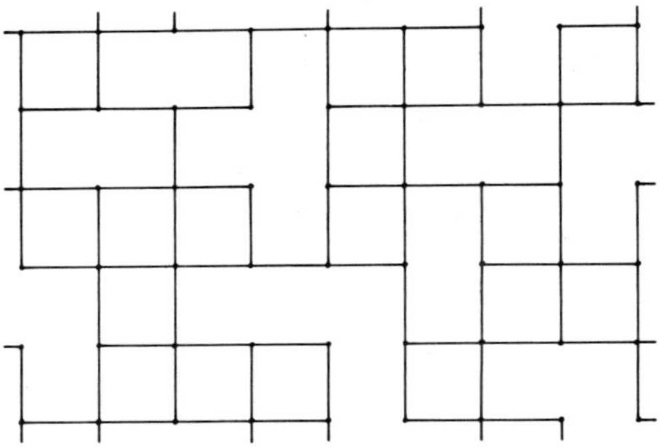

Figure 11-5. Portion of model for diffuse neural network of coelenterates.

series of elements of integral length laid out on a two-dimensional grid. At each intersection point, there is a two-way synapse between vertical and horizontal elements. Horizontal elements do not synapse on other horizontal elements, and vertical elements do not synapse on the other vertical elements. If a horizontal element ends at one vertical grid line, another horizontal element begins not at this grid line but at the next vertical grid line, and similarly for vertical elements. A representative example of the topology characteristic of these models is shown in Figure 11-5. In the model a decision whether or not to fire an action potential in response to input is made at every synapse. A certain fraction of the synapses are declared to transmit every pulse that arrives, whereas the remaining fraction transmit or do not transmit according to the three-state variable neuromime with fatigue described earlier by Reiss in Chapter 8. The simulated experiments with these models then involved prescribing certain stimulation patterns in some regions of the network and observing the spread of the resulting activity.

Representative simulations are those presented by Fehmi and Bullock (1967) who used a 100×100 matrix containing thereby about 10,000 junctions. In this simulation, cells occurred with frequency distribution of 4/32, 7/32, 10/32, 732, and 4/32 of the total number at 2, 3, 4, 5, and 6 units of length, respectively. These networks consistently and repeatedly provided different degrees of spread for different temporal patterns of stimulation. The authors suggest that this represents in general terms the ability of such networks to translate temporally coded information into spatial form. Moreover, for many of the particular networks studied in these simulations, the response spread seemed to be greater for even temporal distributions of stimulation than in response to stimulation patterns containing clustered occurrence of events.

A significant advance in computer efficiency in simulating neuron pools with neuromimes was made by Perkel (1964*a*). The previous simulations had used algorithms which march through time, systematically updating state variables of all cells at all times according to present input and the neuromime rules. Thus, time is discrete in the sense of being partitioned into intervals, but the intervals are sufficiently small to approximate continuous behavior. Perkel borrowed from the computer language SIMSCRIPT an approach based on so-called "interesting times" wherein cells are sampled only at those times that they might be expected to fire: times at which input pulses arrive or at which "spontaneous" firing might occur according to a mean level of background activity and the time of previous firing. Perkel used the neuromime described in Chapter 8 to construct his multiple neuromime-network-simulation model which he called PABLO. There is no need to integrate in this model because the basic equation is a sum and not a differential equation. To simulate a network, one supposes that input events to one cell are determined according to the output-spike times of another cell with appropriate time lags. Connection matrices and conduction time matrices can be specified in any desirable fashion to delineate any particular network.

Both the structure of the networks to be simulated and the physiological properties are assumed to be stochastic. The user specifies mean values, standard deviations, and types of distributions (normal, Poisson, etc.) for network variables such as number of connection, strength of connections, conduction times, and the like. The particular values used in individual cells or connections are chosen from appropriate distributions and fixed once and for all.

Perkel's program was given to workers at the California Institute of Technology who incorporated it on their system and gave it a new name PLEXUS. The system itself is quite elegant, involving interactive computer graphic output and the implied ability of the user to interact with the structure of his network, changing parameters, interconnections, input functions and parameters, and the like as the simulations proceed. As far as we know, however, publications describing application of this system to particular networks have not appeared. Perkel himself, however, has been continuing his network studies, some of which we will describe in later chapters.

MacGregor (1974) and his colleagues at the University of Colorado have developed several simulation techniques using multiple neuromimes to describe the activity of neuron pools. The work is based to a large extent on the use of the four-state variable neuromime described in Chapter 8. This approach invoves the marching-through-time method wherein state variables for all cells are updated at each small time interval to approximate continuous variables. For this approach, the differential equations are

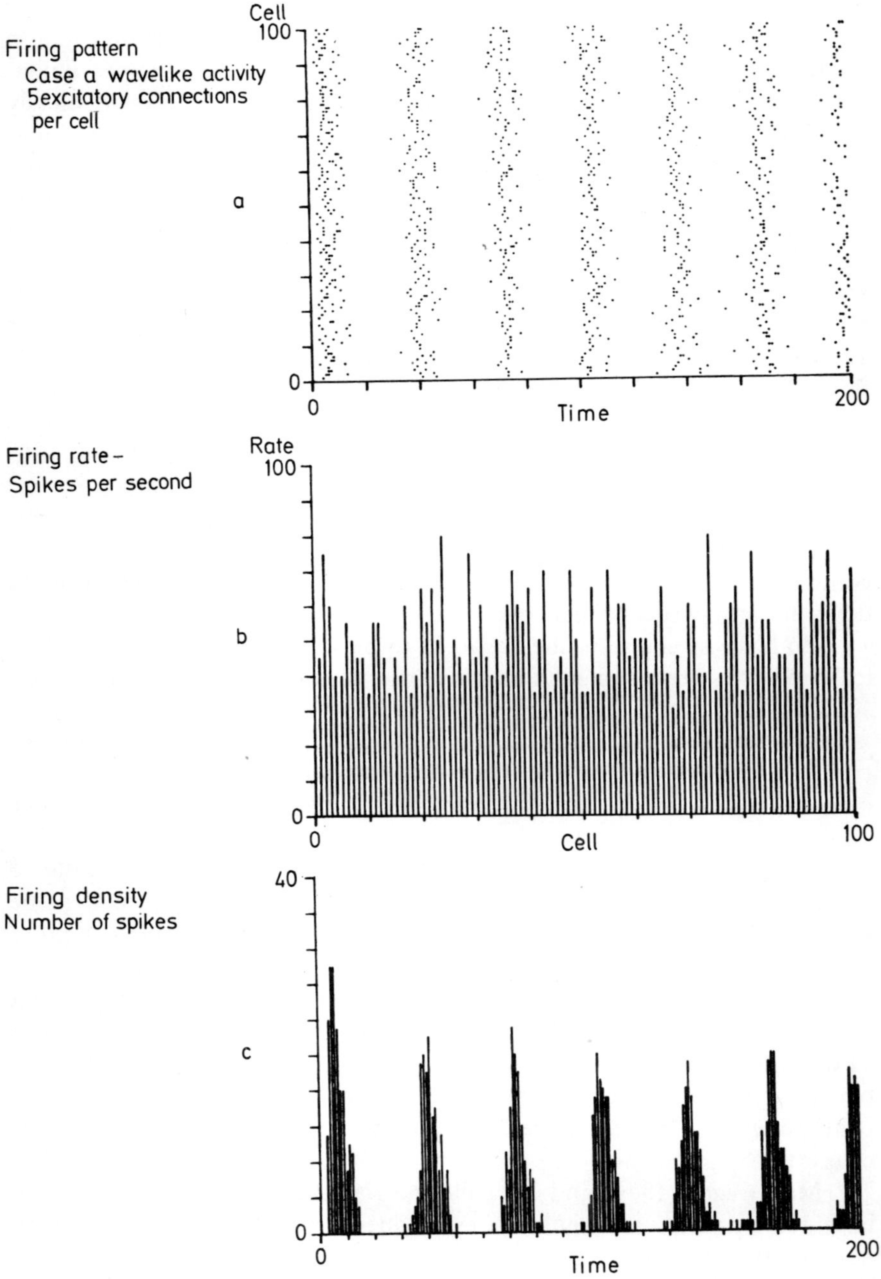

Figure 11-6. Representative examples of coordinated rhythmic oscillation in a neuron pool. (a) Spatiotemporal occurrence of individual spikes; (b) distribution of firing rates; (c) number of cells active versus time.

rewritten in finite difference form as

$$E = e^{-[1+g_K+\mathrm{GE}+\mathrm{GI}]/5}E + \frac{1 - e^{-[1+g_K+\mathrm{GE}+\mathrm{GI}]/5}(\mathrm{SC} + 7\mathrm{GE} - \mathrm{GI} - g_K)}{(1 + g_K + \mathrm{GE} + \mathrm{GI})}$$

$$g_K = e^{-1/\tau}g_K + bS \tag{11-5}$$

$$\theta = e^{-1/\tau}\theta + (1 - e^{-1/\tau})(1 + cE)$$

$$S = \begin{cases} 1 & \text{if } E \geqslant \theta \\ 0 & \text{otherwise} \end{cases}$$

Time is represented at discrete intervals. The quantities on the left side of the equation represent updated values from their previous values used in the right side of the equations. At any time when E is greater than θ, signaling an event in a particular cell, increments of excitation or inhibition are added into the appropriate SC, GE, or GI matrices corresponding to the output connections of that particular cell. A matrix SC (I, J), for example, represents the activating current that will arrive at cell I, J time units from now. At the end of each run through all cells at a given time, all these connection matrices are shifted to the left over J by one time unit. Arbitrary network configurations can be simulated according to how one specifies the interconnection rules used in projecting output from each cell. This digital multiple neuromime network model is generally available for use in conventional batch processing. A representative pattern obtained from a simulated pool of mutually exciting neurons and a representative sample of unit activity from a constituent neuron are shown in Figure 11-6.

A second digital-computer realization of this network model incorporates the "interesting times" method used by Perkel. Here, the four-state variable neuromime is reduced to a three-state variable neuromime:

$$\frac{dE}{dt} = -E - g_K(E + 1) + [\mathrm{SC} - \mathrm{GE}(E - 7) - \mathrm{GI}(E + 1)]$$

$$\frac{dg_K}{dt} = -\frac{g_K}{\tau} + bS + cE \tag{11-6}$$

$$S = \begin{cases} 1 & \text{if } E > 1 \\ 0 & \text{if } E \leqslant 1 \end{cases}$$

This system of equations has a general behavior very similar to that of the original neuromime. From this system, an approximate solution to the course of membrane potential and potassium conductance after the firing of a spike in the presence of only a steady stimulating current SC can be obtained. One can then predict the next interesting time when a cell might fire according to the fixed background level of SC and also determine the modifications in the potassium conductance and membrane potentials at intervening interesting times corresponding to activation by a synaptic impulse or externally applied input. The second approach is not an entirely accurate representation of the first differential equations, but, on the other hand, it is a reasonable representation of the dynamic activity of individual neurons. It also greatly reduces operating costs when the rate of occurrence of interesting times is significantly less than one per time interval per cell.

In addition to the digital-computer simulation programs, we are also constructing a general-purpose electronic simulator for the activities of neuron pools. The simulator consists of hundreds of small electronic modules representing cells and axons, a 4×4 ft powered patch board into which the modules may be plugged, and various interconnecting wires. The electronic cell modules are packaged in a cube approximately 1 inch on a side. The axon modules are about the same volume but have slightly elongated shapes. The main concepts are illustrated in Figures 11-7 and 11-8. The researcher stands before the patch board which is placed on the wall and physically constructs the network he wishes to simulate by inserting cells and axons in appropriate arrangements on the board. Synaptic connections between two cells are realized by insulated wires which fit readily over a pin on the axon of the sending cell and inserted via a sharp elongated pin on the end of the wire into a conducting sponge on the surface of the receiving cell. The sponge is about 1 inch square and can accommodate approximately 50 input pins, each of which represents a separate input synapse. Resistors imbedded within the connecting wires determine various synaptic strengths.

The system is activated by various pulse generators which can also be directly connected to the conducting-sponge receiving surface of the cells with sharp pins. Light bulbs attached on top of the individual cells flash whenever the cell fires an action potential. Red lights are on excitatory cells and green lights are on inhibitory cells. The researcher can thereby observe directly various patterns of activity in the flashing lights of the pool. Moreover, he might represent extracellular microelectrode investigations by recording only the output pulse trains of individual cells. The simulated transmembrane potential of individual units is available on a tap on each unit. Thus, intracellular microelectric recording can be simulated by recording this voltage. The combined information from all cells can be recorded and handled in any desired fashion.

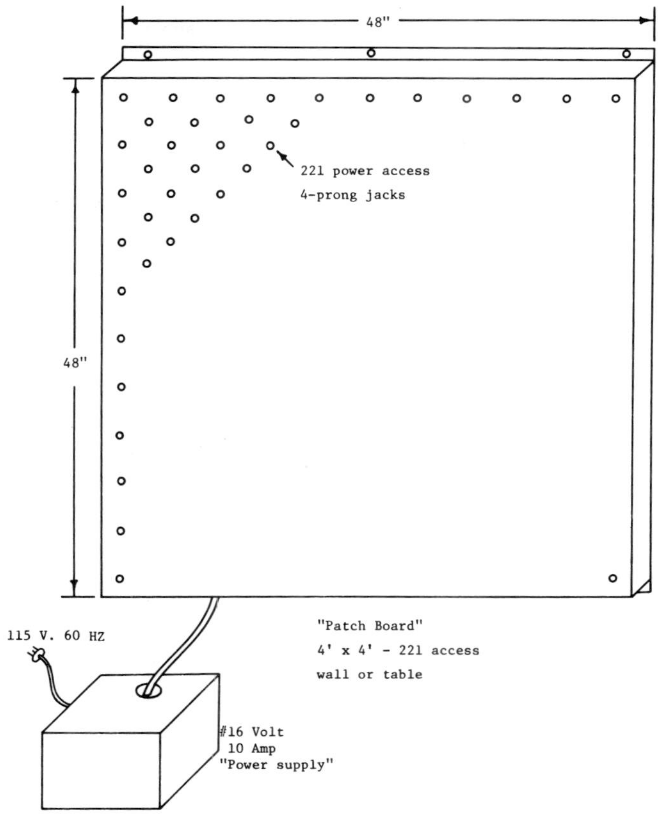

Figure 11-7. Patchboard for modular electronic network simulator.

The individual cell modules contain the circuitry discussed in Chapter 8 for the four-state variable neuromime. The four adaptation and accommodation parameters γ, τ, b, and c [see equation 11-5)] can be adjusted to create different types of cells separately for each cell. The axon elements produce delayed output pulses in an adjoining cell unit when activated by a pulse. Each axon element can produce pulses with three different time delays. The first time delay can be set anywhere from zero to 1 msec; the second delay is this value plus one more millisecond, and the third delay is the original delay plus two more milliseconds. Individual axon elements can be stacked together to provide any desired time delay. The model is slowed down in time by a factor of 8, so that the delays described above require 8 msec of real time, and a model unit fires once per second to represent a neuron firing at eight per second. This was done to enhance the effective-

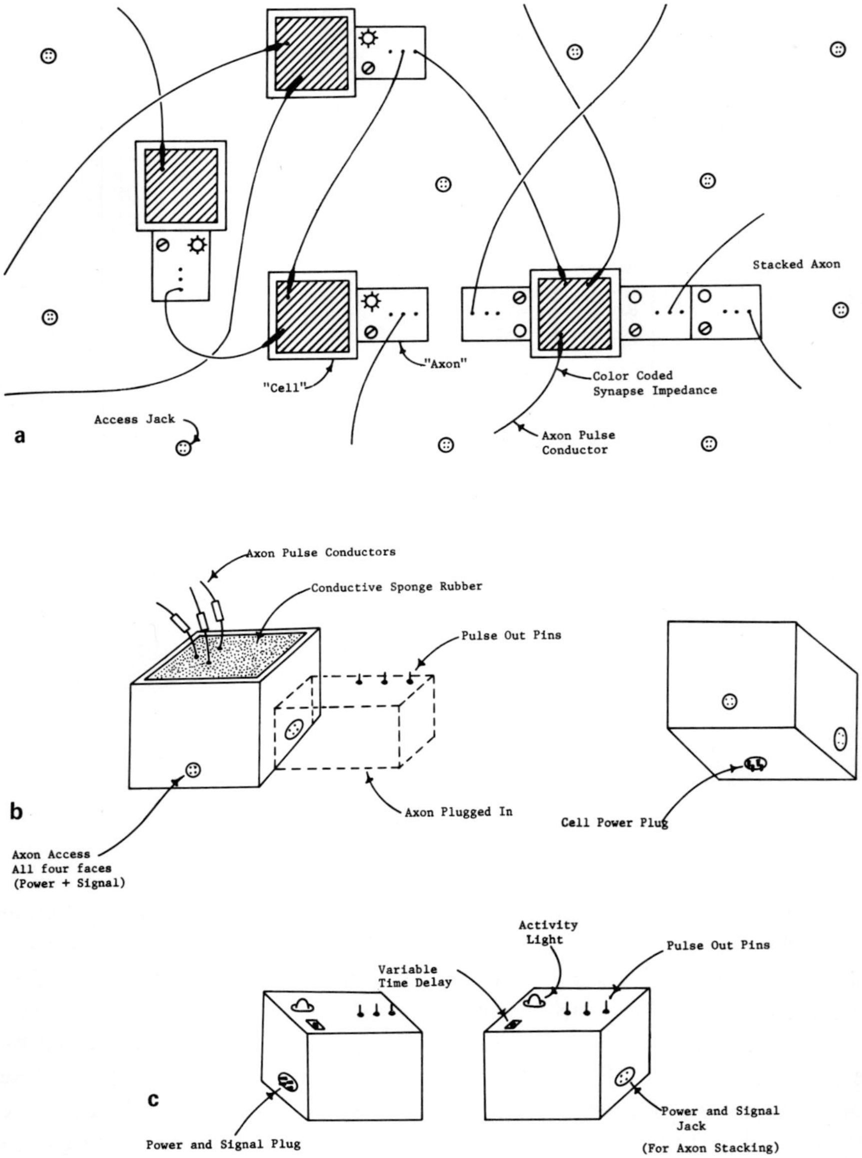

Figure 11-8. (a) Patching configuration and modules for electronic network simulator; (b) "cell" (active module); (c) "axon" (time-delay module).

Figure 11-9. Small network simulated with first-generation electronic model.

ness of the display of flashing lights. We are currently building 100-cell unit modules and 400-axon modules. In an earlier pilot study, we built 12 cells and 36 connecting elements. These are illustrated in Figure 11-9.

The general-purpose digital computer simulation program has been used to explore the activity of diffusely connected neuron populations. The standard ingredient in these simulations was a parent population of 50 cells and 50 afferent fibers which activated them. Each afferent fiber activated 25 parent population cells chosen at random by a random-number generator. The strengths of the individual synapses between afferent fibers and parent population cells were chosen randomly from a triangular distribution. The mean (and mode) value of this distribution produced an EPSP of 0.2 of threshold when the receiving cell was at resting potential. The range of the distribution was from 0.05 of threshold to 0.7 of threshold. In addition to the afferent fibers and population cells, some simulation cases used systems of self-exciting collaterals from the parent population. When these were used, a diffuse connection scheme utilizing 25 synaptic terminals per parent population fiber was effected. Also, a second population of 50 inhibitory cells was someimes used. When used, this population was activated by the output fibers of the parent population, again with the diffuse interconnection scheme involving 25 synaptic terminals per fiber and the distribution of PSP strengths given above. Sometimes, however, the inhibitory population was excited by a system of afferent fibers instead of (or in addition to) fibers from the parent population. In these cases, the incoming fibers were sometimes shared with the parent population and sometimes were entirely separate. In all cases connections were diffuse, involving 25 synapses per fiber from each of 50 fibers. The inhibitory population effected inhibitory connections to the parent population. This connectivity interface utilized ten synaptic terminals per inhibitory fiber. The strengths of the individual inhibitory synapses were chosen from a triangular distribution whose mean (and mode) rendered an IPSP of −0.25 of threshold when the receiving cell was at resting potential. A time delay representing conduction time plus synaptic delay was assigned to each fiber in the afferent fiber system, in the conduction system from parent population to inhibitory population, and in the conduction system from inhibitory population to parent population. These time delays were chosen at random for the individual fibers from a truncated quasi-Gaussian distribution with mean 3 msec and range 1 to 5 msec.

The primary source of activity was a random pattern of pulses distributed over the 50 afferent fibers. An overall mean rate of activity, typically, say, 16 per second per fiber, was prescribed. The actual firing rate in any single fiber was then chosen from an exponential distribution with this mean. Moreover, the times of individual pulses in a given afferent fiber were drawn from an exponential distribution of interspike intervals with mean value corresponding to the mean rate of the particular fiber. Altogether, about 100 simulation cases representing 14 different structural configurations were run. The standard simulation length was 6 sec of model time. In the mechanics of the simulations, the inhibitory population was always present and almost always activated by the parent population, but sometimes not connected back to the parent population. The result of this is that input–output relations for single diffuse transfer were obtained for a wide variety of input patterns which are different from the random pattern presented to the parent population.

The main characteristics of simple diffuse transfer are illustrated in Figure 11-10. Each of the six figures in Figure 11-10 shows a 1-sec sample of the firing pattern in a population of cells by representing each individual action potential by a dot appropriately placed in a plane of cell number versus time. The figure shows input and corresponding output population patterns for each of three different input patterns. The first main characteristic illustrated by these data is that the degree of clustering of action potentials (that is, the occurrence of simultaneous or nearly simultaneous action potentials in several to many cells of the population) is always more

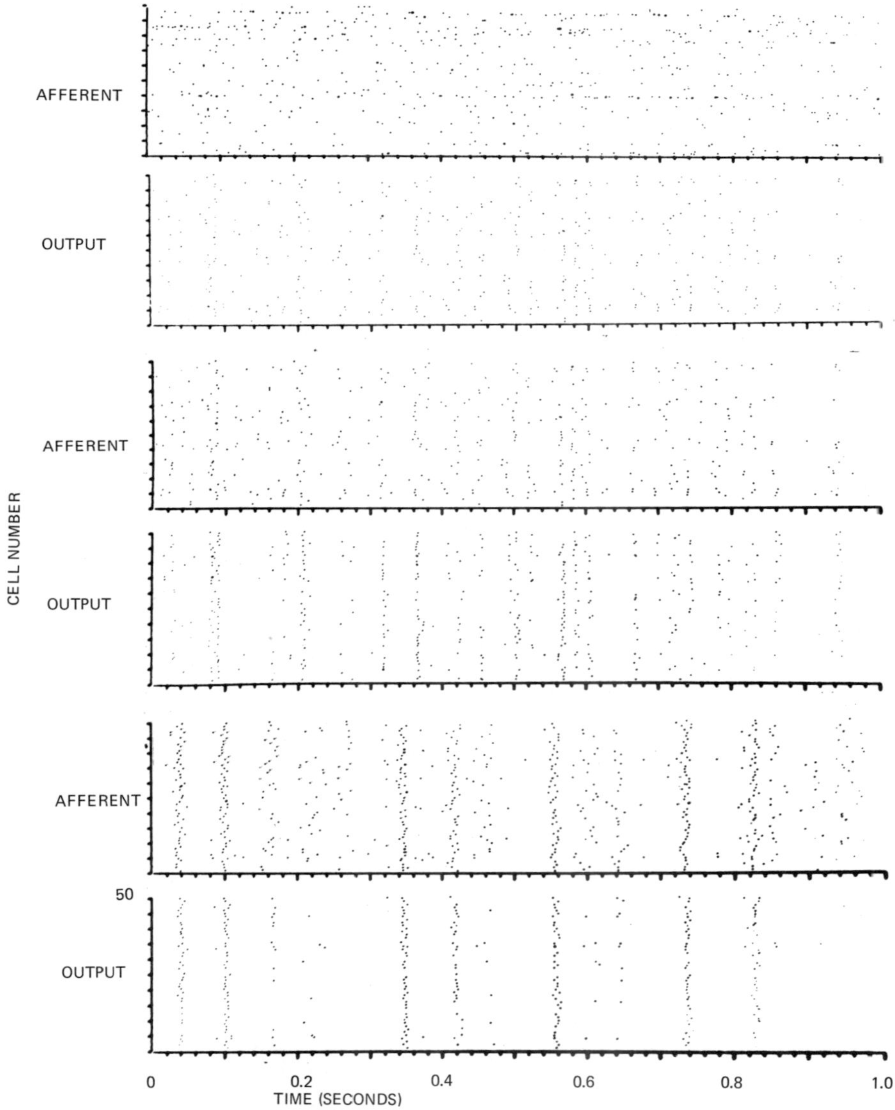

Figure 11-10. Afferent and output population patterns for simple diffuse transfer: (a) random afferent pattern; (b) slightly clustered afferent pattern; (c) highly clustered afferent pattern.

marked in the output pattern than in the input pattern. The second main characteristic indicated in Figure 11-10 is that clusters of spikes in the output pattern are generally generated by clusters in the input pattern.

There are only a few exceptions to this observation, and it holds true for even very weak clusters. These two characteristics are tantamount to the suggestion that the simple diffuse transfer mechanisms tend to selectively pass on synchronized clusters of activity (and accentuate them if necessary) while remaining generally much less responsive to ongoing random bombardment. In this sense, one can conceive of the diffuse transfer structure as a filter, responding to clusters of simultaneous action potentials among its constituent input fibers and cells. This picture is particularly driven home by the observation that the second-stage output, obtained when two simple diffuse transfer structures are arranged in series, invariably consists of sporadic series of synchronized clusters in response to virtually any input pattern to the first stage. A representative example is shown in Figure 11-10b where the upper figure is the output of the first stage subjected to the random input shown in the upper figure of Fig. 11-10a, and the lower figure in Fig. 11-10b represents the output of the second stage.

In the dynamics of these structures, the strength of the individual clusters seems to be secondary to the occurrence of the clusters. That is, if a very weak cluster occurs and is passed into a system characterized by simple diffuse transfer, the cluster will attain a moderate strength and will be passed on indefinitely through the system. If a strong cluster occurs, it will be somewhat weakened but passed on in moderate strength indefinitely. If a cluster does not occur, there is generally no cluster in the output. However, although this is almost always true, there are some exceptions. For example, if a single cell in the input population is fired at a very high rate, there is an increased tendency to scattered sporadic clusters which is not always obviously attributable to weak clusters in the input pattern (and, of course, if one adjusts parameters in a somewhat artificial manner, it is possible to have a single input fiber drive clusters involving all the cells to which it connects).

The main dynamic characteristic of diffusely connected self-exciting populations is their extremely marked tendency to fire synchronized clusters of spikes. For example, when such populations are subjected to random afferent bombardment, they tend to generate rhythmic series of recurring synchronized clusters, an example of which is illustrated in Figure 11-11. The rate of cluster occurrence depends primarily upon the refractory properties of the constituent cells and is modulated by the overall level of afferent input, as indicated in Figure 11-12. Moreover, the series of clusters is rather regular at all rates. The standard deviation in these simulations, for example, ranged from two to one, with smaller values occurring at higher input rates.*

*In these populations the self-exciting collaterals and the diffuse interconnections both bring about synchronized firing of clusters in groups of individual cells. The cells which fire tend to go through similar refractory trajectories and to recover from refractoriness at about the same

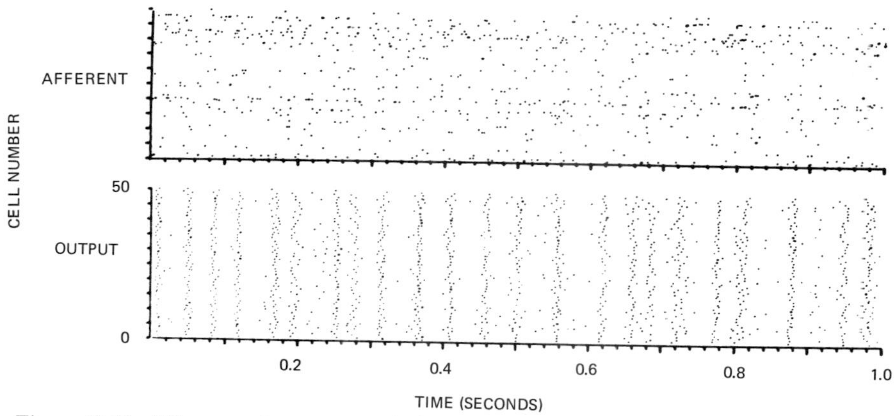

Figure 11-11. Afferent and output population patterns for diffuse transfer with diffuse self-exciting collaterals.

The only noticeable effect of afferent inhibition on simple diffuse transfer is to diminish the transfer rate. As the rate of activity in afferent inhibitory fibers impinging on a self-exciting population is increased, the series of rhythmic clusters associated with the self-exciting population tends to become more sporadic and less frequent until eventually, at very high rates of afferent inhibition, it disappears entirely.

---

time. This latter fact allows subsequent synchronized firing in the same cells and provides the basis for determining the rate of cluster occurrence. It is often suggested that self-exciting populations of neurons exhibit sustained reverberations of activity and might provide a basis for short-term memory through self-exciting chains of interconnected cells. These possibilities may occur, but if they do, they are in a different portion of the parameter space from the regions explored in these simulations. One might expect such activity in self-exciting populations that have a smaller percentage of self-exciting terminals for total cell population, and stronger individual synapses. These possibilities need to be explored. In the diffusely connected pools explored here, self-exciting chains of reverberations do not occur because refractoriness occurs virtually simultaneously in large numbers of constituent cells due to synchronized clustering.

The dynamics of these diffusely connected self-exciting nets is better thought of in terms of "a natural frequency" of recurring synchronized clusters rather than in terms of reverberations through self-exciting chains. The natural frequency of such a pool would be the average cluster rate obtained when the cell is subjected to random background bombardment, approximately 30/sec as shown in Figure 11-11. When such nets are driven at progressively higher rates of shocks superimposed upon random background activity, one finds that when the individual shocks with synchronized clusters have reached rates up to about twice the natural frequency, misses begin to occur. Similarly, if repetitive shocks are applied at rates lower than the natural frequency, one often finds misses depending upon whether the individual shocks occur during the refractory troughs.

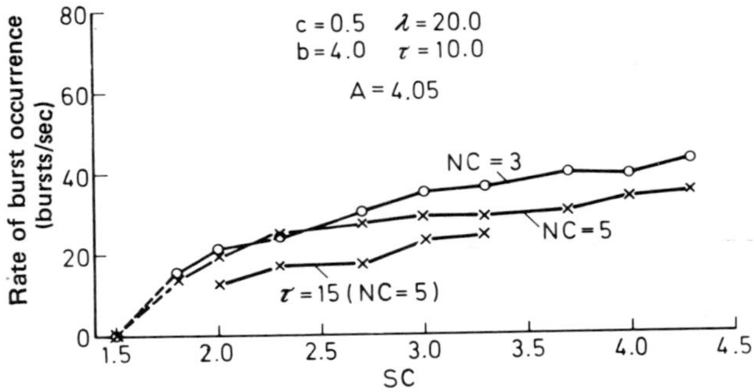

Figure 11-12. Rate of burst occurrence.

Recurrent inhibition has a very weak effect upon simple diffuse transfer. The overall transfer rate of individual spikes is diminished and there is a slight (but only very slight) increased tendency to cluster.† There is, however, a more marked effect of recurrent inhibition on self-exciting populations. First, the overall rate of synchronized clusters is up significantly by about 35%. The individual clusters are sharpened in the sense of being diminished in duration. The net result of these two occurrences is that the overall transfer rate of individual action potentials is down some 25%. Otherwise, with respect to overall regularity of rhythmic cluster series and other features, the dynamic activity of diffusely connected self-exciting populations with recurrent inhibition is very similar to those without recurrent inhibition.‡

†It has sometimes been suggested that recurrent inhibition is analogous to the negative feedback system with time delay in engineering systems analysis, and might thereby tend to bring about rhythmic firing in neuron populations. This is not true for the diffusely connected systems explored here. Moreover, one might suggest that because recurrent inhibition causes a substantial number of parent population cells to be inhibited simultaneously and thereby recover approximately simultaneously, many of the cells might tend to fire approximately simultaneously. These points are critical with respect to certain theories concerning the origin of the alpha and the theta rhythm, as will be seen in Chapter 15. The simulations used here incorporate a moderate level of inhibition and IPSPs which die out accoring to the membrane time constant (5 msec in these simulations). Thus, although we do not discover rhythmic activity in these systems, this does not necessarily contradict its possible occurence in systems which incorporate much stronger and prolonged inhibition as in the thalamus and hippocampus.

‡This surprising influence of recurrent inhibition on the cluster rate of self-exciting populations can be explained as follows. when a cluster of parent population cells fires synchronously, a widespread recurrent inhibition blocks firing in a substantial number of other population cells. These cells are blocked from firing by the inhibition but are not refractory. They thus go into a recovery phase similar to those of cells that fired; however, they recover sooner and are available for excitation sooner than if they had been in the cluster that did indeed fire. This basis, played upon by the synchronizing mechanisms of diffuse connectivity and self-exciting collaterals, brings about, as a net result, a higher rate of synchronized clusters.

The main dynamic characteristics of diffusely connected neuron pools as simulated here is their propensity to produce, and to respond selectively to, synchronized clusters of action potentials occurring simultaneously or nearly simultaneously in a substantial number of constituent cells or fibers; in particular, the simple diffuse transfer structure acts as a filter, responding selectively to synchronized clusters of action potentials in subsets of the input fibers, and signals their occurrence with corresponding clusters in its output pattern. The structure of diffusely connected self-exciting collaterals has shown itself extremely prone to the production of such synchronized clusters. Once these dynamic signatures are grasped, the underlying mechanisms are easy to understand. Simple diffuse transfer is selectively sensitive to clusters because of conventional spatial summation. Individual cells require the simultaneous arrival of several to many EPSPs to trigger firing. On the other hand, the production of synchronized clusters in simple diffusely connected pools reflects a high degree of overlapping of input terminals. In such an arrangement, when one population cell has received input sufficient to excite it, other constituent population cells are likely to have shared at least some fraction of that same input and thereby be more likely to fire at about that time than they might otherwise be. The cluster-producing properties of the self-exciting structure are extremely clear. When one population cell fires in this system, it directly excites other cells of the system, encouraging them to fire with very short time delays. If the afferent input to the system is diffusely connected, then many of these receiving cells tend to be ready to fire as described above since they share some of the input which triggered firing in the initially considered cell.

The question arises as to the extent to which these features might occur in larger pools or pools involving different parameters from those simulated here. The key parameters are n and n/N, where N is the number of constituent cells (assumed in this discussion to be equal to the number of afferent fibers), and n is the average number of terminals per receiving cell (here equal to the average number of terminals per input fiber). The sensitivity to clustering depends primarily on n, whereas the production of clusters depends primarily on the ratio n/N. If the average number of afferent terminals per receiving cell, n, is large, selective sensitivity to clustering within the input fibers incident upon a particular cell is possible merely by reducing the strength of the individual EPSPs to slightly more than $1/n$ of threshold. The ratio n/N can serve as a quantitative measure of diffuseness of connections. If this ratio is high, there is, correspondingly, a high degree of overlapping or sharing of input terminals among constituent cells. These systems tend to produce synchronized clusters by the mechanisms described above. Low values of this ratio correspond to more discretely connected systems which allow for more finely discriminative dynamics. The tendency of self-exciting collateral systems to produce synchronized clusters, however, increases with the number and strength of the self-exciting connections and with the ratio n/N. Thus, the reduced ability to generate clusters in nets with low n/N ratios can be compensated for by appropriately designed self-exciting collateral systems.

These observations provide a basis for a generalized theory of information coding in primitive nervous systems. Thus, one can speculate that early in phylogenesis nervous systems evolved the capability to respond selectively to synchronized clusters (simple diffuse transfer structures), evolved structures to produce such clusters (self-exciting collateral systems, and to a lesser extent simple diffuse transfer structures), and, correspondingly, evolved higher-order configurations to make use of such clusters. In this theory the synchronized cluster emerges as a critical dynamic entity which links system dynamics to structure, on the one hand, and to experimental electrophysiological tests, on the other hand, and which serves as a key element for generating models of information processing in a functional context. For example, one can picture primitive nervous systems as containing series of pools each of which acts as a unit or relay in passing on synchronized clusters. Information could be coded in the temporal structure of series of synchronized clusters in individual pools or in confluent fiber systems, much in the same way we picture information to be coded in the spike trains of individual cells. Such an arrangement would ensure transmission and effective activation of subsequent pools and effector systems because of the large number of simultaneous impulses involved. It would also help stabilize systems against spurious activity in small numbers of units. Various invertebrate locomotor systems are likely candidates for specific realization of this theoretical scheme. Diffusely connected components within more highly developed nervous systems would also exhibit dynamics characterized at least in part by synchronized clusters. One hypothetical example is a model for the origin of synchronized brain waves based on synchronization-sensitive recurrent inhibition, discussed in Chapter 15.

This theoretical orientation provides a basis for generating hypotheses concerning structure–function relationships in diffusely connected nervous components. It also provides clear guidelines for experimental testings of such models. Perhaps the best possible approach at present is to examine cross correlations and firing patterns in several to many cells recorded simultaneously from the same system. Although existing electrophysiological data are as yet inadequate to reliably reflect on the extent of occurrences of synchronized clusters in real systems, and we are only beginning to look purposefully for such clusters in our own research, preliminary explorations of such firing patterns in groups of up to four simultaneously recorded cells in individual ventral ganglia of the earthworm have revealed rhythmic series of synchronized clusters (at about 30 per second) indistinguishable from those produced by our diffusely connected self-exciting simulated pools.

The general area of neuron pools is in a period of exciting development. Many ideas and tools exist, but many more need to be developed. We should hope to see many contributions and valuable insights in the foreseeable future.

12

Models of Large Networks: Analytic Approaches

Chapter 8 on neuromimes described models for simulating the activities of single or small numbers of neurons, say, less than about 20, while Chapter 11 considered techniques for analyzing up to hundreds or thousands of interconnected cells. In this and the next chapter we consider techniques for predicting the activity of vast numbers of interconnected neurons, say, from tens of thousands to millions or more. We will find that the various existing techniques fall into two categories: those that are essentially analytic in character, and those that are explicitly or primarily appropriate for realization on large-scale, high-speed computers. Corresponding to these two categories, there are two main germinal contributions which have greatly influenced most subsequent work in their respective categories. The primary starting point for the analytic studies is a paper published in 1956 by Beurle, while that for the computer-oriented approaches is a classic paper by McCulloch and Pitts which appeared in 1943. In this chapter we take up the analytic approaches to large neural networks which stem largely from the work of Beurle.

Various investigators have recognized for many years that it might be unnecessary and even inadvisable to try to predict or follow the activity of every individual neuron in a complex network. One might consider, for example, the total number of cells in the network firing at a given time as the significant variable of the network. A more refined extension of this concept is to consider the relative proportion of cells which are firing in a small control region as a function of the position of the control region and of time. Here the control region is conceived to be differentially small compared to the overall size of the network but still is sufficiently large to contain large numbers of neurons. This general orientation can be characterized as the field theoretic approach.

One of the earliest theoreticians advocating this point of view was Beurle who published in 1956 an extremely intriguing paper entitled "Properties of a Mass of Cells Capable of Regenerating Pulses." Beurle's main concepts as discussed in this paper are, first, a function $F(x, t)$ which represents the number of excited cells at position x and at time t, and, second, the distinction between "used" and "sensitive" cells. Used cells are cells which fired at the preceding time interval and which are, therefore, according to his assumption of absolute refractoriness for one time interval, incapable of firing at the present time interval. Sensitive cells are cells which did not fire at the last time interval and are therefore available for activation at the present time interval. Note that Beurle has implicity assumed here a partitioning of the time continuum into discrete intervals, each of which has the length of one interneuronal delay time τ. All delay times are assumed to have this value. Three main quantitative relationships comprise the basis of Beurle's method for framing the network problem in mathematical terms. The first of these is an expression representing the mean rate of pulses arriving at a given position x at time $t + \tau$. This expression is given as

$$\text{Mean rate of arrival of impulses at } x \text{ and } t+\tau = \int_{-\infty}^{\infty} F(X, t)\xi(x - X)dX \tag{12-1}$$

Here, $\xi(x - X)$ gives the mean number of connections between cells separated by the distance $x - X$. The second is the expression given by Beurle for the mean value of integrated excitation at point x at time $t + \tau$:

$$\text{Mean value of integrated excitation at } x \text{ and } t + \tau = \overline{N}(x, t)$$
$$= \int_{-\infty}^{0} \int_{-\infty}^{\infty} F(X, T)\xi(x - X)\chi(t - T)dx\,dT \tag{12-2}$$

Here, the function $\chi(\tau)$ represents the portion of the excitation produced by a given primary cell which remains at time τ after it has become active. Finally, Beurle uses the Poisson approximation to the binomial distribution to get an expression for the proportion of cells which have a total excitation equal to $q - 1$ at time $t + \tau$ in terms of the mean value of integrated excitation $\overline{N}$ at $t + \tau$. This expression is

$$\text{Proportion of cells at } x \text{ which have total excitation of } q-1 \text{ at } t+\tau = p_{q-1} = e^{-N}\frac{N^{q-1}}{(q-1)!} \tag{12-3}$$

This expression represents, according to the binominal theorem, an approximation to the number of ways of choosing $q - 1$ elements from N elements when N is very large. Here, the assumption is that the detailed connections of the network are such that this result for a totally random selection of connections will be applicable.

This general methodology is applied by Beurle to two particular cases. In the first case Beurle considers continuous random activity in a fixed neural network. The main equation for this case as given by Beurle is

$$F(x, t + \tau) = RP_{q-1} \int_{-\infty}^{\infty} F(X, t)\xi(x - X)\,dX \tag{12-4}$$

This equation states that the number of cells active at position x at time $t + \tau$ is equal to the total number of pulses that arrive at that position at that time times the number of nonrefractory cells which need just one more input to reach threshold. There, R is the proportion of cells which are nonrefractory, and q is taken as the threshold. Beurle then makes two conditions necessary to sustain constant activity at a given level. These are given as

$$\begin{aligned} F(t + \tau) &= F(t) \\ 1 - R &= Fr \end{aligned} \tag{12-5}$$

The first equation states that the number of active cells is constant from one time to the next; the second condition is that the rate at which used cells recover should equal the rate at which sensitive cells become active. Here, r is the recovery period. When these conditions are applied to the basic regenerative relationship given in equation (12-4), the following condition for sustained steady firing is obtained:

$$e^{-Fks} \frac{(Fks)^{q-1}}{(q-1)!}\left(k - Fks\frac{r}{s}\right) = 1 \tag{12-6}$$

Here, F is the value of sustained activity, k is the integral of ξ over its entire range, and s is the time interval over which excitation is integrated by the cells. Thus, for example, the total integrated excitation N is equal to Fks. Beurle goes on to discuss this relationship in terms of cortical circuitry. He suggests, among other things, that a block of cells having these properties cannot support uniform continuous random activity if threshold q is above 5.

Beurle also extends this methodology in an attempt to derive a basic differential equation for the propagation of waves through nervous tissue. The basic assumptions used in this attempt are

$$\begin{aligned} dF/dt &= R\phi - F \\ dR/dt &= -F \end{aligned} \tag{12-7}$$

Here, F and R are again the number of cells that are active and the number of sensitive cells, both at position x and time t, respectively, and ϕ is the probability that a sensitive cell will be excited above its threshold in unit time; ϕ is also a function of both position and time. The first of these equations states simply that the rate of change in the number of cells active at a given time is equal to the number that become active at that time minus the number that are active at that time. The second equation represents the observation that each time a cell becomes active the number of sensitive cells is reduced by one. Combining these two basic physical assumptions, Beurle arrives at the wave equation

$$\frac{d^2R}{dt^2} + \left\{1 - R\frac{\phi}{F}\right\}\frac{dR}{dt} = 0 \tag{12-8}$$

Generally, the solution to equation (12-8) is extremely difficult to obtain, particularly since the function ϕ is difficult to find. The quantity ϕ is generally a function of F, x, and t. It implicitly contains space differentials. Thus, equation (12-8) is certainly a partial differential equation and not an ordinary differential equation with respect to time which its form might suggest. Beurle examines solutions for the case where ϕ/F is a constant, m. He shows for these cases that a solution in the form of a traveling wave might be represented as

$$\begin{aligned} R &= \frac{1}{m} - \frac{2B}{m}\tanh B(t - t_0) \\ F &= \frac{2B^2}{m}\frac{1}{\cosh^2 B(t - t_0)} \end{aligned} \tag{12-9}$$

Here B is a propagation constant which determines the rate of rise and fall of activity. Beurle also notes that the wave equation is nonlinear, so that superposition does not apply.

Griffith (1963) is another investigator who used the field theoretic approach to large neural networks. Griffith defines two classes of quantities: first, field quantities which satisfy a given differential equation, and, second, sources and sinks which represent driving forces on the field quantities. To represent nervous tissue, Griffith takes a function $\psi(x_i, t)$ to represent the overall "excitation" of the nervous network and a function $F(x_i, t)$ to represent the activity at the soma of the neurons. The value of F can be considered proportional to, say, the mean firing rate of the neuron. In this model, to use Griffith's words, "Excitation is regarded as being carried by a continual shuttling between sources and field, F creates ψ, and so on." In obtaining his mathematical formulation of this system, Griffith starts with the two main equations

$$H\psi = kF$$
$$F(x_i, t) = f[\phi(x_i, t)] \tag{12-10}$$

where

$$\phi(x_i, t) = \int_{-\infty}^{\infty} \psi(x_i, t + \epsilon) i(\epsilon)\, d\epsilon \qquad \begin{cases} i(\epsilon) = 0, & \epsilon > -\tau \\ i(\epsilon) \geqslant 0, & \epsilon < -\tau \end{cases}$$

In the first equation, H is an undefined operator and k is a constant; i.e., there is some specific equation for the field excitation variable ψ in terms of the internal sources of activation F. This equation includes the assumption that superposition holds for the system. The second equations show how the general excitation ψ influences single neuronal excitation F. The remaining problem is to work out the appropriate form of the operator H (i.e., the differential equation satisfied by ψ) and the actual functional form for F. Griffith uses various simplifications, including the main assumption that a second-order operator is sufficient, and assumes that the system is left invariant under all translations and rotations to finally arrive at the main equation

$$\nabla^2\psi = \frac{1}{4}\beta^2\psi + \left(\frac{\beta}{v} - 4\pi A_1\right)\frac{\partial\psi}{\partial t} + \left(\frac{1}{v^2} - 4\pi A_2\right)\frac{\partial^2\psi}{\partial t^2} - 4\pi F(\psi) \tag{12-11}$$

In this equation, v is the velocity of propagation of excitation, and β, A_1, and A_2 are constants. Equation (12-11) then is Griffith's main field equation for excitation in neuronal networks. In a subsequent paper published in 1965 Griffith applied this equation system to the question of stability in a randomly connected net. In this work he assumed that the conduction velocity of the neuronal excitation is infinitely large and, moreover, neglected the second time derivative, stating that its retention according to the Taylor series argument is somewhat arbitrary. This results in the equation

$$\nabla^2\psi = \frac{1}{4}\beta^2\psi + \gamma\frac{\partial\psi}{\partial t} - 4\pi F(\psi) \tag{12-12}$$

To solve this problem, one needs to specify β, γ, and the form of the function F. Griffith was able to show that this equation system exhibits sustained oscillations through various degrees of activity. This conclusion was used to help resolve a paradox wherein previous continuous models of network oscillations showed an inability to sustain intermediate activity whereas natural brains normally operate at an intermediate level of sustained activity.

Representative contemporary research on large neural networks has been made over the last few years by Harth *et al.* (1970) at Syracuse University in New York. This work has proceeded on both analytical and computational grounds. Harth declares that although in obtaining various analytical expressions he makes certain simplifying assumptions to make the mathematics less cumbersome, the analytical models are shown to match reasonably well with computational output wherein more rigorous and complicated situations are retained.

We will consider first the analytical development as pursued primarily by Harth *et al.* (1970). Harth considers a group of A neurons. A certain proportion h are assumed to be inhibitory. The others are excitatory. Time is composed of discrete intervals of length τ. In each interval, each neuron may be "on" or "off." If a cell is "on" during a particular time interval, it activates other cells in the net with increments of activity of amplitude k during the next time interval. The number of output connections per cell is μ^+ for excitatory cells, and μ^- for inhibitory cells. A particular cell is activated during an interval if the number of excitatory inputs minus the number of inhibitory inputs is greater than the quantity θ, which is called the threshold. If a cell is activated during an interval, it is not allowed to fire in the next interval. In the first stages of development of this model, Harth used the following simplifying assumptions: (1) All conduction times between cells are one time unit, (2) PSPs sum and decay instantly (i.e., there is no temporal summation), and (3) the refractory period is absolute and quite brief (i.e., there is no relative refractoriness).

Harth then proceeds to develop the implications of these assumptions for internal reverberations within otherwise unactivated networks. In particular, he proceeds to derive an expression for the expected number of cells activated at time $n + 1$ in terms of the number of cells activated at time n and in terms of various network parameters in this work. The main output variable is the fraction of cells activated at any particular time, denoted by α_n. In deriving this expression, Harth proceeds as follows. The main and most powerful assumption in this derivation is

$$p_l = [\alpha_n(1 - h)A\mu^+]_l\left(\frac{1}{A}\right)^l\left(1 - \frac{1}{A}\right)^{\alpha_n(1-h)A\mu^+ - l} \qquad (12\text{-}13)$$

This equation gives the probability p_l that a neuron received l EPSPs at the time $n + 1$ in terms of the various network parameters. This is based on the observation that the quantity $\alpha_n(1 - h)A\mu^+$ represents the total number of EPSPs appearing in the net at time $n + 1$. Since the connections are assumed to be random, each of these EPSPs has the probability $1/A$ of landing on a particular neuron. Equation (12-13) then is simply the binomial distribution representing the number of ways of choosing l EPSPs from the total number of EPSPs, where the probability that a single cell receives a given EPSP is simply $1/A$. Note that in writing this expression it has been implicitly assumed that coherence effects may be disregarded; i.e., the net structure is effectively rerandomized at each step instead of having fixed connections. This assumption, together with the later assumption that activity goes from expectation value to expectation value thereby neglecting fluctuations, leads Harth to label his analytic model the *neuron gas model.*

If one now assumes that the total number of EPSPs is much larger than l, then the Poisson approximation for the binomial distribution is valid, and one may write

$$\begin{aligned} p_l &\simeq [\alpha_n(1-h)\mu^+]^l \exp[-\alpha_n(1-h)\mu^+]/l! \\ q_m &\simeq [\alpha_n h\mu^-]^m \exp[-\alpha_n h\mu^-]/m! \end{aligned} \tag{12-14}$$

The companion expression for q_m included in equation (12-14) is the expression for the probability that a given neuron receives m inhibitory PSPs at time $n + 1$. It is derived precisely analogously to the expression for p_l.

The next step in this derivation is to obtain the probability $P(\alpha_n)$ that a neuron receives a combination of PSPs such that its threshold is exceeded at time $n + 1$:

$$P(\alpha_n) = \sum_{m=0}^{m_{\max}} \sum_{l=\eta}^{l_{\max}} p_l q_m \tag{12-15}$$

Here, η is the minimum number of excitatory inputs necessary to trigger a neuron which has just received m inhibitory inputs. It is given by $\eta = u[(\theta - mk^-)/k^+]$. The function $u[x]$ is defined as the smallest integer which is equal to or greater than x. The upper limits $l_{\max}$ and $m_{\max}$ are the total number of active excitatory and inhibitory connections, respectively: $l_{\max} = A\alpha_n(1-h)\mu^+$ and $m_{\max} = A\alpha_n h\mu^-$. Finally, one can write the expectation value of α_{n+1} in terms of $P(\alpha_n)$ as

$$\langle \alpha_{n+1} \rangle \simeq (1-\alpha_n)P(\alpha_n) \tag{12-16}$$

This equation states that the probability that a cell is activated at time $n + 1$ is equal to the probability that it receives suprathreshold input at that time times the probability that it is not refractory at that time. Combining the steps indicated above leads one to the final expression given by these authors and reproduced here as

$$\begin{aligned} \langle \alpha_{n+1} \rangle \simeq (1-\alpha_n)\exp(-\alpha_n h\mu^-) \sum_{m=0}^{M} (\alpha_n h\mu^-)^m/m! \\ \times \{1 - \exp[-\alpha_n(1-h)\mu^+]\} \sum_{l=0}^{\eta'-1} [\alpha_n(1-h)\mu^+]^l/l! \end{aligned} \tag{12-17}$$

In the final equation the upper limit m has been replaced by M, which is an integer sufficiently large so that the addition of higher terms may be neglected.

Equation (12-17) then is the main theoretical result presented in early papers by Harth and Anninos. Its basic behavior is shown in Figure 12-1. The distinguishing characteristic of the curve is its exhibition of a positive slope for small values of α_n and a negative slope for higher values of α_n. The positive slope for lower values of α_n means that increasing the number of cells active at a given time increases the total number of PSPs in the net and thereby increases the expected number of cells to be activated at time $n + 1$, α_{n+1}. However, the curve exhibits a negative slope at higher values of α_n because of the assumed refractory properties of the cells; i.e., as α_n becomes progressively closer to unity, the proportion of cells which will be refractory at time $n + 1$ is correspondingly increased. In the limit when α_n is equal to unity, all the cells will be refractory at the next time unit, and consequently α_{n+1} must be zero for this case. The parameter η measures the minimum number of excitatory inputs necessary to trigger a neuron in the absence of any inhibitory inputs. It is, therefore, an inverse measure of sensitivity. Clearly then, networks which contain cells with lower values of η exhibit curves with steeper slopes than those which contain cells with higher values of η.

On the basis of the recurrence relation given in equation (12-7) and illustrated in Figure 12-1, these authors define three classes of networks according to their ability to sustain reverberatory activity. Class A nets are networks which produce sustained activity for any initial activity α_n less than unity. The curve for $\eta = 1$ in Figure 12-1 is a representative example. Class B denotes those nets which have a threshold α_n greater than zero for being triggered into sustained activity, as illustrated by the case shown for $\eta = 2$. Class C nets are those nets which show monotonically decreasing activity for all initial activities. Such an example is illustrated by the curves for $\eta = 3$, 4, and 5.

Harth and Anninos show that this model can be extended to include afferent excitation and inhibition according to the expression

$$\langle \alpha_{n+1} \rangle = (1 - \alpha_n)\exp\{-[\mu_0^- h_0 + \mu_0^+(1 - h_0)]\sigma A_0/A - \alpha_n \mu^- h\}$$

$$\times \sum_{I'=0}^{I'_{\max}} \sum_{l'=0}^{l'_{\max}} \sum_{I=0}^{I_{\max}} (\sigma\mu_0^- h_0 A_0/A)^{I'} [\sigma\mu_0^+(1 - h_0)A_0/A]^{l'} (\alpha_n \mu^- h)^I / I'!\, l'!\, I'!$$

$$\times \left\{1 - \exp[-\alpha_n \mu^+(1 - h)] \sum_{l=0}^{\eta'-1} [\alpha_n \mu^+(1 - h)]^l / l!\right\} \qquad (12\text{-}18)$$

$$\eta' = \mu[(\theta + I\kappa^- - l'\kappa_0^+)/\kappa^+]$$

$$I'_{\max} = \sigma\mu_0^- A_0 h_0 \qquad l'_{\max} = \sigma\mu_0^+ A_0(1 - h_0) \qquad I_{\max} = \alpha_n \mu^- A h$$

This model supposes that there is a total number of A_0 afferent fibers of

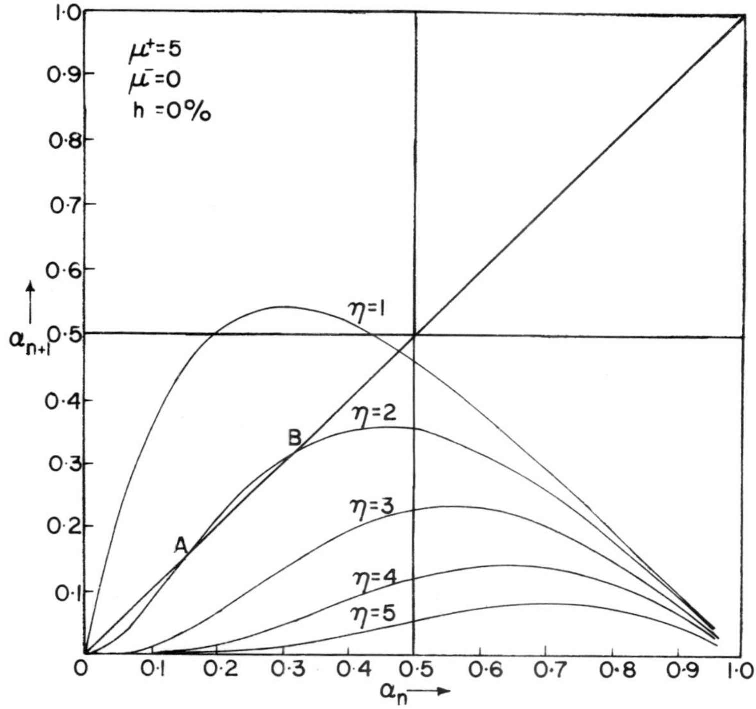

Figure 12-1. Sequential regenerative transfer relationship for a large network of automata units presented by Harth and Anninos.

which h_0 are inhibitory. Each excitatory fiber makes μ_0^+ synaptic connections on different neurons, and each inhibitory fiber makes μ_0^- efferent connections on different neurons. The quantities κ_0^+ and κ_0^- are the average coupling coefficients of the synapses made by the afferent excitatory and inhibitory fibers, respectively; σ is the fraction of afferent fibers carrying action potentials at a given time interval. In the original papers, Anninos *et al.* (1970) discuss phase diagrams based on equations (12-18). An example is shown in Figure 12-2. Each curve may be thought to divide the plane into two areas, one to the left and one to the right of the curve. The right region is one of rising activity, while the left is one of falling activity. Thus, a single-valued curve like $\eta = 3$ represents steady-state activities since for a given σ a momentary departure would tend to restore the system to the point on the curve or at least cause an oscillation about the value at that point. However, multivalued curves like those for $\eta = 1$ and $\eta = 2$ may be divided into stable and unstable portions. Such networks exhibit hysteresis loops when subjected to slow changes in level of afferent activation. If σ is between σ_0 and σ_{crit} as indicated in Figure 12-2, then a large but brief change in σ would elicit a strong and sustained increase in response activity

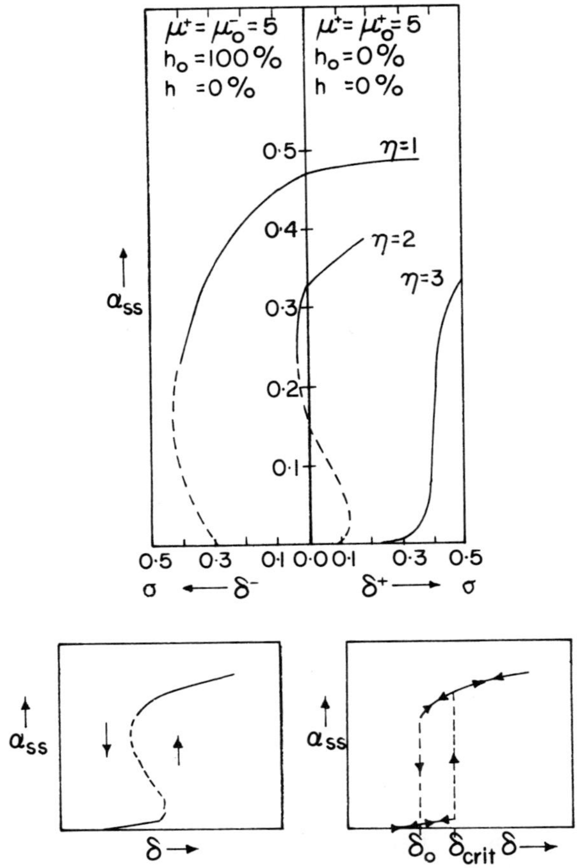

Figure 12-2. Phase diagrams for Harth–Anninos automata-theoretic model of neural networks.

in the network. Anninos *et al.* (1970) suggest that such network activity might be representative of responses in the cerebral cortex and in spinal motoneuron pools in response to activation of the brain stem reticular formation. Other speculative possible applications are mentioned.

In later work Wong and Harth (1973) attempted to extend this analytical model to take into account various combinations of refractory periods, summation times, and effective delays which are salient limiting assumptions in the original development. If cells are assumed to be refractory for k time intervals instead of simply one time interval, then the fractional number of neurons available for firing at time $n + 1$ is given by

$$\begin{matrix}\text{Fractional number of neurons} \\ \text{available for firing at } n+1\end{matrix} = 1 - \sum_{i=1}^{k} \alpha_{n+1-i} \tag{12-19}$$

This expression can be substituted directly into equation (12-16) in place of the quantity $1 - \alpha_n$. A similar extension can allow for fibers having different effective delay times. Let μ_m be the number of afferent fibers from a neuron with effective delay m. Then the average number of active afferents per neuron at time $n + 1$, λ_{n+1}, is given by

$$\lambda_{n+1} = \sum_{m=m_1}^{m_2} \alpha_{n+1-m}\mu_m \tag{12-20}$$

Here, m_1 and m_2 are lower and upper limits on m. With these modifications, the expected value for the number of neurons firing at time $n + 1$ in a completely excitatory network subjected to no afferent stimulation is given by

$$\langle\alpha_{n+1}\rangle = \left(1 - \sum_{i=1}^{k} \alpha_{n+1-i}\right)\left(1 - \sum_{l=0}^{n-1} e^{-\lambda_{n+1}}(\lambda_{n+1})^l/l!\right) \tag{12-21}$$

Suppose now that the summation time is L, and assume an exponential decay of the PSP with the time constant γ. The average total PSP per neuron at time $n + 1$ is now given by

$$\lambda'_{n+1} = \sum_{m=m_1}^{m_2} \mu_m \sum_{j=0}^{L} \alpha_{n+1-(m+j)} e^{-j\nu} \tag{12-22}$$

With these modifications, Wong and Harth have effectively increased the number of time intervals with which the net must retain memory in order to determine activity at time $n + 1$; i.e., one can no longer predict activity at time $n + 1$ from activity at n alone. To properly display such activity, the two-dimensional graph of $\langle\alpha_{n+1}\rangle$ versus α_n is now replaced by a recursion surface in a space $\alpha_{n+1}, \alpha_n, \ldots, \alpha_{n+1-k}$, where k is the order of the net. Wong and Harth discuss several interesting ways to display activity for such higher-order networks in the original papers, to which the interested reader is referred.

Recently, Wilson and Cowan (1973) have presented a rather elegant development of field equations for activity in neuronal networks, building largely upon the work and methodology of Beurle. These authors, like Beurle, use two main-field variables to describe the activity of a neural network. The first is a function $E(x, t)$ which represents the proportion of excitatory cells becoming active per unit time at the instant t and at the point x. The function $R(x, t)$ represents the number of so-called "sensitive" neurons at x and t, i.e., the number of cells which are not in a refractory period and are, therefore, able to fire.

The main tool derived by Wilson and Cowan is an equation which gives the proportion of cells excited at time $t + \tau$ and at position x in terms of the proportion of cells excited at previous times, network parameters, and input functions. This equation is given here as

$$E(\chi, t + \tau)\rho_e \delta\chi\delta t = [1 - \int_{t-r_e}^{1} E(\chi, T)\, dT] \cdot \rho_e \delta\chi$$
$$\times L_e \Big\{ \int_{-\infty}^{t} \Big[\int_{-\infty}^{\infty} \rho_e E\Big(X, T - \frac{|\chi - X|}{v_e}\Big) \beta_{ee}(\chi - X)\, dX$$
$$- \int_{-\infty}^{\infty} \rho_i I\Big(X, T - \frac{|\chi - X|}{v_i}\Big) \beta_{ie}(\chi - X)\, dX \pm P(z, T)\Big] \alpha(t - T)\, dT \Big\} \delta t$$

In this expression ρ_e is the surface density of excitatory neurons, L_e is the expected proportion of excitatory neurons that receive at least threshold excitation per unit of time, and the β's are interconnection functions. The function $I(x, t)$ is the proportion of inhibitory cells becoming active at point x and time t; P represents afferent excitation applied from outside the net; $\alpha(t)$ is an excitatory postsynaptic potential. There are five main quantitative statements which were used in deriving this equation. First, the mean rate of arrival of pulses on excitatory cells at x and time t from other excitatory cells is given as

$$\text{Mean rate of arrival of excitatory pulses at } x \text{ and time } t = \rho_e E\Big(X, t - \frac{|x - X|}{v_e}\Big) \beta_{ee}(x - X)\, dX \tag{12-24}$$

Second, the mean value of "integrated excitation" for excitatory neurons at x is given as

$$\overline{N}_e(x, t) = \int_{-\infty}^{t} \Big[\int_{-\infty}^{\infty} \rho_e E\Big(X, T - \frac{|x - X|}{v_e}\Big) \beta_{ee}(x - X)\, dX$$
$$- \int_{-\infty}^{\infty} \rho_i I\Big(X, T - \frac{|x - X|}{v_i}\Big) \beta_{ie}(x - X)\, dX \pm P(x, T)\Big] \alpha(t - T)\, dT \tag{12-25}$$

Third, the expected proportion of excitatory neurons that receive at least threshold excitation per unit time is given as

$$L_e(\overline{N}_e) = \int_0^{\overline{N}_e} G(\theta_e)\, d\theta_e \tag{12-26}$$

G is a distribution function of excitatory neuronal thresholds. Fourth, the number of excitatory neurons which are sensitive, i.e., not refractory, at time t is given by

$$R_e(x,t) = \left[1 - \int_{t-r_e}^{1} E(x,T)\,dT\right]\rho_e\,\delta x \tag{12-27}$$

Finally, the main physical assumption in this approach, as in Beurle's, is the statement that the number of cells which become excited at $t + \tau$ is equal to the product of the number of cells which are sensitive at time t times the number of cells which have input exceeding their threshold at time t. This expression is given as

$$E(x,t+\tau)\rho_e\,\delta x\delta t = R_e \cdot L_e(\overline{N}_e)\delta t \tag{12-28}$$

Combining these five quantitative concepts and terms results in equation (12-23), from which one can obtain the temporal evolution of the proportion of excitatory cells activated given appropriate connectivity functions and input functions. Wilson and Cowan also present an analogous companion expression for $I(x,t)$.

Wilson and Cowan also present a simplified version of this general equation which they claim is a valid approximation for a wide variety of neuronal activities of interest. First, they assume that in the cortical networks with which they are concerned, the times required for conduction of impulses along axons and collaterals are very small compared to membrane time constants, so that conduction velocities are effectively infinite. They also assume that they can write the expected value for E in terms of an integral of $E(x,t)$ as

$$\langle E(x,1)\rangle = \frac{1}{\mu}\int_{-\infty}^{1} E(x,T)\exp\left[-\frac{(t-T)}{\mu}\right]dT \tag{12-29}$$

$E(x,t)$ can then be written in terms of its expected value as

$$E(x,t) = \mu\frac{\partial}{\partial t}\langle E(x,t)\rangle + \langle E(x,t)\rangle \tag{12-30}$$

If the refractory period is assumed to be much less than the membrane time constant, as is the synaptic delay, then one can reduce equation (12-30) to the following:

$$\begin{aligned}\mu\frac{\partial}{\partial t}\langle E(x,t)\rangle &= -\langle E(x,t)\rangle + [1 - r_e\langle E(x,t)\rangle] \\ &\times L_e\{\alpha\mu[\rho_e\langle E(x,t)\rangle \otimes \beta_{ee}(x) - \rho_i\langle I(x,t)\rangle \otimes \beta_{ie}(x) \pm \langle P(x,t)\rangle]\}\end{aligned} \tag{12-31}$$

where ⊗ represents spatial convolution. This simplified equation is a partial differential equation of the first order in time for the spatial and temporal development of the expected value of the excitation E. The connection functions β_{ee} and β_{ie} occur in convolutionlike manner. This equation is a so-called coarse-grained version of equation (12-23), a technique which was first applied to problems in statistical physics by Kirkwood. In deriving these equations for the dynamics of neuronal tissue, Wilson and Cowan (1973) have made certain assumptions concerning the character of that neuronal tissue. They have been concerned primarily with cortical networks and have assumed that these may be considered to be essentially two-dimensional in a functional sense. They view the cortex as a laterally organized net of columnar functional units. The overall activity of neurons within a column is taken as functionally significant rather than the responses of any single neuron in a column. They have also supposed that although the main activity variable E is a random variable, one might suppose that the probability distribution associated with it is sharply peaked about its mean value so that the stochastic character of the dynamics might be neglected. Also, the equations are derived for the limited one-dimensional case. The generalization to two dimensions is said to be straightforward. Moreover, these authors consider only afferent activities that vary in one spatial dimension. Thus, because of the assumed homogeneity and isotropy of the model, the neural responses also vary only in the same dimension.

The fundamental equations developed in this paper are inherently nonlinear and generate complex nonlinear phenomena. Analytic solutions to such equations are usually impossible to obtain, and, moreover, one generally is unable to map the parameter ranges associated with certain observed response properties. Thus, although equations (12-23) and (12-31) might represent a reasonable representation of neuronal net activity in terms of the assumptions made by Wilson and Cowan, their usefulness in predicting actual dynamic activity is severely limited. However, these authors have provided some exploration of their properties. For example, numerical solutions of the equations have illustrated three qualitatively different dynamical modes: active, transient responses followed by relaxation to the resting state; spatially localized limit cycles; and spatially inhomogeneous stable, steady states that reflect attributes of prior stimuli. These dynamic properties are discussed in terms of sensory-information processing, oscillations, waves and rhythms, and possible mechanisms for short-term memory. The interested reader is referred to the publication of Wilson and Cowan (1973) for details.

In the 1940s Norbert Weiner suggested that the correct tool to spell out the dynamics of neural networks would be a form of statistical mechanics. Weiner was influenced by the marked theoretical and aesthetic success of classical statistical mechanics in describing the interaction of molecules in a gas. Certainly, on the surface at least, one might draw a parallel between many millions of moving particles in a gas interacting and colliding according to nonlinear rules, on the one hand, and many millions of interconnected neurons bombarding one another with electrical pulses, on the other hand. In both cases, one is concerned not so much with the details of interaction at any one or set of molecules or neurons but rather with gross measures of overall activity which one might relate to directly observable macroscropic properties in the former or functional characteristics in the latter. Despite the obvious merit of this suggestion, only one investigator, J. D. Cowan, of the University of Chicago, has actively pursued it. Cowan (1968) describes his model and metholodology as follows:

> In an attempt toward a formulation that does make contact with experimental variables, I have recently developed a statistical mechanics of the activity of nervous nets. The "equations of motion" which I have obtained are nonlinear differential equations that represent the changes in nervous activity that result from interactions within and between assemblies of nervous nets. In a certain special case these equations represent neural oscillators and can be transformed into the well-known Hamiltonian mechanics of the physicist. From this the transition to statistical mechanics is fairly standard. The utility of this approach is that the statistics that it generates are related to certain ensemble averages—formal analogues of temperature, energy, and entropy—which appear to give useful descriptions of the changes that occur in the activity of nets following interaction. There is some reason to believe that the mathematical model so constructed will predict some of the statistics associated with microelectrode data, and perhaps some of the statistical parameters of macroelectrode recordings from nervous tissue. An example may make this clearer. The equations of motion are as follows:
>
> $$\frac{dX_r}{dT} = (\gamma_r + \beta_r^{-1} \sum_s \alpha_{sr} X_s) X_r (1 - X_r)$$
>
> $$r = 1, \ldots, N, \qquad s = 1, \ldots, N$$
>
> X_r is a dynamical variable measuring the sensitivity of the rth neuron in a net. It measures the fraction of time in a long time interval during which the neuron is not refractory and can be fired. γ_r is a growth coefficient, a function of the various electrical parameters of the neural membrane, as is β_r. However, γ_r is also a function of control stimuli, i.e., the input to the neuron enters nonlinearly through γ_r. The coupling coefficient of the interaction between the rth and sth neurons, α_{sr}, is given by
>
> $$\alpha_{sr} = \begin{cases} r^{-1}(-E_m)\delta_e g_{sr} b_{sr} & \text{(excitatory synapse)} \\ r^{-1}(E_i - E_m)\delta_i g_{sr} b_{sr} & \text{(inhibitory synapse)} \end{cases}$$
>
> The coefficient r is the refractory period of the neuron. E_m is the resting potential of the cell membrane, E_i the equilibrium potential for an IPSP, δ_e and δ_i are the operating times for synaptic transmission, g_{sr} is the conductance of the srth synapse, and b_{sr} is an attenuation factor that serves to represent electrotonic decrement in the flow of excitation from dendritic to somatic regions of the cell. The variable T is dimensionless, and is in seconds/τ_{sr}, where τ_{sr} is the mean intercellular transit time for neural activity. Incorporated in τ_{sr} are delays due to axonal conduction, synaptic transmission, and passive electrotonic flow in dendrites.
>
> These equations are obtained by combining Eccles' equivalent circuit for neural SD-membrane with an approximation to Rall's equations for the electrotonic spread of

postsynaptic potentials from dendrites to soma plus the assumption that the mean rate of neural spike emissions is given by the function $\Phi(x)$, where x is the mean current intensity built up in the cell membrane. The equations are only a heuristic step toward an exact treatment of the activity of nets, and their validity can best be judged by the accuracy of the predictions which can be deduced from them.

To illustrate this methodology, Cowan looks at a very simple specific example. This example consists of a single two-cell loop wherein one connection is excitatory and the other inhibitory. Each of these cells is supposed to represent a network of cells, and cells within a given network are not coupled. Under these assumptions the general equations reduce to the following:

$$\begin{aligned} \frac{dv_r}{dT} &= \sum_s \lambda_{sr} \frac{\partial H}{\partial v_r} \\ v_r &= \ln[(X_r/q_r)(1 - X_r)^{-1}] \\ \lambda_{sr} &= \alpha_{sr}/\beta_s \beta_r \\ H &= \sum_r \beta_r[1 + q_r \exp(v_r) - q_r v_r] \end{aligned} \tag{12-32}$$

Here the variable q_r is the stationary state of X_r, the state for which $dX_r/dt = 0$. The Hamiltonian H is the formal analog of energy and here is a function of certain logarithms of the probabilities of spike emission of neurons in the net. In this case, H can be thought of as a positive constant of the motion of a system of neural oscillators. If one considers a single oscillator to be continually bombarded by other circuits, then the average behavior of the oscillator can be viewed as if it were a member of a very large assembly. To a first-order approximation, there is an invariant measure of the form

$$\frac{\exp(-\alpha H)}{\Pi_r \int_{-\infty}^{+\infty} \exp(-\alpha H_r)\, dv_r} \tag{12-33}$$

The interval distributions of neuronal firing patterns are given by

$$\begin{aligned} p(\tau_r)dt_r &= \frac{r^{-1}(1 - r/\tau_r)^{p-1}(r/\tau_r)^{q+1}\, d\tau_r}{B[p, q]} \\ p + q &= \alpha\beta_r \end{aligned} \tag{12-34}$$

This particular example was chosen by Cowan primarily to illustrate the mode of thought and methodology employed in his approach. He does suggest, however, that even this very simplified application results in predictions that are at least consistent with what one observes in macroelectrode recordings. Moreover, the predicted interval histogram has sufficient paramenters to match a wide variety of data. In this particular example the variables in the net have been effectively uncoupled.

Cowan suggests that the next step in the development of this theoretical approach is to introduce thermodynamic functions and to investigate the resulting statistical thermoneurodynamics. He points out that in classical statistical mechanics the main macroscopic variable is the thermodynamic variable temperature. The analog in the present approach is the constant α in expression (12-33). Cowan calls this measure the amplitude of fluctation of neural activity and notes that it depends on the first two moments of the distribution of neural sensitivities,

$$\langle X_r \rangle = q_r \quad \text{and} \quad \langle (X_r - \langle X_r \rangle)^2 \rangle = \langle (X_r - q_r)^2 \rangle$$

A slightly different analytic approach to describing the dynamic activity of large neuronal networks is based on extensions of conventional systems-analysis techniques.

The systems-theoretic approach to modeling neural networks is characterized by the use of continuous rather than discrete variables to describe the dynamic activity of the system, and the description of these variables in terms of systems of interrelated ordinary differential equations. Typically, one unknown variable and one dynamic equation is introduced for each element in the system. Therefore, if one knows appropriate initial conditions and external input functions one can, in principle, solve all the equations for all the unknowns as functions of time. The parallel between this description of a neuron network and the corresponding description of the dynamics of interacting groups of mechanical particles is obvious.

In numerous publications, Stephen Grossberg (1969*a*,*b*; 1970), of the Massachusetts Institute of Technology, has presented a coherent theory for neural networks from the systems-theoretic point of view. The basic assumption made by Grossberg for the dynamic activity and individual neurons is shown here as

$$\begin{aligned}\frac{dx_i(t)}{dt} &= -\alpha_i x_i + \sum_{m=1}^{n} B_m[x_m(t-\tau_{mi}) - \Gamma_{mi}]^+ y_{mi}(t) \\ &\quad - \sum_{m=1}^{n} [x_m(t-\tau_{mi}) - \Gamma_{mi}]^- q_{mi} + I_i(t) \qquad (12\text{-}35)\end{aligned}$$

This equation describes the temporal development of the generator potential in the ith cell, x_i, as a function of the externally applied input and internal interconnections of the network. Note that this equation can be thought of as an extension of the basic membrane equation wherein the last three terms of the right represent simply the stimulating current incident on the membrane patch. Here α_i is the reciprocal of the membrane time constant. The quantity I_i is the input applied to the ith cell from sources external to the network. The first summation on the right side is the total input to the ith cell from excitatory synapses originating from presynaptic fibers within the network. Correspondingly, the second summation on the right is the total input at inhibitory synapses from presynaptic fibers originating within the network. In this model, Grossberg has neglected the nonlinearities associated with the synaptic changes in membrane conductance and supposes that all arriving input may be treated as simply stimulating current. It is important to pay particular attention to the manner in which Grossberg effects excitation from cell m to cell i. The bracketed function in equation (12-35) is zero if its argument is negative, and equal to the argument otherwise. Thus, the output emanating from cell m to cell i is nonzero only if the generator potential x_m is greater than its threshold Γ_{mi}. A critical assumption in this model is that the influence on cell i of activity in cell m is proportional to the difference between x_m and the threshold of cell m. Thus, this equation system might be thought to represent a case where only the mean firing rate over some interval is significant. The possibility of information coding in the fine temporal structure of spikes is ignored. Thus, one might suppose that the output firing rate in cell m is approximately proportional to the difference between its generator potential and its threshold. This mean firing rate in turn effects a continuously graded and proportional input influence on cell i. The quantity τ_{mi} in the argument of x_m accounts for the time delay involved in conducting a pulse from the soma of cell m to the soma of cell i. It includes time associated with conduction along the axon of cell m, synaptic delay time as particles diffuse across the synaptic cleft between cell m and a dendrite in cell i, and the time required for electrotonic conduction along the dendrites of cell i: What arrives at cell i at t depends on what x_m was at time $t - \tau_{mi}$. The constants y_{mi} and q_{mi} represent the strength of excitatory and inhibitory synapses, respectively, between cells m and i. The constant β_m represents the relative strength of excitatory synapses as compared to that of inhibitory synapses.

Grossberg has been primarily concerned with the way in which such interrelated groups of neurons might "learn." He has proposed that the equations

$$\begin{aligned} y_{jk}(t) &= p_{jk}\, z_{jk}(t)\left[\sum_{m=1}^{n} p_{jm}\, z_{jm}(t)\right]^{-1} \\ \dot{z}_{jk}(t) &= \begin{cases} -u_{jk}\, Z_{jk}(t) + v_{jk}[x_j(t-\tau_{jk}) - \Gamma_{jk}]^{+}\, x_k(t) & \text{if } p_{jk} > 0 \\ 0 & \text{if } p_{jk} = 0 \end{cases} \end{aligned} \tag{12-36}$$

used in conjunction with equation (12-35) represent a reasonable quantification of a way in which synaptic interconnections might change with activity, thereby encoding memory of particular temporal patterns in the anatomical structure of the network. Here the interconnection matrix p_{jk} is semistochastic in the sense that all p_{jk} are positive and $\sum_{m=1}^{n} p_{jm} = 0$ or 1. Grossberg has described numerous implications of this general equation system when applied to various particular cases. He calls his networks "imbedding fields" to reflect their property of coding their history of activity within the actual structure of the network. Grossberg's approach is to obtain a mathematical theorem from the equation system as applied to a particular case, and then to draw qualitative conclusions about the behavior of the network from the theorem. A representative example of this approach is that of the so-called "outstar" shown in Figure 12-3. For this network the general equation system reduces to the following:

$$\begin{aligned}
\dot{x}_1(t) &= -\alpha x_1(t) + I_1(t) \\
\dot{x}_i(t) &= -\alpha x_i(t) + \beta x_1(t-r) y_{1i}(t) + I_1(t) \\
y_{1i}(t) &= z_{1i}(t)\left[\sum_{m=2}^{n} z_{1m}(t)\right]^{-1} \\
\dot{z}_{1i}(t) &= -u z_{1i}(t) + \beta x_1(t-r) x_i(t)
\end{aligned} \tag{12-37}$$

In this example, Grossberg has neglected the nonlinear threshold mechanism. This equation system allows the following theorem as proved by Grossberg (1969*b*).

Theorem 1. Let $G^{(1)}, \ldots, G^{(N)}, \ldots$, be any sequence of outstars with equal but otherwise arbitrary initial data. Let $I_i^{(N)}$ be any inputs of the form

$$I_1^{(N)}(t) = I_1(t)\chi(t - U_1(N)) + h_N(t)$$

and

$$I_i^{(N)}(t) = \theta_i I(t)\chi(t - U(N)) \qquad i = 2, \ldots, n$$

where

(a) $\{\theta_i : i = 2, \ldots, n\}$ is a fixed but arbitrary probability distribution (i.e., $\theta \geqslant 0$ and $\sum_{i=2}^{n} \theta_i = 1$);

(b) $U_1(N)$ and $U(N)$ are any nonnegative and strictly increasing functions of the integer $N \geqslant 1$;

(c) the inputs $I_i(t)$ and $I(t)$ are constrained only by the existence of positive constants c and T_0 such that

$$\int_0^t e^{-\alpha(t-v)} I_1(v)\, dv \geqq c$$

and

$$\int_0^t e^{-\alpha(t-v)} I(v)\, dv \geqq c$$

for $t \geqq T_0$;

(d) the input $h_N(t)$ is constrained only by being zero until $t > U(N)$;

and

(e)

$$\chi(t) = \begin{cases} 1, & \text{if } t < 0 \\ 0, & \text{if } t \geqq 0 \end{cases}$$

Then

(A) for every $N \geqslant 1$, the limits $Q_i^{(N)} = \lim_{t\to\infty} X_i^{(N)}(t)$ and $P_{1i}^{(N)} = \lim_{t\to\infty} y_{1i}^{(N)}(t)$ exist and are equal;

(B) for every $N \geqslant 1$ and $t \geqq U(N)$, $X_i^{(N)}(t)$ and $y_{1i}^{(N)}(t)$ are monotonic and are contained in the interval $[m_i^{(N)}, M_i^{(N)}]$, where

$$m_i^{(N)} = \min[X_i^{(N)}(U(N)), y_{1i}^{(N)}(U(N))]$$

$$M_i^{(N)} = \max[X_i^{(N)}(U(N)), y_{1i}^{(N)}(U(N))]$$

and

$$\lim_{N\to\infty} m_i^{(N)} = \lim_{N\to\infty} M_i^{(N)} = \theta_i$$

In particular, by (A) and (B),

$$\lim_{N\to\infty} \lim_{t\to\infty} X_i^{(N)}(t) = \lim_{N\to\infty} \lim_{t\to\infty} y_{1i}^{(N)}(t) = \theta_i \qquad i = 2, \ldots, n$$

(C) for every $N \geqslant 1$ and $i \neq 1$, the functions

$$y_{1i}^{(N)}, f_i^{(N)} = y_{1i}^{(N)} - X_i^{(N)}, \text{ and } g_i^{(N)} = X_i^{(N)} - \theta_i$$

change sign at most once and not at all if $f_1^{(N)} g_i^{(N)}(0) \geqslant 0$. Moreover, $f_i^{(N)}(0) g_i^{(N)}(0) > 0$ implies that $f_i^{(N)}(t) > 0$ and that $y_{1i}^{(N)}(t)$ is monotonic for all $t \geqslant 0$.

In the following application of this theorem to the network indicated in Figure 12-3 it is helpful to note that one is assuming here that a patterned input is presented to the peripheral neurons of the outstar, whereas an arbitrary input is presented to the center neuron of the outstar. The pattern is contained in the set of constants θ_1. The network eventually learns this pattern in the sense that the strengths of connections from the center neuron to the peripheral neurons contains that pattern θ_1 originally presented to the peripheral neurons. Grossberg (1969*b*) then states that his theorem is equivalent to the following qualitative conclusions concerning outstars.

1. "Practice makes perfect."
2. An isolated outstar suffers no memory loss.
3. An isolated outstar remembers without overtly practicing.
4. After moderate amounts of practice, the memory of an outstar spontaneously improves.
5. The act of making a correct recall can be repeated as many times as one wishes without destroying the outstar's memory.
6. All errors can be corrected, although the retraining time might be lengthened due to response interference.

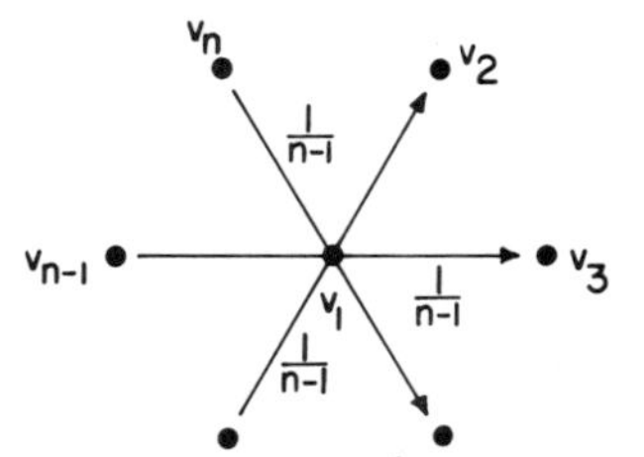

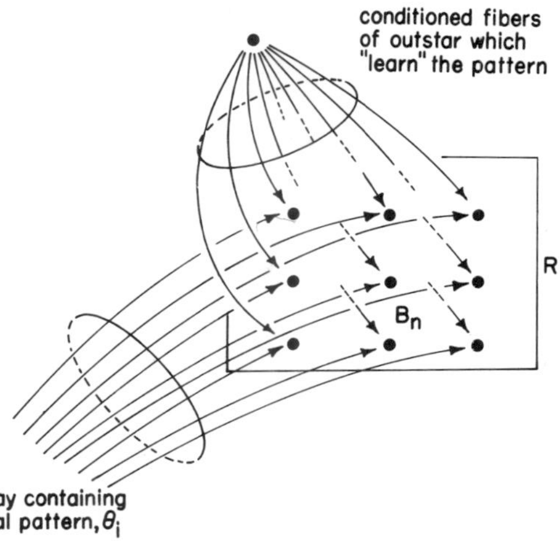

Figure 12-3. Two views of the "outstar" presented by Grossberg.

In the continued development of this example and in many other examples Grossberg's work is rich in qualitative suggestions regarding the functioning of various types of networks. The interested reader is referred to his many publications for further elaboration.

The general field of analytic approaches to large neural networks contains many interesting and potentially fruitful contributions. We have touched on only representative examples in this chapter. However, these approaches have made no verifiable contributions to the understanding of any real nervous network. The work in the area is thus far much too speculative and needs experimental verification. On the one hand, it is not reasonable to subject such theories to test at the single-unit level since the working assumption is that unit properties might be glossed over in some reasonable fashion in obtaining more gross macroscopic measures. On the other hand, the resulting predictions of gross behavior have not been adequately tested or examined in comparison to experimental observations on any particular neuronal network. The statistical-mechanics approach pursued by Cowan is extremely promising and should be further developed. However, we should warn that in the development of classical statistical mechanics of gases the researchers knew and had before them a clear picture of the macroscopic variables whose behavior they were trying to predict as provided by classical macroscopic thermodynamics. This is not yet the case for nervous systems where we generally do not know what overall measures of activity are most appropriate for the system. It may be then that the development of refined statistical mechanics of nervous networks will progress more satisfactorily after this level of understanding has been attained by other means.

13

Models of Large Networks: Computer-Oriented Approaches

This chapter takes up the second mainstream of modeling techniques for large neural networks—those that make extensive use of large-scale digital computers or particularly lend themselves to realization on large-scale computers. A germinal source work which has greatly influenced much subsequent work in this area was published in 1943 by McCulloch and Pitts. This paper has been extremely influential in propagating the so-called automata-theoretic view of neural-network dynamics which has been taken over, or at least influenced, by many analytically oriented researchers as well, as we saw in the previous chapter. In this chapter we discuss this basic source paper, its ramifications, and several contemporary examples of computer-oriented approaches to large neural networks.

The starting point is a classic paper published in 1943 by McCulloch and Pitts entitled "A Logical Calculus of the Ideas Immanent in Nervous Activity." McCulloch and Pitts introduce their paper by framing the then conventional ideas roughly as follows: The nervous system consists of a network of neurons. Each neuron has an axon and a soma. A synapse always proceeds from an axon to a soma. A neuron has a threshold such that if the "excitation" is greater than threshold, a pulse ensues. A pulse, except for its timing, is determined by the neuron, and not by the excitation. The axonal conduction time is not significantly different from one cell to another because long axons conduct rapidly whereas short axons conduct slowly. A neuron is excited when a sufficient number of input pulses arrive within the period of latent addition, which is less than about $\frac{1}{4}$ of a millisecond. (McCulloch and Pitts state that "observed temporal summation of impulses at greater intervals is impossible for single neurons and empirically depends on the structural properties of the net." This, of course, reflects the incomplete neurophysiological knowledge of the time.) During the first part of the pulse, the neuron is absolutely refractory, then it returns to its normal state rather quickly. Inhibitory synapses either raise the threshold for excitation or oppose excitatory synapses.

McCulloch and Pitts (1943) itemize their assumptions formally as follows:

> 1. The activity of the neuron is an "all or none" process.
> 2. A certain fixed number of synapses must be excited within the period of latent addition in order to excite a neuron at any time, and this number is independent of previous activity and position on the neuron.
> 3. The only significant delay within the nervous system is synaptic delay.
> 4. The activity of any inhibitory synapse absolutely prevents excitation of the neuron at that time.
> 5. The structure of the net does not change with time.

In this formulation the state of each constituent neuron is signaled according to a single binary variable which reflects the all-or-none character of the action potential. Time is considered in discrete intervals the length of which corresponds to the time of conduction from one element to another. They argue that interunit conduction times may be considered equal for all cases since long fibers have larger diameters and therefore conduct faster, as compared to shorter fibers which have smaller diameters and therefore conduct more slowly. Thus, at a given time, all cells are in a given state; they exchange information about these states according to the interconnection pattern before the next time interval when they all signal a revised state. McCulloch and Pitts point out that specification at any given time of the activity of all the constituent neurons and of the afferent stimulation determines the state of the system and, moreover, that "specification of the nervous net provides the law of necessary connection whereby one can compute from the description of any state that of the succeeding state." Thus, to activate such a system, one needs merely to specify the output connections of any and all constituent neurons and to specify the temporal character of input for all the cells. One then has a quantitative algorithm whereby the temporal evolution of states of all the constituent neurons may be determined.

The authors then show how such elements can be used to construct various sequential machines to perform combinations of logical operations of any degree of complexity. Embedded in this work is the idea that neuronal activity may be interpretable in terms of propositions. In their terms, "the 'all-or-none' law of nervous activity is sufficient to insure that the activity of any neuron may be represented by a proposition." Thus, as indicated in Figure 13-1, one may readily construct a wide variety of elementary logic circuits from elements obeying the rules described above. In this figure, knobs represent excitatory connections, and loops represent

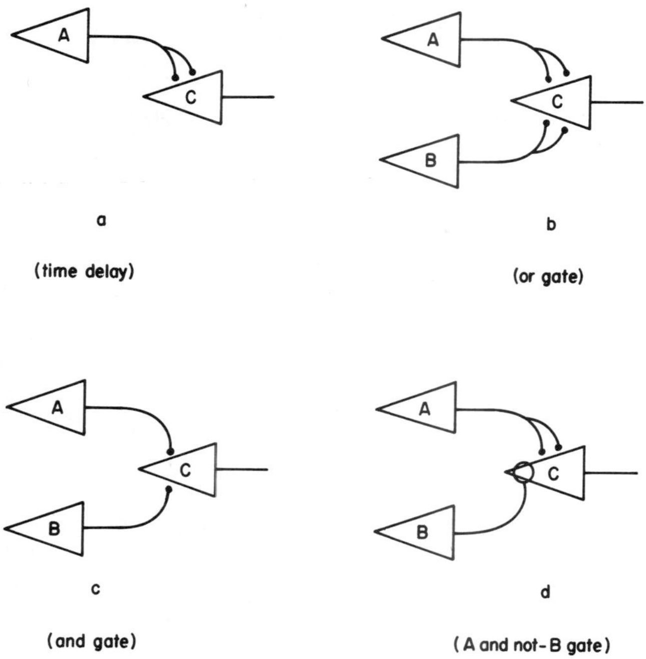

Figure 13-1. Hardware elements for propositional calculus according to McCulloch and Pitts. Knobs represent excitation; loops represent inhibition. Two simultaneous excitatory inputs are required for excitation.

inhibitory connections. In this example, the simultaneous occurrence of two excitatory input events is required to cause activation of the output element, and inhibition blocks any combination of inputs. Simple time delays may be constructed as in Figure 13-1a: Output unit C will fire every time the corresponding input element A does but will invariably effect a time delay corresponding to the synaptic lag. "OR gates" may be constructed as in Figure 13-1b: A single pulse in either of the input units A or B, or in both, is sufficient to cause activation of the output element C. The element in part C can be interpreted as an "AND gate": It is necessary that input pulses arrive simultaneously at both input units A and B to successfully activate the output element C in this element. Using only the rules and assumptions stated by McCulloch and Pitts, logical elements of virtually unlimited variation and complexity can be constructed.

The work of McCulloch and Pitts has been extremely influential by laying the foundation for the entire science of sequential machines, and also for the automata-theoretic approach to neuronal networks. Both these sciences rest on the partitioning of the time continuum into discrete intervals.

On each interval one may identify specific values associated with the variables of the system. A sequential machine is one the values of whose variables on one interval are determined by input on that interval and the values of those variables at previous intervals. Similarly, an automaton model of a neuronal network is one whose output variables are conceived to take on values over certain discrete intervals and to change those values from interval to interval in accordance with interconnection rules that map the influence of state variables of one interval onto corresponding values at subsequent intervals. In practice, this almost invariably means that the values taken on in one interval depend only on those in the previous interval.

Unfortunately, this influence, at least with respect to the modeling of neuronal networks, has been a mixed blessing. On the one hand, the work has generally excited interest in the field and indicated how large-scale computers could be used to simulate vast numbers of interconnected neurons. On the other hand, the work seems to have laid out a large signpost pointing in the wrong direction which has been followed by many subsequent workers. For example, it is not at all clear that it is always or even frequently best to conceive the dynamics of neuronal networks as a sequence of states along a partitioned time continuum. Moreover, McCulloch and Pitts were excessively exuberant in presenting their highly simplified models of network dynamics as a bona fide representation of the activity of real brains. They, indeed, thought that the entirety of psychology had been reduced to the working out of dynamics of appropriate sequential machines. This point of view has unfortunately introduced into the literature of our area a vast number of publications which bear titles including the words "neural networks" or the like and which often contain overreaching claims of wide generality concerning the activity of so-called "neural networks" but which, in point of fact, contain extremely simplified versions of neuronal activity. This body of knowledge can be a misleading trap for the beginning student and will not be discussed here.

Other researchers, however, have been more aware of the limitations of the approach and have provided valuable information concerning the development of the approach and its application to real nervous systems. Caianiello (1961, 1967) and various colleagues have considerably developed the automata-theoretic approach to neural networks. In 1961 Caianiello published a paper entitled "Outline of a Theory of Thought Processes and Thinking Machines." He notes as his guiding principle that the human brain, although tremendous in complexity, yet obeys, if one looks at the operation of individual neurons, dynamical laws that are not necessarily complicated and that these laws are such as to engender in large neuronal assemblies collective modes of behavior to which thought processes are correlated. The primary objective is a convenient formulation of these laws.

The model consists of two sets of equations, the "neuronic" equations and the "mnemonic" equations. The neuronic equations are shown here as

$$u_h(t+\tau) = l[\sum_{k,r} a_{hk}^{(r)} u_k(t - r\tau) - s_h]$$
$$l(x) = \begin{cases} 0 & x < 0 \\ 1 & x \geqslant 0 \end{cases} \qquad (13\text{-}1)$$

In this model, time is divided into discrete intervals of length τ. The variable $u_h(t)$ represents the output spike train of the hth neuron and takes on the value 0 or 1 at each time interval. The network connection coefficients are given by $a_{hk}^{(r)}$, and s_h is the threshold of the hth neuron. The variable r allows Caianiello to use conduction times at any multiple of the elemental delay time τ. In this model, summation of PSPs is instantaneous, as is the decay; inputs must arrive simultaneously to interact.

The "mnemonic" equations are shown as

$$\frac{da_{hk}^{(r)}(t)}{dt} = \{a^{(r)} u_k(t-\tau) u_h(t) - \beta^{(r)} I[a_{hk}^{(r)}(t) - a_{hk}^{(r)}(0)]\} a_{hk}^{(r)}(t) I[A_{hk}^{(r)} - a_{hk}^{(r)}(t)] \qquad (13\text{-}2)$$

These equations determine the means whereby the interconnection coefficients $a_{hk}^{(r)}$ can changes as a function of time according to the history of activity in the network, u_k, and various constants and parameters $a^{(r)}$, $\beta^{(r)}$, and $A_{hk}^{(r)}$. The mnemonic equations are used to outline the way in which the nervous network might "learn" according to its experience. They reflect the common assumption that such learning corresponds to changes in efficiency of individual connections between neurons.

The time scale over which the mnemonic equations behave is assumed to be considerably longer than that over which the neuronic equations behave. To further separate the activities of these two equation systems, Caianiello announced his so-called "adiabatic hypothesis" wherein the $a_{hk}^{(r)}$ are assumed to be constant throughout any particular dynamic-behavior pattern determined by the neuronic equations; i.e., the installation of memory takes a long time and does not disrupt the ongoing activity of the neuronal network. With these three general statements, then, Caianello was able to outline how dynamic activity in his networks of artificial neurons was determined and how his networks might learn; he could also show that these two processes will not disrupt one another.

In this early paper, Caianiello (1961) discussed at length the basis of these equations and speculated on their various applications. In his words:

> A qualitative but rigorous discussion shows that this machine exhibits, as a necessary consequence of the theory, many properties that are typical of the living brain: including need to "sleep," ability spontaneously to form new ideas (patterns) which associate old ones, self-organization towards more reliable operation, and many

others. Future works will deal with the quantitative solution of these equations and with concrete problems of construction—things that appear reasonably feasible. With a transposition of names, this theory could be applied to many sorts of social or, more generally, "collective" problems.

In several papers appearing over the last decade, Caianiello and his colleagues have worked out various implications of this equation system. In a paper appearing in 1968, Caianiello *et al.* characterized their activities as follows: "The essential part of our work consists in taking such a description [neuronic equations, mnemonic equations, and adiabatic learning hypothesis] as the starting point for a mathematical formulation of the problems which must be associated with the neuronic equations in order that they make physical sense. Our tool is ordinary matrix algebra. Our aim is the formulation and solution of some essential theoretical problems, the nature of which seems to discourage different approaches." In this paper, Caianiello and colleagues point out two broad main areas of interest to them.* The first is the attempt to identify various constants of motion for artificial neural nets. For example, when all conduction delays are one time period, the neuronic equations may be rewritten as

$$\mathbf{v}(t) = A l[\mathbf{v}(t - \tau)] - \mathbf{s} \tag{13-3}$$

Here, $\mathbf{v}(t)$ is defined as $\mathbf{g}(t) - \mathbf{s}$, where $\mathbf{g}(t)$ is the generator potentials of the cells and $\mathbf{s}$ is their thresholds. Thus, equation (13-3) derives from the assumption $\mathbf{g}(t) = Al[\mathbf{g}(t - \tau) - \mathbf{s}]$. The quantity A is the connectivity matrix of the network. Then, if $A^T \boldsymbol{\nu} = 0$, where $\boldsymbol{\nu}$ is an arbitrary N-dimensional vector and A^T is the transpose of A, then $\boldsymbol{\nu} \cdot \mathbf{v}$ is a constant of motion. That is, although the individual $v_i(t)$ may fluctuate arbitrarily in the ongoing activity of the network, the quantity $\nu_1 v_1 + \nu_2 v_2 + \cdots + \nu_N v_N$ remains constant. This concept is analogous to the concept of conservation of kinetic energy and momentum in the classical machanics of interacting particles. This particular constant of motion has a direct and satisfactory interpretation. Suppose, for example, that the ν_i correspond to coupling coefficients between the network output fibers and a "summator" of some kind which may or may not be another neuron. Then the quantity $\boldsymbol{\nu} \cdot \mathbf{v}$ measures the total activity incident on the summator which is a weighted measure of the total network activity. There are clearly many places in nervous systems where such a constant measure could be usefully employed.

More generally, Caianiello and colleagues have shown that for a network of N cells, where a_{hk} is of rank K, then there are $n \cdot K$ linear constants of motion. It is their conclusion that such constants of motion represent the natural definition of the collective modes of activity for such a nonlinear network.

*In this and most subsequent developments of this equation system, Caianiello's group assumes all conduction delays to be one time interval τ.

The second main area of interest to them has to do with reverberations. More precisely, they are concerned with conditions under which the system state vector $\mathbf{v}$ goes through certain recurring cycles as demanded by the equation

$$\mathbf{v}_{m+R} = \mathbf{v}_m \tag{13-4}$$

Here $\mathbf{v}_m$ represents the state vector at the time m. The assumption is that since conduction delays are all one time unit and there is no external input to the system or noise within it, then $\mathbf{v}_m$ completely describes the state of the system and defines uniquely the subsequent state $\mathbf{v}_{m+l}$. Thus, if the state $\mathbf{v}_m$ recurs at time $m + R$, the same series of intervening states will be reactivated. This series of states is thus defined as a reverberation. In exploring the conditions under which such reverberations may occur, the authors conclude that in their theory the main role is played by the rank K of the coupling matrix a_{hk}. Most of their subsequent publications are concerned with working out properties of these reverberations according to the rank of various matrices. For example, the following proposition is presented by Accardi (1971):

> *Proposition.* If the matrix A of the coupling coefficients has rank K, after an arbitrary initial transient the number R of admissible states of the network, and *a fortiori* the maximum possible reverberation, is such that
>
> $$R \leqq 2^N - 2^{N-K+1} + 2$$
>
> Thus, by knowing the number of cells in the network and the rank of the connectivity matrix, one can put an upper limit on the period of cyclic activity.

The work of Caianiello and colleagues is mathematically oriented and consists largely of drawing deductions from equation (13-1) as applied to various interconnection matrices a_{hk} under different conditions. Other studies of neural networks using the general approach of automata theory have been presented by Anninos. Anninos has developed digital computer simulation models incorporating the same basic assumptions as used in his early analytic work with Harth and has applied them to several basic problems of coding in neural networks. Anninos' assumptions are that a neuron will fire at a given time interval if the sum of the excitation minus the sum of the inhibition is greater than threshold, that there is instantaneous summation of inputs (there is no temporal summation: all time delays involved in connections between neurons are one time interval), and that there is an absolute refractory period that lasts exactly one time interval. Anninos incorporates a network interconnection scheme which utilizes "design in the large" but "randomness in the small." Thus, the basic connections in Anninos' computer model are represented by a matrix k_{ij} which contains the numbers of interconnected cells and the actual strength of connection associated with each particular connection. These specific values are chosen according to random distributions the overall characteristics of which are specified. For example, Anninos specifies that the total number of output connections from any given neuron will lie between two values, say, μ_0 and μ_n and that the actual number of connections for any given neuron is uniformly distributed between these two values. The actual number existing for a given particular neuron is then chosen to lie between these two values in the computer program with the aid of a random-number generator. Such a value once chosen for a particular cell remains fixed for the duration of the simulation, and similarly for other microscopic connection parameters. Anninos typically represents output in his simulation runs by a microscopic "state vector" $\mathbf{a}(\mathbf{N})$. This state vector has N elements, where N is the total number of cells in the network, each of which has the value 0 or 1 corresponding to whether that particular cell was active or quiescent at that time. The state vector thus prescribes exactly which of the network cells were active at a given time. Its development through time gives a complete description of the network's activities. The total number of cells which are active at any given point in time provides a convenient macroscopic measure of the network's activities.

In an early publication, Anninos was concerned with the properties of cyclic activity by such networks. Thus, he considered various particular networks subjected to randomly chosen initial states or randomly chosen instantaneous input patterns, or both, and observes the development of sustained activity as a function of various network and activation parameters. A typical example is shown in Figure 13-2. This figure shows the development of so-called cycling activity in an artificial nerve net after the occurrence of a transient phase. This particular cyclic mode exhibits five different states which recur indefinitely once initiated. Representative microstate descriptions for these simulations are shown in Figure 13-3. Anninos points out that there is ambiguity in proceeding from macrostate to microstate, as any of a large number of different microstates might correspond to the same total number of cells active at a given time. Anninos reports the simulations for a wide variety of network parameters and initial conditions which develop cyclic modes with numbers of constituent states ranging from 2 to about 50. He concludes that details of microscopic structure do not influence markedly the occurrence of cyclic modes nor their periods. These latter variables are influenced rather by the overall statistical parameters which can influence cyclic behavior dramatically. He also notes that the intensity of input strongly influences the periods of oscillation in these cyclic modes. When investigating the influence of afferent activity upon this behavior, Anninos assumed that the total number of active input pulses per unit time was constant but that the recipient cells were chosen randomly at each time from the total cell population in the network.

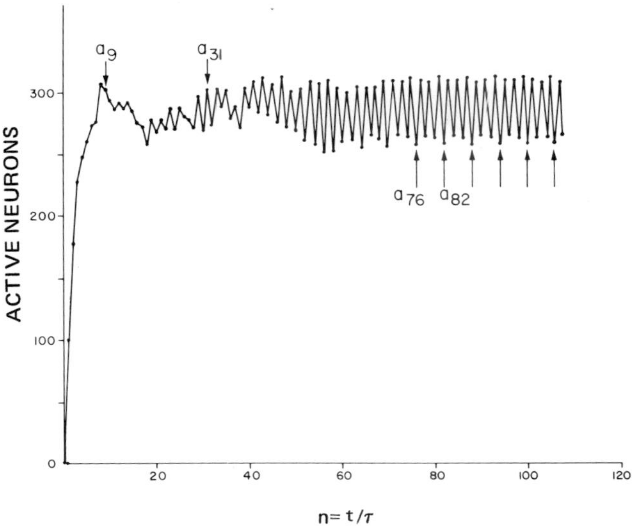

Figure 13-2. Cycling activity in an artificial nerve net. Activity of the net is plotted as a function of time ($n = t/\tau$). Cycling pattern is apparent beginning with $n = 76$. Net parameters: $A = 1000$, $\eta = 1$, $\mu^+ = \mu^- = 4$, and $h = 40\%$. Period of the cyclic mode is indicated by arrows between successive identical states a_{76}, a_{82}, a_{86}, . . . , as discussed in the text. States a_9 and a_{31} are two states with the same number of neurons firing which are not identical.

In another application of the same model, Anninos assumes that the algebraic sum of all PSPs for all cells in such a network might represent a reasonable approximation to the observed properties of the EEG. This model is used to produce such simulated EEG reords for comparison to actually observed ones.

Anninos also applies this model to explore variability in responses of ongoing spontaneous background activity. He notes that so-called evoked responses over some 20 or more individual responses are usually quite constant. Representative responses from Anninos' simulation scheme are shown in Figure 13-4. His main conclusion is that the variability in response is markedly different for two kinds of nets. For networks with high connectivity, there is much variation in individual responses. After a certain latency, cyclic activity appears. This cyclic activity is, in Anninos' words, the network's representation of the stimulus. However, the cyclic modes are considerably different from one presentation of the stimulus to the next and, moreover, are generally quite different from the cyclic modes that existed before presentation of the stimulus. Thus, such networks may be thought to encode the fact that a particular stimulus has been presented by a change in their cyclic modes, but there does not seem to be a clear

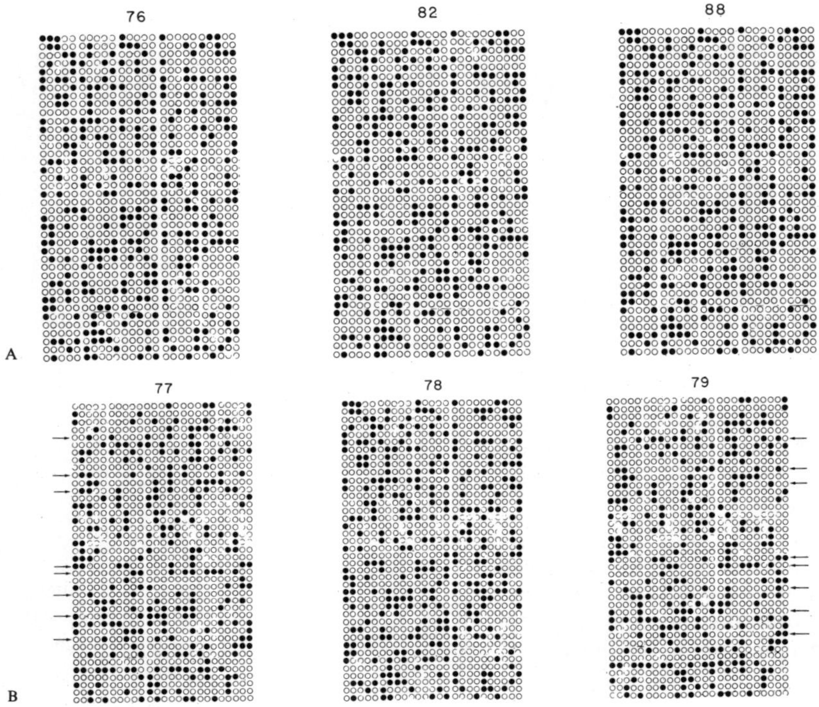

Figure 13-3. Firing pattern in artificial nerve nets. (a). Three state vectors at times $n = 76$, 82, and 88 from the net of Fig. 13.2. Closed and open circles indicate active and inactive neurons, respectively. As may be seen also in Fig. 13.2, those state vectors are identical. (b). Three successive state vectors at times $n = 77$, 78, and 79 for the same net as above. Although these states are already inside the cyclic mode and although states a_{77} and a_{79} appear quite similar, they nevertheless differ in eight specific points, as indicated by the arrows.

relationship between the cyclic modes, on the one hand, and the stimulus, on the other hand. Networks with low connectivity exhibit essentially identical responses to all presentations of stimulation.

In later work, presented up till now only in abstract form, Anninos has explored the relative character of spontaneous activity in artificial networks representative of the reticular formation and the cerebral cortex. Anninos simulated the reticular formation by assuming a Poisson distribution of connections; i.e., the probability of a connection between two cells i and j is exponentially distributed according to the absolute value of $i - j$. Such networks were triggered into sustained activity by the firing of a single constituent cell. Moreover, such sustained activity did not decay but rather remained indefinitely. To simulate the connectivity of the cerebral cortex, Anninos specified a normal distribution of connections; i.e., the probability of connection between two cells i and j was normally distributed according to the absolute value of $i - j$ with a certain mean value μ and standard deviation σ. For such networks, Anninos found bursting, sustained activity only if a large number of cells were activated simultaneously and, moreover, that such bursting activity invariably died out. In another application, Anninos tried to match certain behavioral features of Parkinson's disease with an application of his networks. The basic connection scheme is shown in Figure 13-5. The main cells in the model represent a pool of α-motoneurons. These neurons are excited by excitatory and inhibitory input from higher centers and by excitatory feedback from stretch receptors in the muscles activated by the alpha motoneurons. In actual patients with Parkinson's disease, it is suspected that there is increased muscle tone because of decreased inhibition. Moreover, in response to stretch of the muscle there is so-called cogwheel rigidity characterized by spastic give and take of muscle tone. Anninos reports that in his network simulations of this scheme there is higher-than-average activity in response to a stretch because of the assumed lower-than-normal inhibition. This over-normal response, however, is followed by an under-normal response at the next time unit because of the many neurons that are in a refractory stage. Thus, the main characteristics of his model with a lower level of inhibition is the occurrence of more pronounced oscillations in motoneuron response.

Large-scale digital computers can be appropriately programmed to represent the connections or activity of large neural networks in manners independent of the assumption of automata theory as generally followed by the authors whose works we have discussed thus far. One such use involves the organizing of vast amounts of information relating to a neuronal network prior to actually performing simulations. Suppose, for example, one is interested in the detailed behavior of a network which contains 10,000 neurons, each of which has approximately 100 synapses. In constructing structural or simulation models for such a network, it is not reasonable to specifically designate properties for all the constituent cells and million odd synapses. One can, however, partition the entire set of neurons into classes all of whose members have specific values of dynamic variables drawn from the same distribution and let the computer determine the specific values applicable to each individual constituent cell. If, for example, one assumes a particular neural system of interest to contain, say, five distinct classes of cells each of which can be anatomically and dynamically described by, say, 20 variables measuring mean values and standard deviations for thresholds, number of output connections, size of dendritic trees, etc., then one specifies 100 parameters to characterize these properties for all the cells of the system.

Such a modeling system can be designed to "grow" interconnected networks from a basis of detailed anatomical information by using some

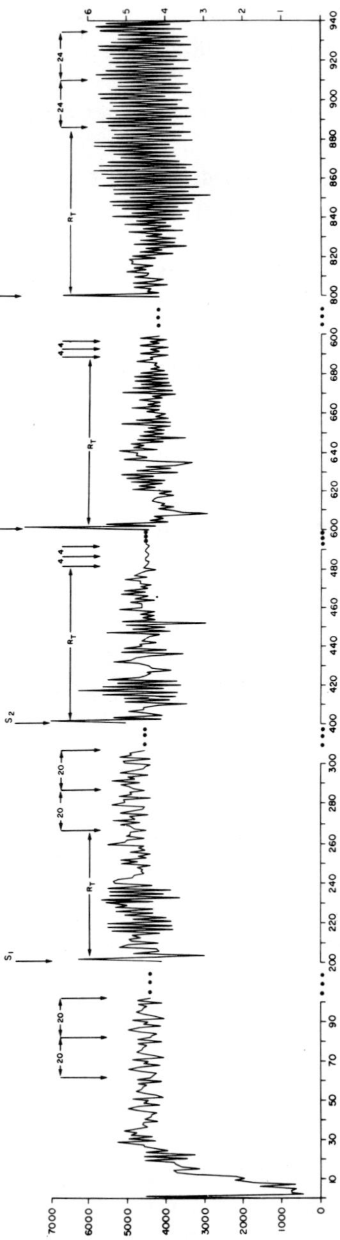

a

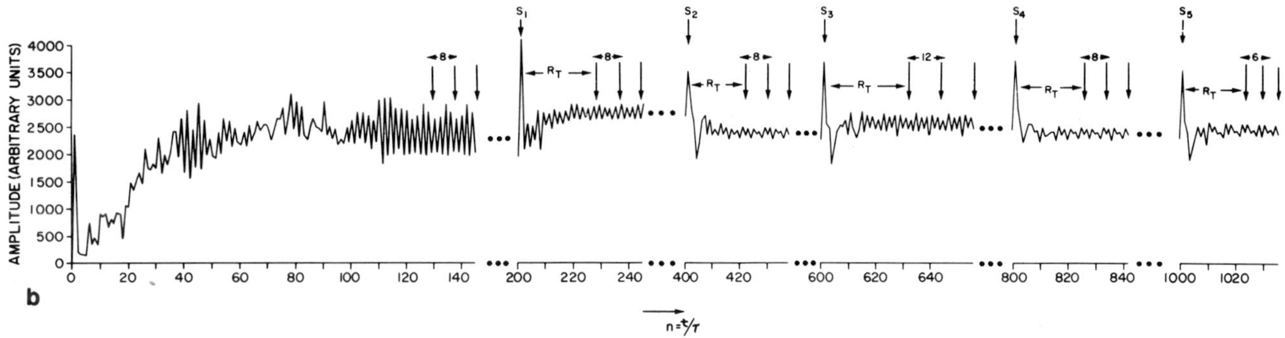

Figure 13-4. (a) Simulated evoked potentials from a neural-net model with parameters $\eta = 1$, $A = 1000$, $\mu^+ = \mu^- = 4$, $h = 40\%$, $\mu_0^+ = 4$, and $\sigma^+ = 0.5$. The abscissa represts the time in units of the synaptic delay τ, and the ordinate represents the amplitude of the potentials in arbitrary units. The S_1, S_2, S_3, and S_4 are the identical stimuli presented at times given at the corresponding arrows. The quantity R_τ represents the latency, i.e., the time during which the system is in a labile state prior to the cyclic activity. The numbers on the left and right sides of S_1 are the periods of the cyclic activity prior and after the presentation of the stimuli, respectively. Here μ_0^+ is the average number of branches from each afferent fiber impinging on the neural net. (b) Another example of simulated evoked potentials for a neural net with parameters $n = 1$, $\mu^+ = \mu^- = 3$, $A = 1000$, $h = 40\%$, $\mu_0^+ = 3$, and $\sigma^+ = 0.020$.

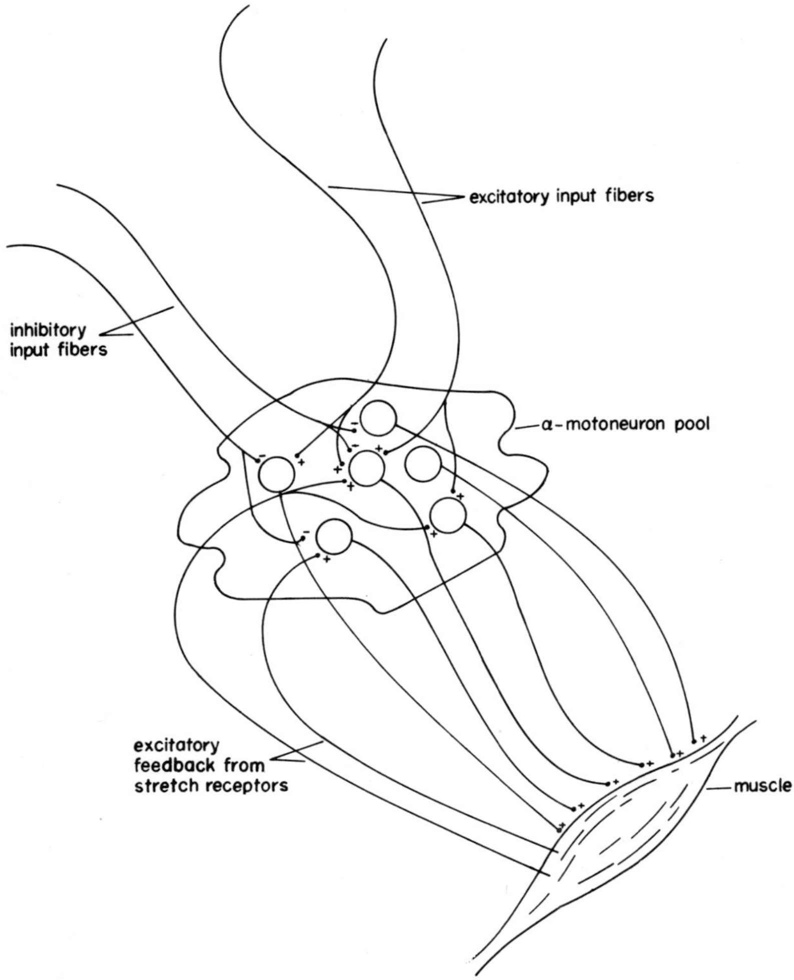

Figure 13-5. Model simulated by Anninos to investigate basic features of Parkinson's disease.

reasonable rule for the development of synapses between cells, such as, for example, assuming that a synapse occurs between cells A and B if the axon terminal of cell A is sufficiently close to the dendritic tree of cell B. Such cells are pictured as existing within some dimensioned space and precisely localized within the space by the computer according to appropriate overall parameters. With such a system, one can then explore in detail anatomical observations on single and small numbers of neurons in relation to their significance within the network as a whole. Such structure models can conceivably be of considerable value in assessing and studying the overall interconnection patterns associated with neuronal networks.

Moreover, one can also clearly include a wealth of information concerning electrophysiological properties for constituent cells. In principle, one can provide certain prescribed stimulation patterns and thereby produce simulated activity in such networks. Response activity in such networks can be stored, analyzed in any of a multitude of ways, and indeed compared to activity from the real networks the model is intended to represent. A difficulty with the latter effort is that very large networks require very large storage space, operating time, and associated costs. Two investigators, D. H. Perkel (unpublished) and Wittie (1974), have initiated such studies. Perkel has begun such a study on the interconnections of the networks of the cerebellar cortex using the wealth of detailed anatomical information available on that structure. Wittie has developed a program called BOSS (brain organization simulation system) which can, in principle, simulate up to 8,000 neurons and 80,000 synapses. These studies are potentially extremely valuable, but as far as we know neither has as yet provided significant information with respect to any real network.

It seems to us that, in general terms, computer-oriented approaches to problems of large neuronal networks are potentially very fruitful. Valid and promising tools exist. However, virtually no solid applications have been established, or general principles discovered, as of yet. The automata-theoretic approach, although promising (now as it has been for 30 years), has led to a large amount of irrelevant work and only a small number of studies relevant to real networks. Cultivation of these tools as applied to specific networks and to general canonical systems is needed in this area of research.

14

Models of Field Potentials, Slow Waves, and the EEG

By far, the larger portion of the book thus far has been concerned with modeling studies which can be related to electrical recordings attributable to single neurons. There exists, however, another class of recordings which measure continuous, graded and more slowly fluctuating waves of activity thought to represent the combined activity of large numbers of neurons or of large regions of neuronal tissue. Such recordings are usually made with macroelectrodes inbedded within nervous tissue, or on the surface of the skull or brain. Even though such recordings were first reported 45 years ago by Berger, contemporary understanding of their physiological basis is remarkedly undeveloped. It is generally assumed that these slow waves are somehow associated with the conglomerate of extracellular field potentials arising from generator potentials in the vast numbers of neurons and glial cells in the general regions recorded from. It is not clear if the characteristic signals of the EEG are determined by physiological properties of individual units, by characteristic patterns of interconnection within local neuronal networks, by interregional circuits, by glial cells, or by some combination of these. Moreover, it is not clear whether or not these fields in turn modulate and influence the activity of neighboring cells such that a truly adequate and comprehensive model would have to include interaction between two distinct but interrelated models of electrical signaling. Moreover, the relationship of unit activity or even multiple unit activity to characteristic features of the EEG has not been established for any significant situation. Quantitative models for any of these processes (which are as yet only understood in outline if at all qualitatively) are virtually nonexistent.

In this chapter we will describe models for describing extracellular field potentials resulting from membrane activity in single cells and then discuss several approaches to modeling the EEG.

14.1 Models of Field Potentials Resulting from Unit Activity

Neurons are embedded in and surrounded by an ionic solution (interstitial or intercellular fluid) which is electrically conducting. This means that the electrical signals which occur across cell membranes produce currents and electrical fields which propagate and decay through the fluid. These volume fields and currents can be measured, and in turn they probably influence activity of other neurons nonsynaptically.

A simple question which relates to the interpretation of wave forms of potentials recorded extracellularly with microelectrodes is: What is the wave form recorded at a site P some distance from a uniform circular cylinder which propagates a square-wave electrical pulse? This situation and the answer to the question are shown in Figure 14-1. A triphasic wave, positive–negative–positive, is recorded when the electrical pulse approaches the recording point from afar, passes at some minimum distance from it, and subsequently recedes into the distance in the opposite direction. A biphasic positive–negative wave is recorded when an impulse approaches the recording site but for one reason or another dies out before reaching it. A biphasic negative–positive wave is recorded when an impulse originates at some distance from the recording site and travels away from it.

These results can be explained qualitatively (and, as desired, quantitatively) by a simple model which attributes the recorded potentials to currents induced by electric dipoles moving through a volume conductor. A classical result of field theory in this situation is that the potential due to a dipole layer of constant moment is proportional to the solid angle subtended by the surface of the layer at the point. Moreover, the potential

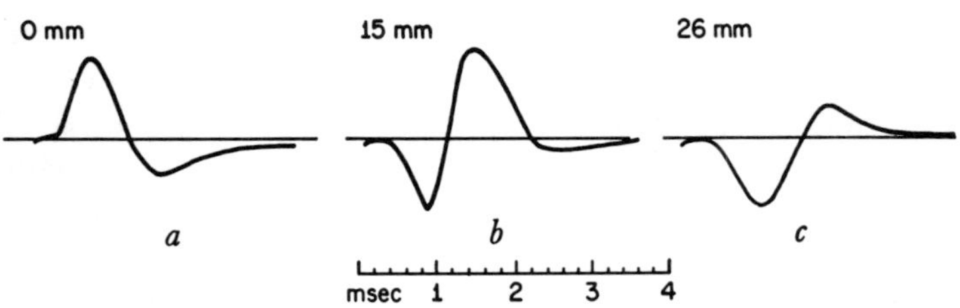

Figure 14-1. Nerve action potentials in a volume conductor. An excised bullfrog nerve is arranged so that it enters a volume conductor at $x = 0$ and exits from it at $x = 26$ mm. The ordinate is the potential of a close electrode with respect to a distant one (negative upward). The abscissa is the time. (a) Potential set up by a nerve volley when recording electrode is at level of nerve entry ($x = 0$) and distant electrode is 3 mm away from nerve ($y = 3$). (b) Potential recorded at $x = 15$ mm, $v = 3$ mm. (c) Potential at $x = 26$ mm, $y = 3$ mm. (From H. D. Patton, Special properties of nerve trunks and tracts, in: *Medical Physiology and Biophysics*, T. C. Ruch and J. F. Fulton, eds., 18th ed., W. B. Saunders, 1960.)

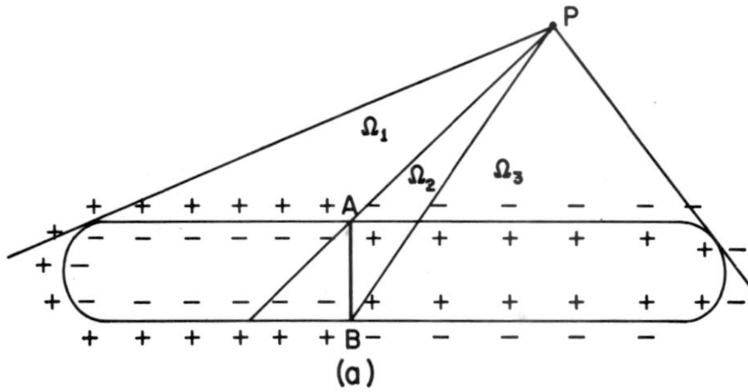

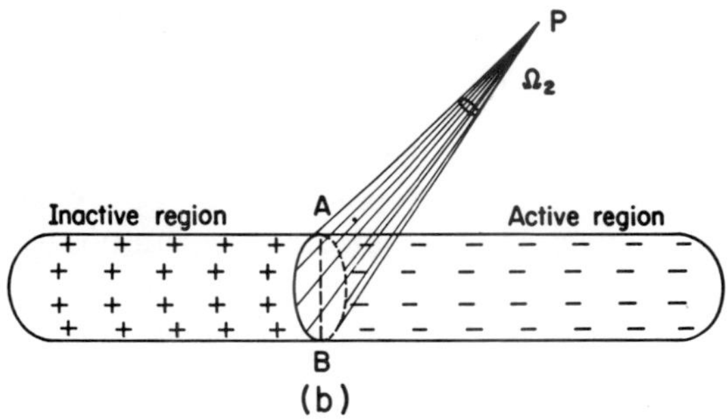

Figure 14-2. Potential at an external point P due to the rising phase of action potential in a cell at some instant. Line AB indicates where the transmembrane potential reverses; for simplicity the reversal is assumed to occur abruptly. (a) Axial section through a cylindrical cell showing cell membrane; diameter has been exaggerated. Total solid angle of cell is divided into three portions Ω_1, Ω_2, and Ω_3 by lines PA and PB. Potential at P due to solid angles Ω_1 and Ω_3 is zero, since the nearer and farther membranes contribute equal but opposite potentials. However, in Ω_2 the nearer membrane is active (outside negative) and contributes a potential at P of the same sign as the more distant, inactive membrane. (b) Diagram to show that under conditions in (a), the potential at P is the same as would be obtained if the membrane charges were placed on the cross section AB. The size of the potential at P at any instant is proportional to the apparent size of the cross section of the nerve at the wavefront. (From H. D. Patton, Special properties of nerve trunks and tracts, in: *Medical Physiology and Biophysics*, T. C. Ruch and J. F. Fulton, eds., 18th ed., W. B. Saunders, 1960.)

set up at a point in a volume conductor by a square-wave action potential is proportional to the product of the height of the transmembrane potential and the solid angle of the wave boundaries as measured at the recording site. The solid angle is a measure of the apparent size of the object as viewed from a particular point. Figure 14-2 shows how the potential arising from the rising phase of a square-wave potential can be calculated according to subtended solid angles. This diagram shows a longitudinal section of an excitable cell with a wave of depolarization near the center. The solid angles Ω_1 and Ω_3 contribute no net potential because the proximal and distal portions of the membrane contribute equal and opposite potentials. However, in the intermediate region determined by angle Ω_2 the proximal membrane is active and contributes a negative potential to P, while the distal membrane is inactive and also contributes a negative potential to P. The nerve impulse is a wave traveling at a constant speed, so Figure 14-2 represents the situation at one instant of time. As the wave moves from right to left, the solid angle first increases and then gradually decreases to zero. The sequence of potential changes recorded at P as the wave front passes can be estimated according to the diagrams shown in Figure 14-3. The triphasic wave (PNP) arises because the waves of depolarization and repolarization are sufficiently close together that both contribute significantly to the potential. To the left of P_1 the wave front dominates and the potential is positive. At P_1 the two solid angles are equal and opposite in sign such that the resulting potential is zero. At P_2 both boundaries contribute a negative potential; at P_3 the departing wave of repolarization dominates and the potential is again positive. The situations where action potentials only approach or only depart from the recording site can be interpreted as components of this total event.

The current status of the field-theoretic approach to determining extracellularly recorded wave potentials in nervous tissue is well represented by recent modeling studies presented by Nicholson (1973). Nicholson is concerned with describing the potential recorded at a single point in a volume conductor resulting from a given arbitrary distribution of current sources. The main assumption is that the extracellular fluid may be described as an anisotropic medium characterized by three main components; i.e., it is understood that vector and tensor components are measured relative to a coordinate system oriented with the principal axes of the stress tensor. It is understood that the current sources determining the field correspond to transmembrane currents associated with electrical events in constituent cells. It is assumed that these are not in turn influenced by the external field produced.. Under these conditions, Nicholson shows that Poisson's equation holds in the form

$$\sigma_x \frac{\partial^2 \phi}{\partial x^2} + \sigma_y \frac{\partial^2 \phi}{\partial y^2} + \sigma_z \frac{\partial^2 \phi}{\partial z^2} = -Im(x,y,z) \tag{14-1}$$

Nicholson's approach is to write Green's integral representation of the Poisson equation, given here as

$$\phi(\xi') = \frac{1}{4\pi} \int_U \frac{I(\xi)}{|\xi' - \xi|} d^3\xi \tag{14-2}$$

$$\xi = x/\sqrt{\sigma_x}, \qquad \eta = y/\sqrt{\sigma_y}, \qquad \zeta = z/\sqrt{\sigma_z}$$

The application of this methodology to particular cases is straightfor-

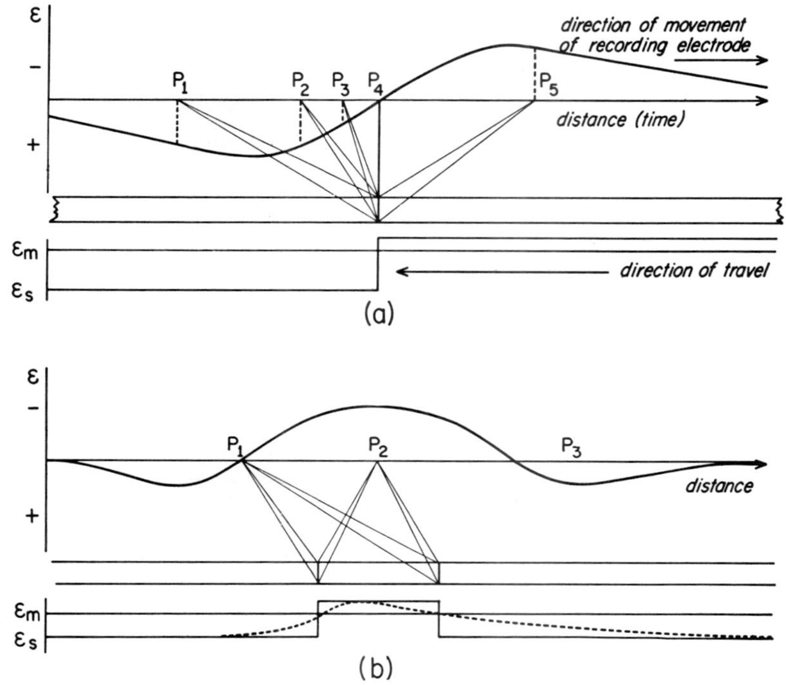

Figure 14-3. Potential generated by propagating wavefronts. (a) Diphasic potential due to a wavefront traveling from right to left at constant velocity. Time sequence of potential changes at a fixed point is shown along line parallel to nerve fiber. Sequence is constructed by supposing that wavefront is stationary and that a recording electrode is moved from left to right along a line parallel to nerve. When the recording electrode is at points P_1, P_2, and P_3, it "sees" the positive side of the wavefront. Dashed line at P has a height proportional to the solid angle of the wavefront from P_1 and thus proportional to the voltage recorded at P_1 As the recording electrode moves toward the right, the solid angle of the wavefront at first increases slowly owing to closer approach, and then decreases rapidly as the electrode sees the wavefront more and more on edge. At P_4 the potential is zero and immediately thereafter becomes negative. The plot is both of ϵ as a function of distance at a fixed time and of ϵ as a function of time at a fixed point. (b) Triphasic potential due to an idealized impulse. Solid angles of waves of both depolarization and repolarization must be added to obtain ϵ. Construction method same as in (a). P is point of zero potential; note that it is to the left of the wavefront. Triphasic (+, −, +) potential is typical of a propagated nerve impulse. Note that the maximum external potential is much smaller than the internal potential. An internal electrode at P_2 would see positive changes over 4π, while outside, even on the surface of the fiber, the solid angle is of the order of $4\pi/100$. (From H. D. Patton, Special properties of nerve trunks and tracts in: *Medical Physiology and Biophysics*, T. C. Ruch and L. F. Fulton, eds., 18th ed., W. B. Saunders, 1960.)

ward but can become quite complicated according to the particular geometry at hand. Nicholson has developed the equation system to model field potentials recorded in the cerebellum and compared the theoretical

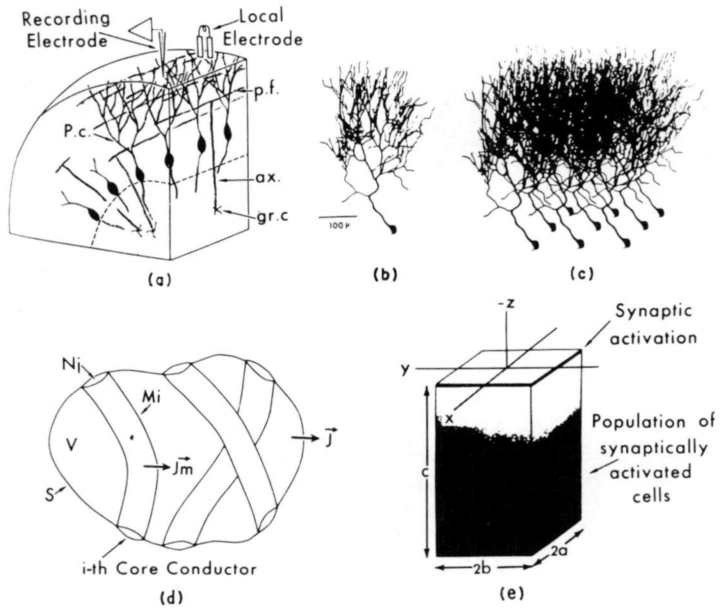

Figure 14-4. Neuronal ensembles and their representations. (a) Simplified diagram of the region of the cerebellum; Purkinje cell (Pc) dendrites lie in planes at right angles to the parallel fibers (pf), which are t-shaped axonal extensions of granule cell (gr. c) axons (ax). A local electrode can stimulate a superficial beam of parallel fibers, and field potentials resulting from the activation of pf–Pc synapses are measured by the micropipette (recording electrode) and amplifier. (b) Accurate drawing of a single Purkinje cell (alligator) showing extensive dendritic ramification. (c) Superimposition of nine Purkinje cells to show the density of dendrites, which approximate a continuous distribution of core conductors. (d) Derivation of the Poisson equation. The arbitrary region is bounded by surface S containing volume V, through which core conductors with surface Mi intersecting S in disk Nj are passing. Current J crosses S and Jm crosses Mi. (e) Representation of the neuronal population for the derivation of the weighting function. The synaptically excited population is confined to a rectangular volume $2a \cdot 2b \cdot c$: A synaptic input causes a distributed current sink in the xy plane at $z = 0$, while the rest of the volume becomes a distributed current source due to the vertical orientation of the core conductors (dendrites). (From Nicholson, 1973.)

predictions with experimental observations. The situation is shown in Figure 14-4. The cerebellum is treated as a two-dimensional region wherein current volume sources are supposed to represent activity from parallel fibers which emerge perpendicularly to the plane. The current source densities in the parallel fibers are assumed to depend only on the value of z. Under these conditions equation (14-2) reduces to the following:

$$\phi(\xi') = \frac{1}{4\pi}\int_0^{\gamma} I(\zeta)\,d\zeta \int_{-\beta}^{\beta}\int_{-\alpha}^{\alpha} \frac{d\eta\, d\xi}{[(\zeta' - \zeta)^2 + (\eta' - \eta)^2 + (\xi' - \xi)^2]^{1/2}} \tag{14-3}$$

Appropriate weighting functions are derived, and predicted fields and currents for various applied current source densities are worked out. Agreement of theory with experimental observations is reasonably good.

14.2. Models of the EEG

One can envisage four levels of sophistication in methodologies for modeling various EEG waves. It seems to us that our eventual understanding of the basis of EEG signals will include considerable development at all four of these levels, whereas so far only the first two levels have been seriously addressed with detailed information available only at the first. The first of these levels is the level of frequency decomposition. This is the most primitive level; quantitative or analytical work here are better characterized as descriptions rather than models. The aim of this work is simply to characterize the nature of the signal or interrelate components of the signal. Conventional spectral analysis wherein an arbitrary analog wave form is decomposed into its various frequency components is the general tool appropriate here and of course has been used in laboratories around the world to characterize EEG signals for decades.

Interesting quantitative work in this area has been presented recently by Nunez (1947*a*,*b*), who is concerned with describing EEG waves with the language of classical field theory for physical waves. Nunez is particularly concerned with the existence and significance of a possible dispersion relationship for brain waves which relates the frequency of a wave to its spatial wave length. Nunez claims to have found experimental evidence for the existence of dispersion relationships for the alpha rhythm and suggests that such dispersion relationships might be quantitative measures of the spatiotemporal synchrony of underlying neural generators. Mathematical analysis of such considerations for brain tissue is more complicated than for conventional homogeneous media in that the existence of relatively long point-to-point connecting fibers requires integral-equation representation rather than the more conventional and easier to work with partial differential equations of classical field theory.

A second level of approach attempts to relate salient electrical features of the EEG to corresponding peculiarities in the electrical activity of single cells. The goal here is not to establish detailed precise quantitative relationships between unit activity and observed field potentials but rather, by using experimentation, simulation, or theorizing, to find constellations of activity in neurons or neuron groups which correlate with significant features of the recorded EEG. Several studies have attempted to relate salient EEG properties to signals in isolated single units. The most salient general theorizing of this type is presented by Adey (1970) with considerable experimental help from Elul (1972). Elul recorded spontaneous EEG-like signals from two separate microelectrodes 30 μm apart and also saw large rhythmic waves about 5 to 15 mV in amplitude very similar to the EEG in cortical neurons. From this, Elul and Adey concluded that the germinal source of the EEG must be as small as single cells and could very well reside in single neurons. Adey suggests that characteristic EEG rhythms such as the alpha rhythm could be mediated by populations of independent neurons all of which exhibit oscillating electrical signals at frequencies determined by the characteristics of the cells. If the cells are all very similar, the characteristic rhythms could be assumed to be drawn from a single distribution with a given mean value. An EEG could be constructed as simply the sum of the signals generated from this constellation of cells. Adey argues that summed EEG signals would have the same characteristic frequency as the frequency of the individual components.

Fujita and Sato (1964) also recorded slow rhythmic waves in the generator potentials of pyramidal cells of the hippocampus which appeared to coincide with the theta rhythm characteristic of that structure. Petsch *et al.* (1962) on the other hand, found pacemaker cells in the medial septum which fire bursts of spikes associated with each wave of the theta rhythm and established that the occurrence of the theta rhythm in the hippocampus was dependent upon activation of these cells. In earlier work these investigators have suggested that the processes of inactivation in hippocampal pyramidal cells (wherein a depolarizing wave follows activation and brings about a complicated burst of spikes, with subsequent absence of firing for tens to hundreds of milliseconds) could be responsible for the theta rhythm in the hippocampus. This is a case where a specific physiological mechanism of a single cell type has been suggested to be the causal factor in an EEG rhythm. The suggestion, though intriguing, is incomplete as an adequate model for the theta rhythm.

An alternative orientation within the same general level of approach looks for features of activity which are characteristic of networks of large numbers of interconnected neurons which might relate to observed characteristics of the EEG. We have already seen several examples in each of our chapters on models of neural networks. For example, the studies of Freeman on neuron pools entail measurements of pool activity which purport to measure the combined activity of aggregates of cells simultaneously. Such a measure does not consider details of propagation of signals through volume conductors and consequent distortion of shape but nonetheless should conceivably bear some relationship to observed recordings. Freeman's work is discussed in Chapter 12 and the interested reader is advised to refer to Freeman's original publications for comparison of his modeling predictions to experimental data obtained from the olfactory bulb. Also, the field equations of Beurle, Griffith, and Cowan described in Chapter 12 can, at least approximately, relate to EEG signals. Finally, as we saw in Chapter 13, Anninos produced some generator potentials from computer simulation for comparison with EEG signals. Andersen and Andersson (1968) have recently proposed that the alpha rhythm of the EEG reflects the characteristics of a thalamocortical circuit and particularly the properties of recurrent inhibition within the thalamus. This model is discussed in Chapter 15 and is probably the most salient example of a model which attempts to relate EEG characteristics to specific physiological properties of a neuronal circuit.

The third level in modeling EEGs would consist of using an extended field-theoretic approach along the line suggested by Nicholson. In such an approach one would assume a given current distribution and appropriate geometry, and take into account the spatial and temporal distortion produced on the assumed distributions to predict potential wave forms observed at some given recording point. Difficulties would be that the approach is extremely complicated in terms of adequately representing and simplifying the geometry. Moreover, one would not know precisely what current distribution to use, so that reasonable although perhaps initially arbitrary assumptions would have to be made. This level of approach would depend on relatively reliable information gained from the second level of analysis described above. The limitations of this level is that it would not take into account the potential effect of such generated fields on the assumed current sources themselves.

The fourth level of approach would assume a considerable degree of development in each of the preceding three levels and would be the level wherein the possible coupling of field activity to cellular activity would be qualitatively and quantitatively represented resulting in a model system with two interacting and perhaps coequal components, cellular signals and field signals.

15

Models of Specific Neural Networks

This chapter describes various selected models for certain prominent networks of the vertebrate brain. We have not attempted to provide an exhaustive or comprehensive coverage of vertebrate networks or even of models for those networks considered. However, the collection of models contained here can be considered representative of the state of the art in this area generally.

15.1. Cerebral Cortex

The cerebral cortex has excited interest among many students of the brain as being more than any other part the "organ of mind." It seems to be that portion of the brain wherein the internal representation of objects and concepts occurs, and where thoughts represented in words are effected. The cortex is more pronounced in primates than in other mammals and more pronounced in man than in other primates. Moreover, within man the cortex contains the vast majority of the nerve cells of the nervous system—some 90%. As yet we do not have any but the most speculative knowledge about how the cortex works and virtually no sound ideas relating function to structure. Models of the cortex have been generally based not on its structure, but rather on what one wants the cortex to do.

A salient early model of this variety was presented by Pitts and McCulloch (1947) in a paper entitled "How We Know Universals. The Perception of Auditory and Visual Forms." In this paper Pitts and McCulloch make suggestions as to how the cerebral cortex might abstract generalities from given patterns of activity. They suggest that numerous nets serve to classify information according to common characteristics. In vision, for example, the equivalence of apparitions related by similarity and congruency like those of a single thing seen from various places is strongly registered. Their basis for analyzing the encoding of such equivalence classes is based on the idea of a group of transformations such that any member distribution which is a part of the generality generates all the other member distributions. Thus, except for order, the set of transformations applied on member distribution a_i will result in the same set of new distributions as results from the transformations operating on member distribution a_j, where a_i and a_j are any member distributions of the generality. They suggest that nervous networks might perform such abstractions of universals with four main component parts: first, a basic display network in which to display the original given distribution of activity ψ. Second, $N - 1$ display nets to display all the transforms of the original distribution ψ, including appropriate interconnections from the original network to effect these transformations. Third, N identical functional computational networks which would assign a unique number to the entire pattern of distribution in each of the basic display nets. Fourth, a master storage/display manifold which would receive and sum the output from all the functional computational networks. The master storage/display manifold, then, signals the appropriate universal involved in the original distribution according to the number that it displays. Different universals would be represented with different numbers in the master storage/display manifold.

It is clear that in this arrangement the net result in the master storage/display depends only on the set of distributions and not on their order. Thus, any member of the appropriate generality, by definition of the transformation relations, generates the same set of distributions in the display manifolds, albeit in different order, and therefore the same number in the master storage/display manifold. Moreover, different generalities lead to different numbers in the master display manifold.

Pitts and McCulloch give three examples where such abstractions may occur in the human nervous system. Here we will discuss one, relating to the perception of a given musical chord regardless of pitch in the auditory cortex. Pitts and McCulloch suppose that the auditory mechanism which recognizes chord and timbre independent of pitch is situated in Heschl's gyrus on the superior surface of the temporal lobe, and that position on the cortex within this gyrus corresponds to the pitch of tones. In this arrangement octaves span equal cortical distances. In this case, the fundamental display network is a one-dimensional strip, and ψ, the original distribution, is a function of the single variable x which measures position along it. The group of transforms is the group of uniform translations of $\psi(x, t)$ of excitation along the strip. This group of transformations is determined by adding various constants to the coordinate x. For example, the dominant

7th chord, played in, for example, the key of C, would use positions of $x =$ 1, 5, 8, 11, and 13 (corresponding to C, E, G, B^{b}, and C), whereas the dominant 7th, in, for example, the key of D, would correspond to $x = 3$, 7, 10, 13, 15 (D, $F^{\#}$, A, C, and D). These chords are in the same equivalence class in that either one is contained in the set of distributions generated by uniform translations of the other. Pitts and McCulloch note that this particular transformation might be effected by simply slanting the afferent fibers in parallel as indicated in Figure 15-1.

They note that one problem in this arrangement is the necessity of having the N-fold repetition of the functional computational circuit. In one modification of this arrangement, one such net could do the work provided it operated on the various distributions sequentially. To effect this, they suppose that the N display manifolds are connected as before, but that their thresholds are raised so that their specific afferents alone are no longer able to excite them. In this case there are "adjuvant" fibers to ramify throughout each one so that when active they remedy the deficiency in summation and permit the display manifold to respond to their afferent stimulation as before. They let all the neurons with the same coordinate x on the N different manifolds send axons to the similarly located neuron on another recipient sheet exactly like the others and suppose that any one of these input neurons can excite this neuron. They note that, under these conditions, if the adjuvant neurons are excited in a regular cycle so that every one of the manifolds in turn, and only one at a time, receives the increment of summation it requires for activity, then all of the transforms will be displayed successively on the receiving sheet. Under these conditions a single functional computational network will now suffice to produce all the values in turn as the time scanning presents all the distributions in the course of the cycle. Thus, in this arrangement, each of the $N - 1$ manifolds is "read out" in turn. They suggest that the alpha rhythm might relate to this temporal scanning of the cortex.

There is no more reason to believe that this model represents the way in which our brains extract universalities now than there was when it was presented 30 years ago. On the other hand, the model does focus attention on the important characteristic of feature extraction in cortical function and how it might work.

Another collection of conceptual and computational modeling efforts is concerned with the overall organization of cerebral networks and how certain subsystems of functionally interrelated neurons might evolve. The main theoreticians responsible for this point of view are Hebb (1949) and Lashley (1960). Lashley proposes a theory for overall organization of cerebral tissue based on his so-called "trace systems." The theory is based mainly on inferences from psychological observations and on many years of research in physiological psychology. The theory is described very well in Lashley's own words (Lashley, 1960) as follows:

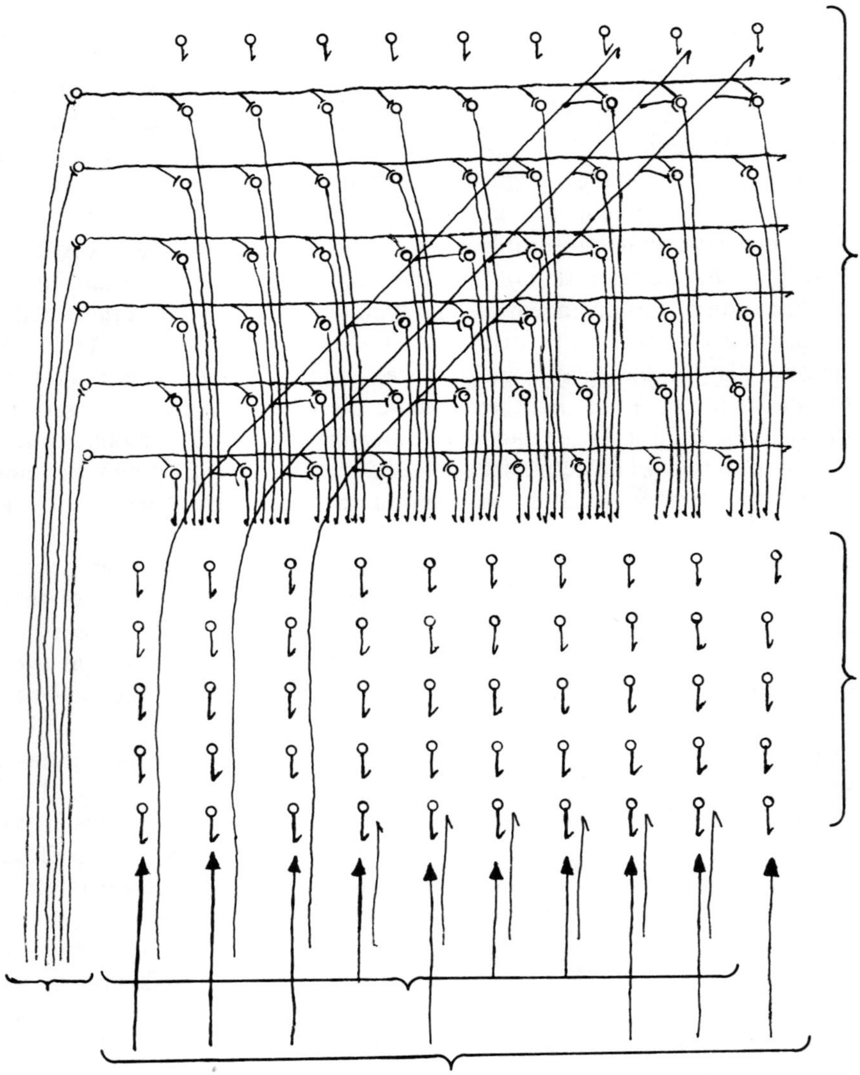

Figure 15-1. Impulses of some chord enter slantwise along the specific afferents, marked by pluses, and ascend until they reach the level $\mathfrak{M}_a$ in the columns of the receptive layer activated at the moment by the nonspecific afferents. These provide summation adequate to permit the impulses to enter that level but no other. From there the impulses descend along columns to the depth. The level in the column, facilitated by the nonspecific afferents, moves repetitively up and down, so that the excitement delivered to the depths moves uniformly back and forth as if the sounds moved up and down together in pitch, preserving intervals. In the deep columns various combinations are made of the excitation and are averaged during a cycle of scansion to produce results depending only on the chord. (From Pitts and McCulloch, 1947.)

1. The billions of neurons in the cerebral network are organized into a large number of systems. Each system consists of the traces of a number of habits or memories. Knowledge of the moves and games of chess would constitute one such system; memories of neural anatomy, another; and so on through all of the individual's varied interests. The traces or engramata in any system are more closely connected with one another than with other systems. The systems are not anatomically separate, and the same neurons in different permutations may participate in many systems. For brevity I shall call these "trace systems."

2. Such a trace system may be thrown into a state of tonic activity by an external stimulus which activates one set of traces within it. In the tonic state the traces of the system are readily excitable and available to recall. Other systems are in abeyance. Thus, when one plays chess, the Evans gambit or Philodor's defense may be readily recalled. But if the player is interrupted by the question, "Who won the pennant last year?," he will take some time to "collect his thoughts"; that is, to organize the baseball trace system. Such a general tonic activation seems a plausible explanation of concentration upon a particular subject, limiting associations to that subject. Questions of preparatory adjustment, organization of purposive activities, and the like can be formulated in such terms. An activated trace system would limit associations of the flow of ideas to the topics included in that group of traces. As I have stressed elsewhere (10), grammatical structure and other ordered activities imply some sorting or arranging mechanism, active before the words reach overt speech or silent thought. The relations in thought structure are antecedent to consciousness. The tonic background might provide the basis for this preorganization. This leads to the problem of the neurology of logic, of which a bit more, later.

3. A system in tonic activity dominates the brain field, limiting the organization of other systems. It is relatively impervious to unrelated excitations. An intense stimulus, or an emotionally charged one, such as the sound of one's name, may break in, but the great mass of afferent excitations is excluded. This blocking might be either an active inhibition or the preemption of neurons which might otherwise be included in the blocked system. The phenomena of attention demand some such hypothesis.

4. The neurons in a trace system, under tonic activation, exert some mutual facilitation. The tonic state of the whole system is thus built up and maintained. This was the only sort of explanation that I could find for the reduction in efficiency of behavior in proportion to the extent of brain damage. The asumption is consistent with Woodworth's notion (19) that long practiced habits may be transformed into drives; that activities which have acquired many associations may become obsessive.

5. The level of tonus in the partially activated system may vary. Circuits which have just been fully activated may retain a high level of subthreshold activity and thus contribute to the temporal organization evident in the memory span.

6. Fixation in memory is generally possible only when the remembered material forms part of such a dominant system. Learning has been classified in two types; mechanical, produced by repetition; and rational, which is not dependent on repetition. We remember the content of a book, not in the author's words but in meanings which fit into previous knowledge of the subject. During the reading the meanings are not necessarily formulated clearly in verbal or other thought forms, but they may be so formulated later. That is, associations may be formed during reading with traces in the system which are not activated above tonic levels during learning.

This assumption would provide an explanation of some types of confusion in which the patient may show disconnected activities or disordered thinking, with no later memory of what occurred during the state.

Such a system of low-level tonic excitation in a system of memory traces would provide a basis for many of the characteristics of mental organization. The circuits of the trace system which are actively firing at levels sufficient to excite other traces would constitute the content of experience, limited and changing from moment to moment. The background of tonic activity would determine the direction of attention and of the flow of thought, restricting it to related associations. It would provide the binding force that holds together the temporal sequences through memory span and more permanent associations. Reduction in the tonic level of the system, as in sleep under anesthetics, or violent invasion, as in a convulsive seizure, would destroy the organization necessary for memory and the continuity of the conscious state. Partial incomplete disorganization in sleep would permit interaction of different partial systems with the bizarre contaminations that occur in dreams.

These assumptions concerning cerebral organization are, of course, purely speculative and mainly inferences from psychological events. There is no present direct evidence from physiology in support of them. However, they are not inconsistent with what is now known of the physiology of the brain, and I believe that some such mechanism is implied by our present knowledge of the structure and activities of the cerebral cortex. . . .

It is probable that every neuron in the cerebral cortex, and indeed in the whole nervous system, is subject to a continuous bombardment by nerve impulses. This follows from consideration of cell number, of frequency of sensory input, and of the continuity of the cerebral network. The observations of Gray Walter on synchronous firing of many areas of the cortex with driven alpha rhythms show how widespread the excitations may be.

Not all and perhaps not even a large proportion of the neurons are fired by the bombardment. It requires at least two and perhaps many more impulses from end buds to fire the cell. But partial depolarization with increase in excitability may result from subthreshold volleys and, because of decremental conduction in the cell membrane, may be selective for special connections. When one set of circuits is actively fired, it should be expected that some neurons in associative connection with it would receive too few impulses to fire them, but sufficient to prime them for response to impulses from other circuits. Sherrington's studies of overlap in motor pools revealed a condition of increased excitatory state which could provide a mechanism for the tonic state of associated neurons at higher levels. This would result in a sort of priming of an entire associative or trace system.

I believe that what is known of the learning process justifies these assumptions concerning the role of the tonic background in the formation of associations. The importance of repetition in learning has been greatly exaggerated. Thorndike, who was the most eminent American student of learning and for many years an advocate of the theory of learning by repetition, finally denied its importance. He substituted for repetition a somewhat vague concept of "belongingness" which implies, so far as I can interpret it, some unconscious relational structure or unconscious associations.

An organization of associated neurons excited at subthreshold levels seems not only a reasonable but an almost necessary consequence of the structure and known physiological properties of the cerebral cortex. As I have indicated, the interplay of such systems would provide a mechanism for a large proportion of the selective activities of mind.

Lashley's model is a generalized and embellished statement of a theory suggested a decade earlier by Hebb in a classic work, "The Organization of Behavior." In this work Hebb suggests that the cortex has two main kinds of organization: first, an intrinsic hypersynchrony and, second, the activity of so-called "cell assemblies." The cell assembly, according to Hebb, is a collection of cells the activity of which is coordinated and which grows in coordination by repeated stimulation. It is an elemental internal representation of an idea or a concept. The cell assembly generally becomes part of a collection of assemblies which mutually facilitate or inhibit each other. This organization is called by Hebb "the phase sequence." The perception of any pattern involves such a sequence. Such patterns also correspond to attention, habits, and the like. In this system, as in Lashley's, the collection of cell assemblies is diffuse, containing a large number of widely separated cells which are related functionally. Different cell assemblies are supposed to interpenetrate anatomically.

In his monograph Hebb elaborates how such cell assemblies and phase sequences might account for various elemental psychological phenomenon. However, the main spark of interest among modelers was generated by Hebb's suggestion as to how such cell assemblies might come into being. Thus, Hebb suggested that such cell assemblies evolved slowly through experience as the strengths of individual synapses between cells increased with use. Thus, the main neurophysiological postulate underlying Hebb's position is "when an axon of cell A is near enough to excite a cell B and repeatedly or persistently takes part in firing it, some growth process or metabolic change takes place in one or both cells such that A's efficiency as one of the cells firing B is increased." As we have discussed in Chapter 11, several investigators in the 1950s and 1960s undertook digital-computer simulation of numbers of connected neurons, using the rule that the strength of interconnections increases with use, to see whether or not they might evolve such organized cell assemblies from originally randomly interconnected networks.

A very imaginative model of cortical organization has been presented by Marr (1969, 1970). Marr's model, as are many, is based on proposals concerning the function of the cortex, but Marr goes beyond other such modelers in working out specific predictions which relate his concepts to the structure of cortical circuitry. Marr's model is presented on three levels: a general conceptual level, probabilistic mathematical statements, and statements concerning specific neuronal networks. Marr's concern, very much like that of Pitts and McCulloch, is with generalizations based on experience. Thus, Marr proposes that the basic operations of the cortex involve the use of experience to form "classificatory units" with which to interpret subsequent experience. He suggests that such units are formed whenever either something occurs frequently in the brain's experience, or enough redundancy appears in the form of clusters of slightly differing inputs. Marr arrives at this suggestion on the basis of his so-called fundamental hypothesis: "There are instances where a particular collection of intrinsic properties (i.e., properties already diagnosed from sensory information") tends to be grouped such that if some are present, most are, then other useful properties are likely to exist which generalize over such instances. Further, properties are often grouped in this way." Marr, then, is mainly concerned with the mechanics of how such classificatory units might function in terms of elementary neurophysiological rules and how such units might relate to the structure of the cerebral cortex.

Marr notes that if such classificatory units do indeed represent the main function of the cerebral cortex in human beings, then there are two main characteristic problems which they handle readily. He labels these the diagnosis problem and the interpretation problem. Consider first the diagnosis problem. Suppose there is an established class unit Ω in the brain which has been formed through previous exposure to many somehow similar events E_i. We suppose now that some new event E is presented to the brain which may be presented only one time and which may be similar but nonetheless is different from the other E_i and may indeed be unique. The diagnostic problem then is: How is the probability of the event E belonging in the class Ω determined? The second and related interpretation problem concerns the question of how an event E can be placed in a class Ω when the event contains only partial information concerning all the possible dimensions of the class Ω. Marr proceeds to work out mechanics whereby his classificatory units might solve these problems readily. To help solve these problems, Marr imagines a so-called diagnostic space wherein input events E may be placed. A set of previous E_i's, which are members of a certain established class Ω, may be localized as occupying a certain locus of points within that space. A given new event E may also be placed somewhere in the space. The likelihood then that the new event E belongs in the class Ω may be approximated on the basis of its proximity to the locus of previous events E_i which did indeed belong to the class ω. Marr (1970) then suggests, in a crucial link in his theorizing, that this mode of operation has the following implications for cortical structure:

> P1. Evidence is stored in the form of conditional probabilities at modifiable synapses between "evidence function" cells and a so-called "output cell" for Ω (eventually identified with a cortical pyramidal cell).
>
> P2. Evidence is consulted by applying an input event E which causes evidence cells relevant to E to fire. The output cell then has active afferent synapses only from the relevant evidence cells.
>
> P3. The weighting function comes about because nearby events will use overlapping evidence cells, just as very similar mossy fiber inputs are translated into firing in overlapping collections of cerebellar granule cells. The exact size of subset detector cells used for collecting evidence depends upon the category of Ω: recognition of speech may, for example, require a generally higher subset size than the 4 or 5 used in the cerebellar cortex.

The interpretation problem is concerned primarily with instances where a decision has to be made on the basis of incomplete information. Marr uses the example of a recognition system set up to recognize a particular face which is presented in a pencil sketch of that face which can be recognized even though much information is missing. Marr attacks this problem again by imagining a diagnostic space and, moreover, supposing that such a diagnostic space is defined by N fibers. He pictures the event supplying incomplete information E as providing signals on only the first R of these fibers. One then imagines a trace through the diagnostic space occupying the given values for the first R fibers and having any acceptable values assigned to the last $N - R$ fibers. Such a set of traces Marr calls "completions" of the original event E. To help effect possible modes of resolutions of these two main problems, Marr (1970) presents the following two main theorems:

Diagnosis Theorem. Let Ω be a binary-valued random variable, and let $p_1, \ldots, p_k$ be independent estimates of the probability p that $\Omega = 1$. Then the maximum-likelihood estimate for p is

$$p_0 = (1/k) \sum_i pi \qquad (M\text{-}1)$$

Interpretation Theorem. Let X be a subevent of X with completions $E_1, \ldots, E_k$. Then

$$P(\Omega|X) = \sum_{i=1}^{K} P(\Omega|E_i)P(E_i|X) \qquad (M\text{-}2)$$

and is unique.

These theorems then provide specific guidance for Marr as he constructs his hypothetical neural models to perform these operations.

Marr pictures the cerebral cortex as containing a set of output cells, each of which corresponds to a given classificatory unit. Thus, if a given input event is presented to the brain, all those classificatory units to which it belongs are thrown into action such that the resulting distribution of activity over these cells signals the set of classes to which it belongs. The basic model for cortical circuitry which Marr deduces from this theorizing is shown in Figure 15-2. In this figure the cell labeled Ω is the main output cell whose firing defines whether or not the event is in that class. The cells c_i correspond to the dimensions of the diagnostic space. Thus, each one of the c_i provide weighted information corresponding to whether or not the event E is in the class Ω. The rate of activity in the c_i cells is monotonically related to the probability p_i given in the diagnosis theorem [equation (M-1)]. The synapses from the c_i cells to the Ω cells are modifiable, so that the entire classificatory unit is built up according to the experience of the activity in the unit.* These evidence cells, the c_i, then are excitatory and their contributions to the Ω cells are summed. The strength of the synapse from

*These synapses may be what Marr calls Hebb synapses or Brindley synapses. The Hebb synapse is facilitated by simultaneous pre- and postsynaptic activation. The Brindley synapses consist of a Hebb modifiable synapse together with another unmodifiable excitatory component.

Figure 15-2. The basic neural model for diagnosis and interpretation. The evidence cells $c_1, \ldots, c_7$ are codon cells with Brindley afferent synapses. The *G*-cell controls the codon cell threshold: it uses negative feedback through its ascending dendrite to keep the number of codon cells active roughly constant. Its descending dendrite samples the input fibers directly, thus providing a fast pathway through which an initial estimate is made.

each c_i cell to each Ω cell corresponds to the probability that an event is in Ω given that it is in c_i. The *S* cells are inhibitory cells spread over the dendritic tree to perform, according to Marr, subtraction, while the *D* cells are inhibitory cells acting on the soma which perform, according to Marr, division. The significance of the latter two classes of cells is discussed in Marr's paper.

The total amount of activity arriving from all the c_i cells can accurately reflect the probability that *E* is in Ω only when the system is balanced to allow for events with incomplete information. A problem is encountered in this system when there is incomplete information and only a subset of the evidence cells c_i are activated. To get around this problem, Marr modifies the set of evidence cells c_i such that the total number of cells active is constant regardless of the degree of completeness of the activating event. He effects this by the inhibitory feedback cell *G* which performs precisely this function. The resulting system is called by Marr a "Condon" system.

Marr then goes on to interpret the structure of the cerebral cortex in terms of this model. These predictions are given by Marr (1970) as follows:

> Three stars*** indicates a prediction which, if shown to be false, would disprove the theory; two stars** indicates that a prediction is strongly suggested, but that remnants of the theory would survive its disproof; one star* indicates that a prediction

is clear, but that its disproof would not be a serious embarrassment, since other factors may be involved; and no stars indicates a prediction which is strictly outside the range of the theory, but about which the theory provides a strong hint.

Each Martinotti cell should have many inputs,*** mainly from intercortical association fibers,** which should terminate by means of excitatory synapses.*** Each should also have inhibitory inputs,*** subtractive in effect** and therefore widely distributed over the dendrites.** These should be driven by local cells*** with locally arborizing axons,*** designed to keep the amount of Martinotti cell activity evoked by different inputs roughly constant.**

Excitatory Martinotti cell afferent synapses are probably modifiable,* and if they are modifiable, they are probably Brindley synapses,* becoming modifiable only during the correct phases of sleep.* If location selection proceeds as in §5.1.1, and if these synapses are modifiable, then they are modifiable only during the correct phases of sleep.*** Martinotti cell dendrites are probably independent.

The output from these cells is excitatory,*** and goes to output (pyramidal) cells*** through modifiable synapses,*** to three** kinds of inhibitory cells*** through unmodifiable synapses,*** and to output selector cells** through unmodifiable synapses.

Cerebral granule cells fall broadly into the same class as Martinotti cells, and the predictions concerning them are the same, with the following exceptions. Their input is mainly more direct than that of the Martinotti cells, and should (because of their smaller size) come from thalamocortical rather than corticocortical projections. They probably do not have modifiable afferent synapses. In the sensory projection areas, where afferents are known to terminate in layer IV, these afferents should form the main source of excitatory synapses on the granule cells.*

The pyramidal cells of layers III and V, and probably also those of layer II, are interpreted as output cells, in the sense of the theory. On the assumption that this is correct, they receive two kinds of excitatory synapses,** and two kinds of inhibitory synapses.** The majority of afferent synapses comes from Martinotti and granule cells,** almost all such cells making not more than one synapse with any given pyramidal cell.** These synapses are either Hebb or Brindley type modifiable synapses.*** The strength of the synapse from the codon cell c to the output cell Ω stabilizes at the value $P(\Omega|c)$.** (This receives only two stars, since there may be a workable all-or-none approximation to this value.) These synapses should be modifiable during the course of ordinary waking life,*** and probably during sleep as well.* All other afferent synapses described here are unmodifiable.***

If the dendrite is large, there exists a second excitatory input in the form of a climbing fiber.** If there is no climbing fiber present, the other excitatory afferent synapses must be Brindley synapses.*** The climbing fiber input, if it exists, can produce the conditions for synaptic modification in the whole dendrite simultaneously,*** but it is subsequently not the only input able to do this.*

There are two kinds of inhibitory input to the cell**: one scattered, which has the effect of performing a subtraction,** and one clustered at the soma, performing the division.** At least one of these functions is performed,*** but the all-or-none approximation would require only one. Both essentially estimate the number of afferent synapses from codon cells active at the cell.***

The output from these cells is excitatory if it forms the input to a subsequent piece of cortex.** Their axon collaterals synapse with neighboring output and Martinotti cells.

Climbing fibers are present only on output cells.* The climbing fiber at a given pyramidal cell provides an accurate enough pointer for that cell for the spatial recognizer effect to take over and make the cell a receptor for a classificatory unit.***

Climbing fibers are excitatory*** if used for this purpose.

Many of the short axon cells which are not codon or climbing fiber cells are inhibitory.*** The theory distinguishes three principal kinds.** Subtractor cells sample the activity of codon cell axons near local regions of dendrite** and send inhibitory synapses to those regions.** These have a subtractive effect.** Division cells, the basket cells, are inhibitory;** and so are cerebellar Golgi cell analogues, which keep the amount of codon cell activity about constant.**

The granule cell threshold controls receive excitatory** synapses from either the granule cell excitatory afferents or the granule cell axons,*** and perhaps from both.* They send inhibitory synapses to the granules themselves,*** and these synapses are scattered over the granule cell soma and dendrites.** The Martinotti cell threshold controls receive excitatory*** synapses either from the Martinotti afferents or from the Martinotti axons.*** In view of the length of the Martinotti axons, they probably receive from both,** and therefore have an ascending dendritic shaft.** Layers VI and VII contain fusiform cells which could be Martinotti cell threshold controllers.

The axonal and dendritic distributions of the inhibitory cells of the cortex depend on the distributions of the afferents, and of the codon cell axons, in a complicated way.

One might say about this model that it is extremely speculative and that there is considerable danger occurring here in two stages going from proposed function to specific nets as in this model. First, there are many ways of constructing and using the classificatory units such as Marr proposes, different from the diagnostic spaces and mechanics discussed here. Also, there are many ways to represent such diagnostic spaces should they correspond to brain function and nervous structure. Moreover, it is most likely that the cerebral cortex performs more functions than only classifying events; for example, the comparing and relating of classes is a main function probably performed by cortical circuitry. Also, this theory is essentially connectionist in that specific anatomical neurons and sets of neurons are supposed to correspond to specific psychological events. There are, at least according to one current theorist, extreme difficulties with this sort of theory, as we will take up immediately below. Nonetheless, Marr's model is extremely penetrating and provoking. It performs the vitally important task of relating proposed functional characteristics to specific predictions concerning actual circuitry in the underlying networks. It is exactly this task which must be successfully performed if we are to understand the workings of these networks.

E. Roy John, in recent publications, has criticized the so-called "switchboard" theories of cortical function. John (1967) has argued that available evidence is not consistent with the view that encoding in cortical tissue is localized in specific neurons or, moreover, is deterministic. John argues that the coding of the information associated with an event corresponds to the pattern of activity over large neuronal ensembles and, moreover, that such coding is statistical in that the information associated with an event corresponds to the average of activity over many neurons within an ensemble rather than being specifically tied to any single member or group of members. As with the previous modelers we have discussed in this section, John supposes that the establishment of representational systems in the brain involves large numbers of neurons whose activity becomes somehow coordinated. However, John supposes that in the internalization of a given event many such systems are spread throughout the brain, and that these systems develop a common mode of activity which is coherent across the various regions and is specific for various stimulus complexes. New response patterns are based upon the establishment of new temporal patterns of ensemble activity rather than on the elaboration of new pathways or connections. John presents experimental data which suggest that wave forms obtained in macroelectrode recordings from different brain regions tend to attain a similar form for a given stimulus context. Moreover, John notes that there is a portion of the electrical activity in the response of a trained animal that is independent of the afferent stimulus and which might be interpretable as a "readout component." He also notes that although single cells within a region display highly variable responses to conditioned stimuli, ensembles of neurons display essentially invariant temporal patterns; i.e., microelectrode recordings are extremely variable, whereas the macroelectrode recordings are highly constant. John's theory does not deny the significance of structural interconnections between neurons in determining the ensemble modes of activity which he considers to be important. The theory simply asserts that it is the overall patterns of activity which are important rather than the single-unit behavior and, moreover, that informational content is stored in the average of ensemble activity rather than in single-cell activity.

The extremely insightful Nobel laureate Sir John Eccles, although not, as far as we know, advancing any specific model for cortical organization, does present an interesting discussion of the propagation of wave fronts through cortical networks. That discussion is given in his own words as follows (Eccles, 1970):

> It is impossible to conceive the complexity actually obtaining in propagation over neuronal chains, where each neuron is linked to hundreds of other neurons and where the convergence of many impulses within a few milliseconds is necessary in order to evoke a discharge from any one neuron. It is therefore desirable to study the properties of simple models of neuronal networks as was done by Burns, by Fessard, by Cowan and more recently by numerous investigators.
>
> [Figure 15-3a] gives a diagrammatic representation of impulse propagation through a series of neurons arranged formally in the characteristic manner of in-parallel and in-series. This model is of course tremendously simplified, but it does help to illustrate the manner of operation of assemblages of nerve cells in the central nervous system. In the upper diagram two of the cells in column A are assumed to be excited and are in gray. Various axonal branches and their synaptic connections from cells in column A to column B and so on to C and D are shown. By convention in this model it is assumed that summation of two synaptic actions is required to generate the discharge of an impulse from a cell and so its activation is transmitted (see arrows) to synaptic

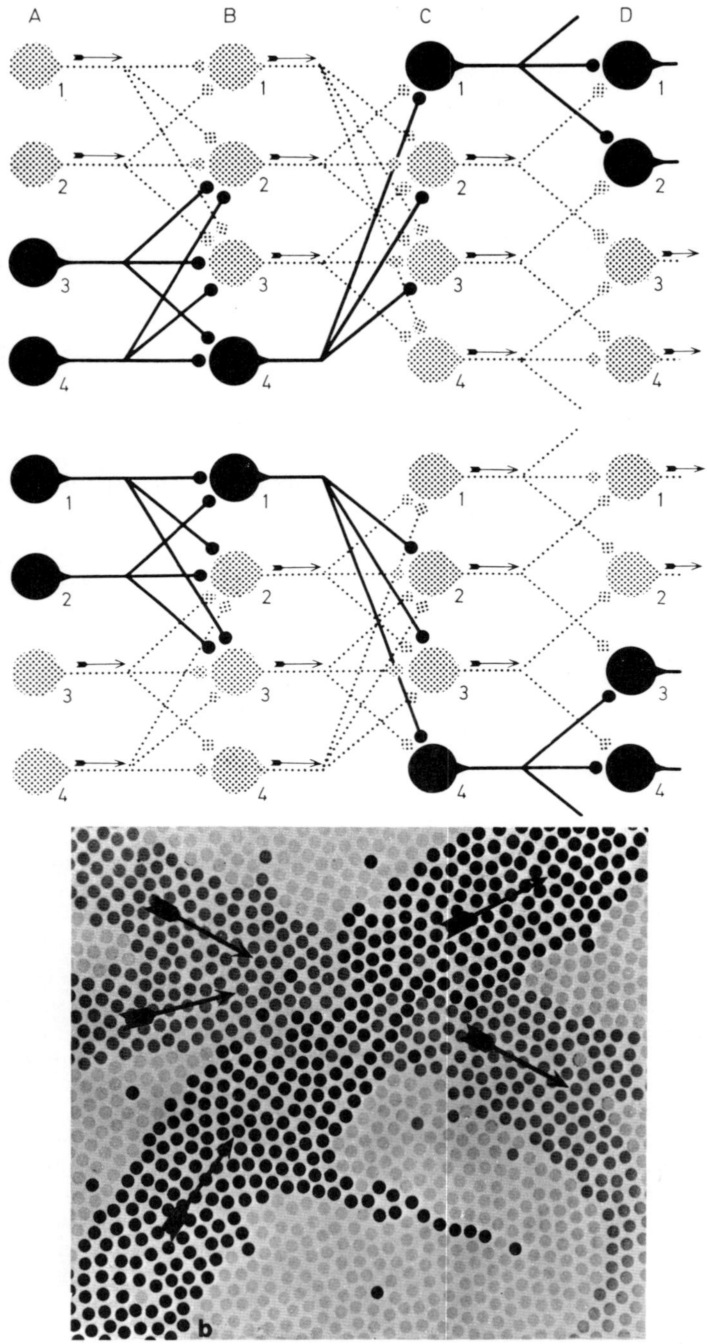

Figure 15-3a,b

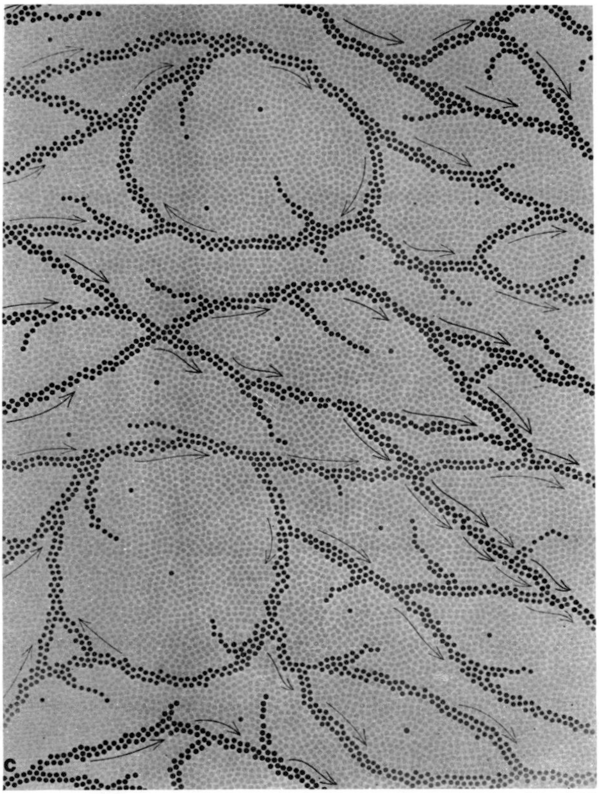

Figure 15-3. (a) Model of a highly schematic neuronal network to illustrate the simplest case of propagation along a multilane pathway. There is exactly the same anatomical network in the upper and lower diagrams. The synaptic connections of the twelve cells in columns A, B, and C are drawn, cells with impulses (note arrows) being shown light gray, while the silent cells are black. The assumption is that a cell fires an impulse if it is excited by two or more synapses (also light gray). Thus an input $A_1 A_2$ results in an output discharge of $D_3 D_4$ (upper diagram), whereas an input of $A_3 A_4$ gives an output of $D_1 D_2$ (lower diagram). Neurons B_2, B_3, C_2, C_3 are activated in the crossing zone for both these inputs. This diagram suffers from the serious defect that it ignores inhibition. (b) In this schematic diagram the nerve cells of the cortex are imagined as laid out as dots in one plane. The multilane traffic in one, evolving a specific neuronal pattern, is shown in black, and in another as dark gray. The light gray cells are not activated by either pattern. Note that at the crossing of these two lanes the same nerve cells should participate in both and so each should be represented as a very dark gray dot, not just dark gray or black as shown. Arrows show direction of propagation. (c) By employing the same conventions as in Fig. 15-3b, an attempt is here made to give a more elaborate representation of the enormous complexity of the spatiotemporal patterns of propagating activity in the cerebral cortex. The arrows show the directions of propagation for two separate evolving patterns (black and dark gray). Note that the black and gray patterns coalesce at two sites and propagate as one advancing wave. This would correspond to some integrative fusion of different sensory inputs to give some perceptual synthesis. (From Eccles, 1970.)

terminals (gray knobs) on the cells of the next serial column. In this way it can be seen that the excitation of cells A_1 and A_2 will result in the output of impulse discharges from cells D_3 and D_4 but not from cells D_1 and D_2. As already mentioned, [Figure 15-3a] is an extremely simplified model because, firstly, it assumes a simple geometrical arrangement with columns of neurons in serial array, whereas there are known to be all manner of synaptic connections bridging such serial order, as for example from cells of column A directly to cells of columns C and D. Also there are many varieties of feedback controls via axon collaterals and the consequent loop operation of large assemblages of cells. Another serious deficiency in [Figure 15-3a] is that no account has been taken of inhibition. Nevertheless [Figure 15-3a] is of value because it gives a simple diagrammatic illustration of the way in which information is propagated from neuron to neuron, summation being required at each synaptic relay. Consequently there is the necessity for the in-parallel arrangement of neurons.

The synaptic connections in the model of [Figure 15-3a] from column A to B to C to D have been designed to illustrate one remarkable and important property of a neuronal network, namely, that two completely different inputs at A (A_1 A_2 in the upper picture or A_3 A_4 in the lower picture) can be transmitted through the same pattern of cell connections (A to B to C to D), crossing each other and emerging as completely different outputs at D (D_3 D_4 in the upper picture and D_1 D_2 in the lower). An interval of some milliseconds between the two wave fronts would eliminate interference by neuronal refractoriness or by summation of synaptic excitations. It will be noted that along the fringe of the advancing wave front in [Figure 15-3a] there are subliminally excited neurons (C_1, D_1, D_2 in the upper picture). Such "fringe neurons" give opportunity for growth of a wave if other influences should also subliminally activate them.

Thus we are introduced to the concept of lability of a wave front. It may be diminished by inhibitory or depressant influences, and so ultimately be extinguished, or it may be enhanced by factors aiding in the activation of "fringe neurons." It will also be appreciated that, if a wave front moves into neuronal pathways having interconnections of suitable configuration, it may bifurcate into two waves propagating independently; while, conversely, two wave fronts propagating at the same time into the same pool of neurons would coalesce and give an onwardly propagating wave having features derived from both waves, with additional features due to the summation.

It will be realized that in the neuronal network of the cerebral cortex the factors involved in transmission of a wave front are vastly more complicated than in the simplified models of [Figure 15-3a]. In the first place, each neuron would make very many synaptic contacts with other neurons, and also receive from many more—probably from hundreds. Perhaps as many as 100 neurons would be involved in each stage of an effectively advancing wave front and not 2 or 3 as in [Figure 15-3a], so this wave would sweep over 100,000 neurons in one second. An advancing wave would also be branching at intervals, often abortively, and coalescing with other waves to give a complex spatiotemporal pattern.

[Figure 15-3b] can be regarded as an enormous development from the simple array of [Figure 15-3a]. In order to illustrate patterned operation of hundreds of neurons there has been suppression of all diagrammatic representation of nerve fibers and synapses (cf. [Figure 15-3a]). In the diagram it is assumed that adjacent cells are synaptically connected in the forward direction indicated by the arrows. Inactive cells are shown in light gray. The dark gray and black arrangements of cells in serial order are meant to illustrate the propagation of impulses along chains of neurons arranged in parallel with branches and convergences. At each stage of serial relay there is assumed to be synaptic summation with the consequent discharge of some cells as in [Figure

> 15-3b]. It will be understood that this is an entirely imaginary diagram of a small fragment of a spatiotemporal pattern of neuronal activation and that it is constructed in order to convey some insight into the extremely complex patterned operation of large neuronal assemblages such as occur in the cerebral cortex. It also illustrates the important property that any one nerve cell can participate in a large number of separate neuronal patterns. Its responses can belong to this or that pattern according to the ensemble of neurons that are activated with it, as will be appreciated from the two models in [Figure 15-3a]. There is now much evidence to indicate that the input of impulses via channels from receptor organs is itself repetitive and evokes long-lasting serial activity in immense numbers of neurons linked up in as yet unknown pathways. . . . Furthermore there is continuous ongoing activity of the cortical pyramidal cells, and this activity can be raised or depressed by sensory inputs.

A more elaborated representation of the propagation of spatiotemporal patterns in cortical networks as discussed by Eccles is shown in Figure 15-3b.

Since cortical circuitry is known only in general outline, models based on structure directly are generally nonexistent. However, a recent model by Dichter and Spencer is at least consistent with what is known of cortical circuitry. These authors suggest two recurrent systems to account for interictal spiking in epileptic seizures. Based on electrical data in seizures from both neocortex and archicortex, they hypothesize a recurrent inhibition with a wide lateral distribution and a recurrent excitation with a more restricted lateral distribution. Computer simulations based on this model show many features similar to those observed in epileptic seizures.

Modeling the cerebral cortex then is a very active area in which many exciting ideas exist. Most current models are based on ideas concerning the overall functioning of the cortex rather than upon its structure. We will need to know a great deal more concerning the anatomy and synaptology of cortical circuits before attempts in the opposite direction, proceeding from structure to operation, are feasible. One suspects that the understanding of cortical operations will be generally resolved within a century but that such unraveling will probably take the better part of it.

15.2. Thalamus and Hippocampus: Brain Rhythms

The two most salient rhythms of the EEG, the alpha and theta waves, have their origin in circuits involving the thalamus and hippocampus, respectively. Since parent population cells in both these regions exibit remarkably strong IPSPs with durations that correspond directly with the periods of the respective waves, it has seemed natural to suspect recurrent inhibition as a main ingredient in the genesis of the rhythms. Recently, models have been presented to explore these possibilities in both structures.

Andersen and various collegues have propounded a theory of recurrent inhibition with particular emphasis on the feature of postinhibitory rebound to describe the genesis of the alpha rhythm in the thalamus. The theory is backed up by a great deal of experimental evidence, much of it obtained directly by Andersen and colleagues. The theory is described in some detail in a book by Andersen and Andersson (1968). A succinct description is provided by Andersen *et al.* (1966) in the following extract taken from a publication describing related computer-simulation studies. The theory is referred to as the "inhibitory phasing theory."

> According to the inhibitory phasing theory projection cells at "rest" have a random probability of discharge. If one cell discharges, it will, by its recurrent axon collaterals, excite one or several inhibitory interneurons, which in turn will produce a long-lasting inhibition in a number of neurons surrounding the initially discharging cell. Following this large inhibition, the cells undergo an inhibitory change, known as the postanodal exaltation, PAE. . . . This postinhibitory increase of the excitability causes a certain number of the surrounding cells to discharge, and the cycle repeats itself. With each spike more and more cells will participate because of the wide distribution of the axons of the inhibitory interneurons. After some cycles of operation, the system will be saturated. With the given degree of postanodal inhibition and the existing distribution of inhibition, no more cells can take part in the rhythmical discharge in any period of time. Disruption of the rhythm occurs because the duration of inhibition is not equal in all cells. In a neuron with short-lasting inhibition, the postinhibitory discharge will occur earlier than in its neighbors. Through its recurrent pathway, it produces inhibition in a number of surrounding cells thus creating a partisan rhythm and starting a breakaway pattern. These small foci of cells beating with a nondominant rhythm will occur in several places, until the whole system again beats at random. The greater the amount of participating cells, the greater is the probability for disruption of the rhythm. The system is therefore self-limiting with regard to the duration of the rhythmic activity.

Andersen *et al.* (1966) then constructed a digital-computer model to work out this hypothesis. Their own description is again quite clear.

> 1. A digital computer has been programmed to simulate a neuronal network consisting of eighty cells with the following characteristics:
>
> (a) All cells in the network had a set random probability of discharge when not affected by other cells. The probability of discharge was chosen as one system parameter (PD).
>
> (b) Subsequent to the firing of any neuron in the network, a certain number of other neurons underwent a change in their probability of discharge, consisting of an inhibitory period followed by a period of increased excitability. The changes of excitability mimic the change in the membrane potential recorded for spontaneously rhythmically active cells in the animal thalamus.
>
> 2. Starting the network, a typical transient response appeared consisting of a burst-like activity. The initial activity was followed, first by random fluctuations of the number of active cells, later by periods of spontaneous rhythmic activity that had several characteristics in common with the type of spontaneous activity seen in the animal thalamus.
>
> 3. The parameters which affected the rhythmic activity of the network most strongly were the degree of the postinhibitory increase of the probability of discharge and the degree of distribution of the inhibition to neighboring cells.
>
> 4. The results are compatible with the inhibitory phasing theory advanced by Andersen and Sears (1964) to explain the occurrence of spontaneous rhythmic activity

in the animal thalamus. However, before sufficient experimental knowledge is available, it is difficult to establish whether the rhythmic activity in the network is of the same type as that of the animal thalamus.

The main result of this model is the finding that a randomly connected pool of mutually inhibiting neurons can be found to fire rhythmically upon random stimulation when the mutual inhibition is followed by a supernormal period. The simulation rhythmicity had a period of 11 to 12 time units. This would suggest that each time unit would correspond to about 8 or 9 msec if the computer results are matched to the 10/sec rhythmicity of the thalamus neurons. This would mean that the inhibition was imposed for the equivalent of some 80 to 90 msec, which corresponds well with experimental observation.

MacGregor, McMullen, and various colleagues are currently developing and evaluating models for rhythmicity in the thalamus and hippocampus. Although this work is in progress and cannot be reported here, an interesting model system has emerged which may have some relevance to either or both of these rhythm-generating structures. The model system posits selective sensitivity of recurrent inhibition to synchronized clusters of action potentials in parent population collaterals as a key ingredient in the genesis of system rhythmicity. This selective sensitivity is a natural property of diffusely connected afferent pathways according to the simulation results discussed in Chapter 11. In this model system, any factor which tends to encourage simultaneous firing of parent population cells tends thereby to encourage synconized rhythmicity in the system, and any factor which tends to discourage simultaneous firing in parent population cells tends to encourage desynchronized activity in the system. For example, one might postulate that any diffuse input system to the parent population or self-exciting collateral system of the parent population tend to encourage the synchronized state, and that any discrete and synaptically strong afferent population impinging on the parent population tends to encourage desynchronized activity by "pushing through" the coordinated recovery implied by simultaneous inhibition and masking any weaker synchronizing influences, thereby discouraging the occurrence of synchronized clusters in the parent population. Such a model system for synchronization-sensitive recurrent inhibition is shown in Figure 15-4.

The dynamic activity of this system can be described qualitatively as follows. First, typical ongoing (unsynchronized) activity can be maintained by a discrete afferent system and involves a minimal level of activity in the inhibitory population. This activity might reflect the processing of information in local somatotopic groups or combinations of such groups provided they make minimal use of large numbers of simultaneous firings in parent population cells, or simply random background activities. Second, anything that initially brings on a synchronized cluster in the parent population leads

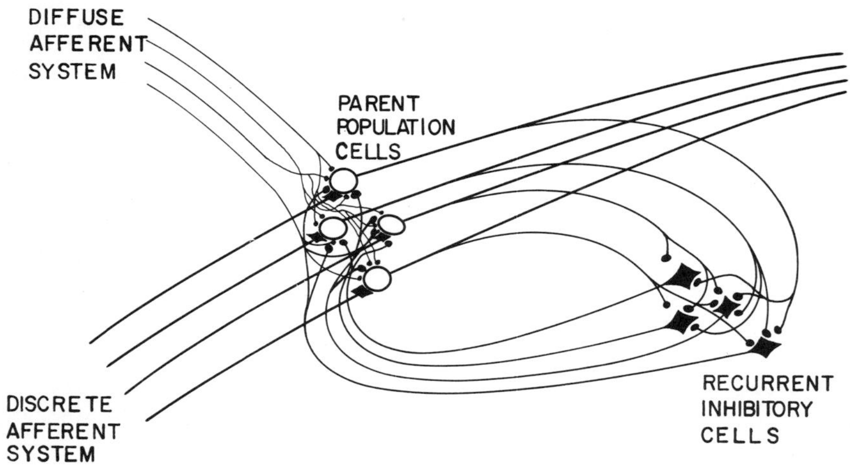

Figure 15-4. General model system.

to massive activation of the recurrent inhibitory system. Such an initial synchronized cluster may be brought about by the chance concurrence of input events within a discrete input system, by the designed coordination of activity within a discrete input system, or by the combination of a sufficiently low level of activity in a discrete system together with activation of a diffuse system or a self-exciting collateral system. Third, the coordinated massive combined IPSP engendered by the recurrent population tends to encourage subsequent synchronization of parent population firing by demanding a more-or-less coordinated recovery from inhibition for all cells so affected. However, whether such a secondary synchronization (and thus, by extension, recurring rhythmic oscillations) does in fact occur depends on the charcter of the input constellation incident upon the parent population cells. If the diffuse afferent system and/or self-exciting collateral mechanisms are dominant, these work with the coordinated recovery to produce recurring synchronized clusters together with the implied recurring activation of the IPSP system so that recurring rhythmic activity ensues. If, on the other hand, a discrete afferent system dominates, then the coordinated recovery is disrupted and subsequent synchronized clusters and associated IPSPs do not occur. Salient examples of these activity patterns obtained from computer-simulation studies are shown in Figures 15-5 and 15-6. Figure 15-5 shows synchronized activity obtained when the diffuse afferent system is activated, whereas Figure 15-6 shows desynchronized activity obtained when the discrete afferent system is activated.

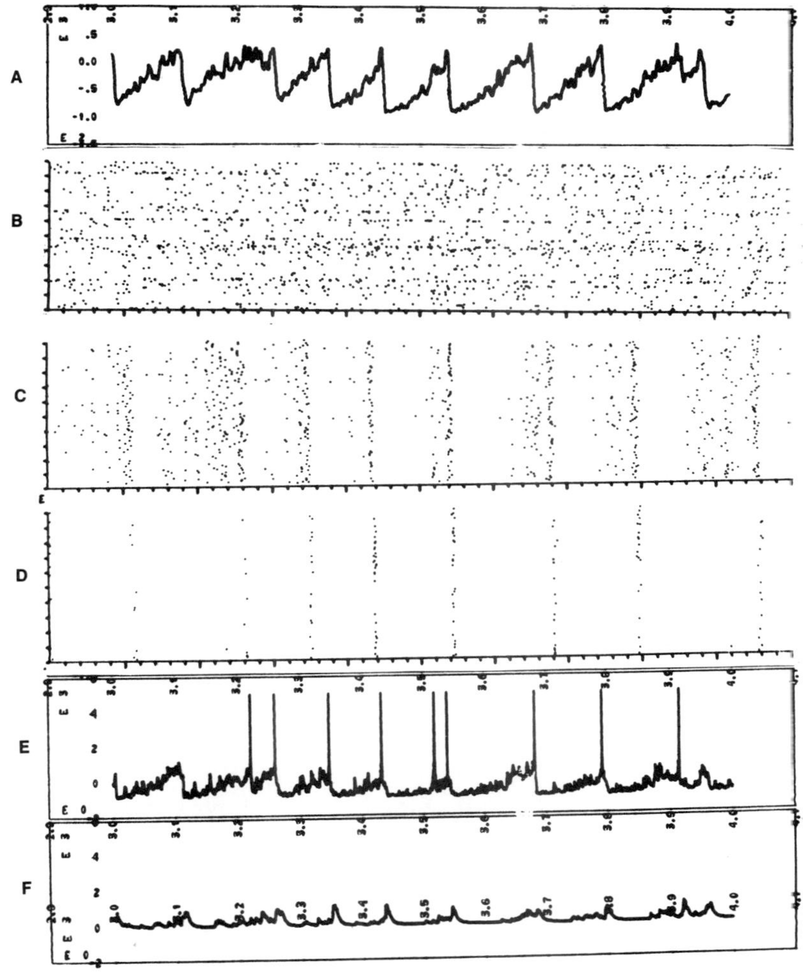

Figure 15-5. Representative computer simulation of general model system for brain waves. (A) Model EEG formed by summing the graded generator potentials of all constituent cells. (B–D) Distribution of action potentials in afferent system, parent population, and inhibitory system, respectively. (E–F) Simulated generator potential for a representative excitatory cell and a representative inhibitory cell, respectively. (Each figure shows a 1-sec sample of activity.) The particular inhibitory cell shown here happened not to fire in the time intervals shown.

Although these considerations seem to us suggestive, many other factors must be taken into account before they can be applied with any confidence to either the thalamus or the hippocampus.

Models for the operation of the hippocampus have also been presented by Marr and by Kilmer.

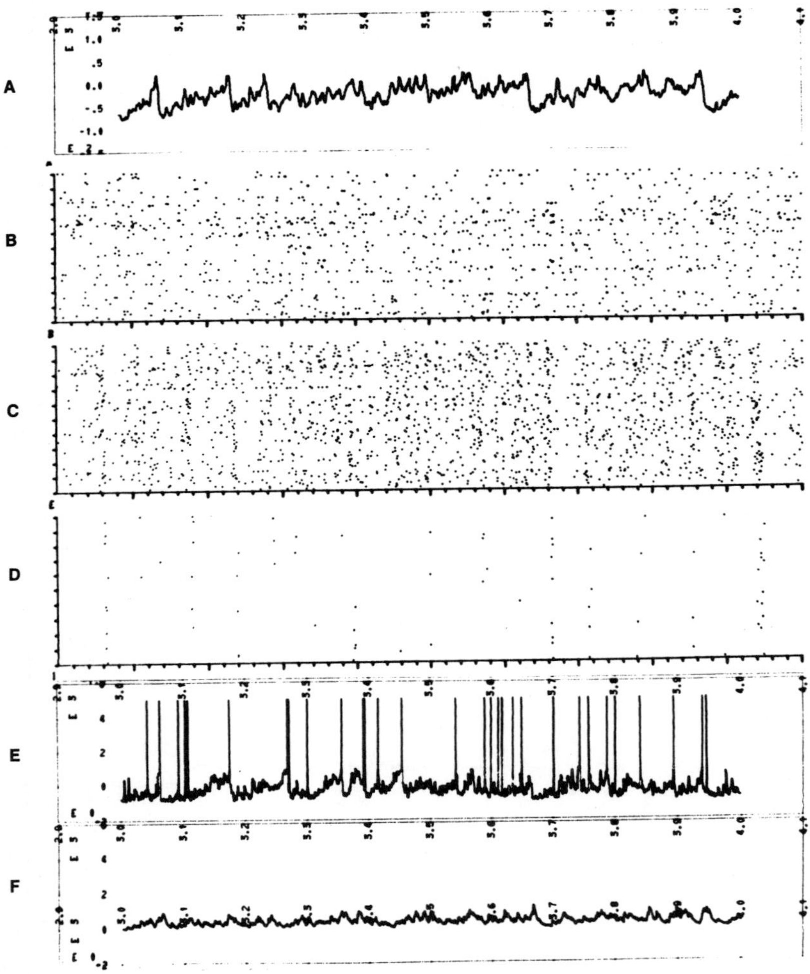

Figure 15-6. Desynchronized activity upon combined discrete and diffuse activation. (B shows activity pattern of discrete input system; diffuse pattern is the same as in B of Figure 15-5.)

15.3. Reticular Formation

A second major central integrative network which has received considerable attention by modelers is the reticular formation of the brain stem. The reticular formation as an anatomical structure extends throughout the core of the brain stem and, moreover, extends caudally throughout the core of the spinal cord and rostrally into the limbic system particularly through the hypothalamus and the medial forebrain bundle, and into the thalamus. It is characterized by cells with rather sparse, unbranching dendrites and many, diffuse connections effected by the many axon collaterals emanating from each axon.

The reticular formation as a unitary functional concept, however, is not very clear. Early theorizing was based largely on the idea of the ARAS—the ascending reticular activating system. This system was thought to correspond to an "arousal" dimension. It was thought to mediate a continuum of states ranging from death through sleep, quiet waking, and on to a highly aroused state. The ARAS was shown to directly influence the state of the cerebral cortex and the nature of brain waves. In the quiet waking state corresponding to low arousal, the EEG shows a steady slow rhythm. In the aroused state the ARAS acts on the cortex to abolish the slow rhythm and bring on low-voltage fast activity. The ARAS acts on all sensory systems to modulate the input to the CNS at any time and sets the tone on the behavior of motoneurons as well. Thus, the reticular formation was thought to be solely responsible for behavioral arousal, to funnel its output rostrally primarily through the midline thalamus, and to respond to sensory input from all modalities homogeneously mixed from primary afferent collaterals. The internal connections were thought to be mediated primarily by short-axon Golgi-type-two cells. The network was thought to be highly multisynaptic, nonspecific, and essentially glandlike or organlike in structural organization and function.

Subsequent work has shown that although the reticular formation is indeed a major integrative center vitally involved in states of consciousness and arousal, this one-dimensional glandlike concept is an oversimplification in many ways. For example, more recent anatomical studies have shown that there are virtually no Golgi-type-two neurons throughout the reticular formation, but rather that each cell has a very long axon with many axon collaterals. Moreover, it has been shown that one can dissociate behavioral arousal from EEG arousal. Thus, if the reticular formation creates arousal, which it does, there must be something else which can perform this function as well. It has also become clear that functions other than arousal have diffuse "nuclei" of sorts embedded in reticular tissue. For example, Casey (1969) has shown the existence of pain centers within the brain stem reticular formation. The so-called pleasure centers in the lateral hypothalamus are embedded in the rostral continuation of the reticular formation. Respiratory centers in the caudal medulla and rostral pons are also structurally indistinguishable from the surrounding reticular formation.

Two more recent models have attempted to incorporate multiplicity of function into reticular organization. Kilmer, working with Blum and McCulloch, has built a model to elaborate their basic hypothesis that the main function of the reticular formation is to commit an animal to one or another mode of behavior. They hypothesize that there should be approximately 25 such modes, including those contained in the following: (1) sleep; (2) eat; (3) drink; (4) fight; (5) flee; (6) hunt (for prey or fodder); (7) search (or explore); (8) urinate; (9) defecate; (10) groom; (11) mate (or sex); (12) give birth (or egg lay); (13) mother the young (including suckling or hatching, retrieval, perineal licking, etc.); (14) build or locate a nest; and (15) special innate forms of behavior such as migrate, hibernate, gnaw, and hoard, depending on the species. They note that a structure which makes mode decisions should have three main characteristics: First, information crucial to certain decisions must be able to penetrate sensory filters whereas irrelevant information should be filtered. Second, the most important mode decisions are often the most urgent and therefore must be responsive in the least amount of time. Third, when a set of information clues is being presented to the organism, those that are the most important must obtain priority and be given authority to influence the state of affairs of the entire system; i.e., information must determine authority. The main operational problem presented by the reticular formation, then, in these authors words, is as follows: "the problem of the RF core is how, in a fraction of a second, can a million or more neurons reach a workable consensus as to the proper mode of total commitment."

To effect this mode of operation Kilmer conceives of a collection of modules each of whose main function is to select a mode based on the input it receives in any particular time period. In this model, time is broken into discrete intervals. The input of each mode comes from external sources and from all the other modules in the system. Also, each module connects to all the other modules, or at least to a subset of them. The model is illustrated in Figure 15-7. The entire system operates as follows. Given an initial set of mode preferences in all the modules and a given state of input parameters at a given time, each module makes a decision concerning the mode the animal should be put in. This decision is communicated to all the other modules to which the module is connected, and the given module in turn receives information from another appropriate subset of modules concerning their states. At this time, then, all the modules again determine their preference for mode based on the updated information. This revision is then communicated to the other modules and the process repeated until a consensus is reached wherein all the modules are signaling the same mode of commitment. Kilmer then constructs computer modules to perform these functions dynamically. In subsequent development of the model he adds capabilities for development, generalization, discrimination, habituation, classical conditioning, extinction, avoidance conditioning, trial and error learning, and the like.

The modeling effort then shows essentially how a given particular concept of functional organization might be realized by a computational device.

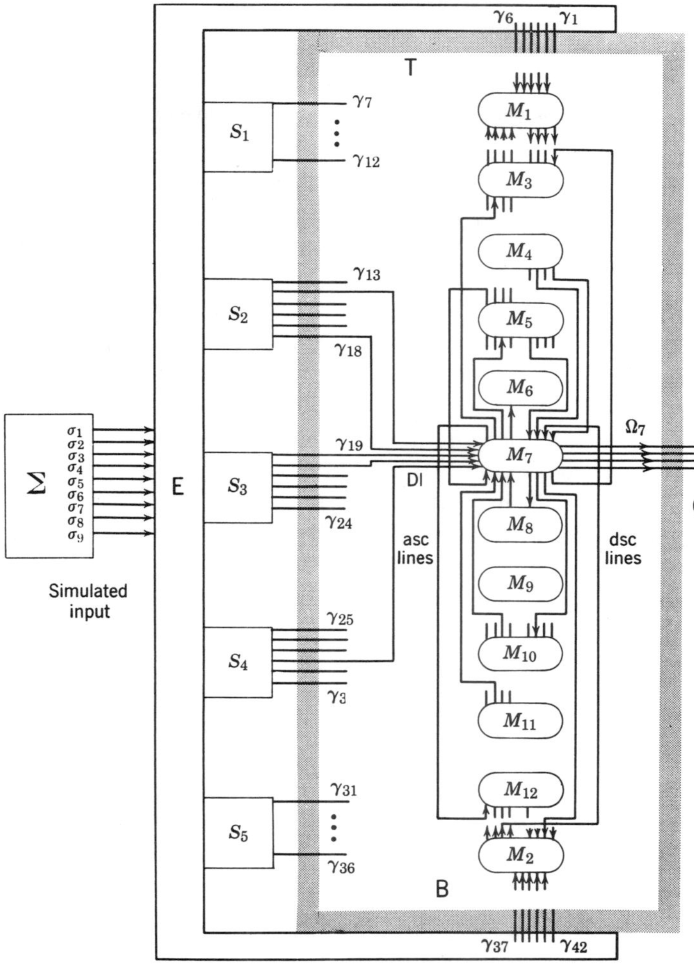

Figure 15-7. S-RETIC simulation model. Connections are shown only about M_7. Each M_i is a caricature of a poker-chip region, or neuropil segment. The S_i are input systems and the Ω_i comprise S-RETIC's output. E generates S_i values by computing functions of the irredundant set of binary variables out of Σ. S-RETIC is said to have converged on a mode if more than half of its M_i indicate that mode with probability (preference) greater than 0.5. (From Kilmer, McCulloch, and Blum, 1968.)

The process of working from function through operation to structure, as in the models of Marr and, in part, Kilmer, is reversed in a modeling effort presented by MacGregor. This is an attempt to make inferences concerning operational dynamics and functional organization from structural information. The model is essentially an elaboration of Schiebel's poker-chip model for the structure of the reticular core. This latter model posits a collection of input fibers arranged in a longitudinal tract and conducting in either direction. Neurons are distributed throughout this fiber tract with their dendritic trees at right angles to the direction of the fibers. The ascending and descending fibers produce collaterals which extend into the network at various places and make synaptic contact with the cells, on the cell bodies or on the dendritic trees. The neurons in turn exhibit axons which join the fiber path, many of which bifurcate and thus extend axonal projections in both ascending and descending directions. These axons also produce collaterals which feed back into the net and make contact on neurons within the net.

By considering the limiting case where the distance between adjacent collaterals is the same for all intercollateral distances, MacGregor suggests the usefulness of the so-called "ladder net" shown in Figure 15-8. In this net the component cells are distributed along a one-dimensional axis, and each component cell connects to all the other cells within the length of its axon. He suggests, first, that the overall functional organization of the reticular formation might be understandable in terms of quasi-independent ladderlike subnets. Each subnet would signal a unique internal state of affairs according to its unique distribution of fibers. Individual nets would be activated preferentially by different patterns of activity in the input fibers. He suggests, second, that, in any case, such a ladder net should be useful in providing some insight into the dynamics of the networks of the reticular formation. MacGregor then goes on to work out predicted dynamical properties of these fiber nets. One main dynamic characteristic of ladder nets is the so-called recruitment fuse. In this response pattern the network, under appropriate stimulation conditions and because of its internal connections, generates a series of spikes occurring sequentially in the network cells until the end of the network is reached. Each cell contributes one spike to such a response. Although such an occurrence would certainly not be noticed in a single microelectrode recording (each cell fires only once in the response), the response pattern is clearly an organic response of the net as a whole. The output spikes from all the

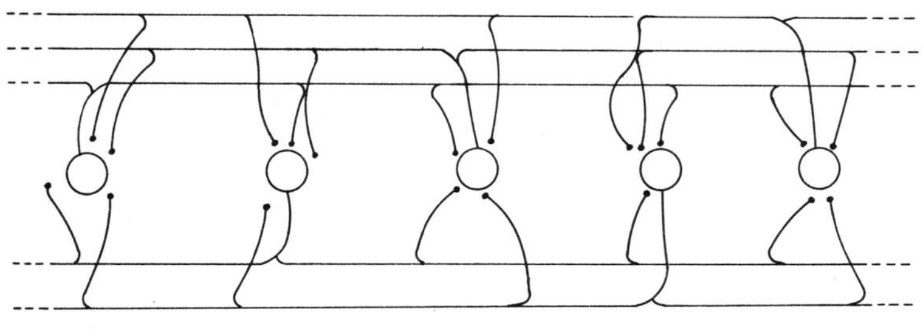

Figure 15-8. The "ladder net."

constituent cells active in the recruitment fuse reach a given output location almost simultaneously and thereby provide a relatively strong output in terms of the number of input pulses presented to a given receptive area. A second main dynamic characteristic of the ladder nets is, first, that the constituent cells tend to fire in bursts of spikes because of the internal reverberations, and, second, that the network cells tend to fire collectively in groups. Finally, it is shown that ladder nets respond preferentially to bypassing fibers which conduct at the same speed as do the fibers of the net. This property could represent a mechanism whereby the reticular networks, or subnetworks in the reticular formation at least, might be preferentially sensitive to slowly conducting pain fibers. This result is of particular interest in view of Bowsher's finding of a relatively sparse number of synaptic contacts from spinoreticular afferents.

15.4. The Retina and Lateral Inhibition

The concept of lateral inhibition is one of the most significant concepts we have concerning the organization of neural networks. Its implications have been worked out in considerable detail and have been shown to be directly applicable to the coding of visual information in the networks of the retina. The network is conceived as lying in a two-dimensional plane. It receives input stimulation at each point in the plane which in turn induces activity at that point. This activity is then passed on to the next stage in the system and also acts to inhibit elements in a local surrounding area. A main property of such a network is spatial sculpturing, as indicated in Figure 15-9. The system emphasizes responses around boundaries in the visual field. For example, a region illuminated by two levels of excitation, *A* and *B*, as indicated in Figure 15-9, leads to a response field containing four response regions: *A* exciting plus *A* inhibiting (I), *A* exciting plus *B* inhibiting (II), *A* inhibiting plus *B* exciting (III), and *B* exciting plus *B* inhibiting (IV). For the case of very bright illumination, *A*, next to very much weaker illumination, *B*, it is easy to see that the most intensely activated area containing *A* excitatory plus *B* inhibitory (II) is adjacent to the most weakly activated area containing *A* inhibitory and *B* excitatory (III). Moreover, these two regions lie precisely on the boundary of the *A* region with the *B* region. A pattern of illumination spread over a two-dimensional plane leads to sculpturing of the pattern in that the boundaries between illumination regions are enhanced and signals within a uniformally illuminated region are depressed.

Moreover, the concept of lateral inhibition can produce a marked sensitivity to temporal contrast as well. For example, if inhibitory signals require a certain time lag to be conducted laterally, then the response at a

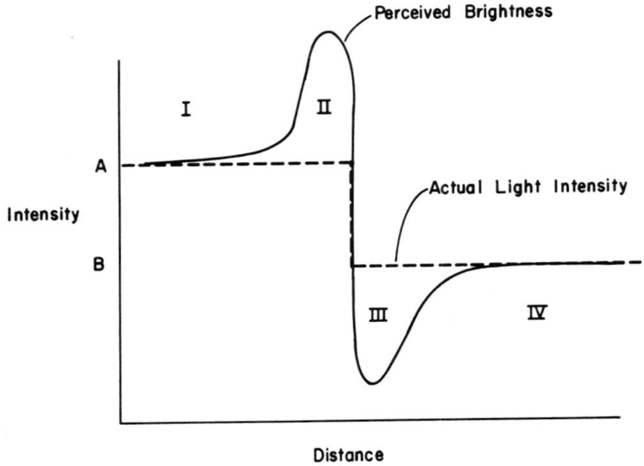

Figure 15-9. Enhancement of contrast boundary by lateral inhibition.

point to a bright uniform step illumination will show an initial response at all points corresponding to the time interval when the direct excitation is active and the lateral inhibition has not yet arrived. This will then be followed by a more surpressed period which reflects the combination of direct excitation with the laterally mediated inhibition. The lateral inhibitory mechanisms are also intimately involved in the dynamics of the opponent process units in color vision.*

There have been many attempts to electronically stimulate or mathematically model the dynamic activity of such laterally inhibitory networks. Runge *et al.* (1968), for example, have constructed an electronic simulation of the avian retina, containing 145 cone cell analogs, 58 horizontal cell analogs, 80 bipolar cell analogs, 88 amacrines, and 7 ganglion cell analogs. A picture of this model and a representative circuit diagram for one of the constituent elements are shown in Figure 15-10. The primary purpose of this modeling effort was to simulate the interactions leading up to specific detector cells at the level of the ganglion cells. Runge *et al.* were able to construct luminosity detectors, general edge detectors, convex edge detectors, vertical edge detectors, directional detectors, and horizontal edge detectors by adjusting parameters.

*Thus, it is clear that at the level of the retinal outer plexiform layer, color information is arranged in two sets of opposing pairs of elements: red and green elements are paired, as are blue and yellow elements. However, receptors are organized accoring to a tricolor scheme: pigments occur in each of three sets, each with maximum sensitivity at a different wavelength.

Varjü (1965) used matrix algebra to describe the interactions of a laterally inhibiting network, and this work has been recently taken up by Morishita and Yajima (1972). This modeling effort uses a continuous-variable model of the neuron as a basic element which consists of a many-input adder, a first-order, low-pass filter, and a diode-type nonlinearity. This is indicated in Figure 15-11. Morishita and Yajima write equations for this system then as follows:

$$\left.\begin{aligned} \tau\frac{dy_i(t)}{dt} + y_i(t) &= x_i(t) - \sum_{\substack{j=1\\ j=i}}^{n} w_{ij}z_j(t) \\ x_i(t) &= \sum_{j=i}^{M} v_{ij}u_j(t) \\ z_i(t) &= y_1(t)\phi[y_i(t)] \end{aligned}\right\} \quad i = 1, 2, \ldots, N \qquad (15\text{-}1)$$

Here, y_i is a continuous variable corresponding to the activity of the cell at location i, and z_i is the corresponding output signal of that cell; z_i is equal

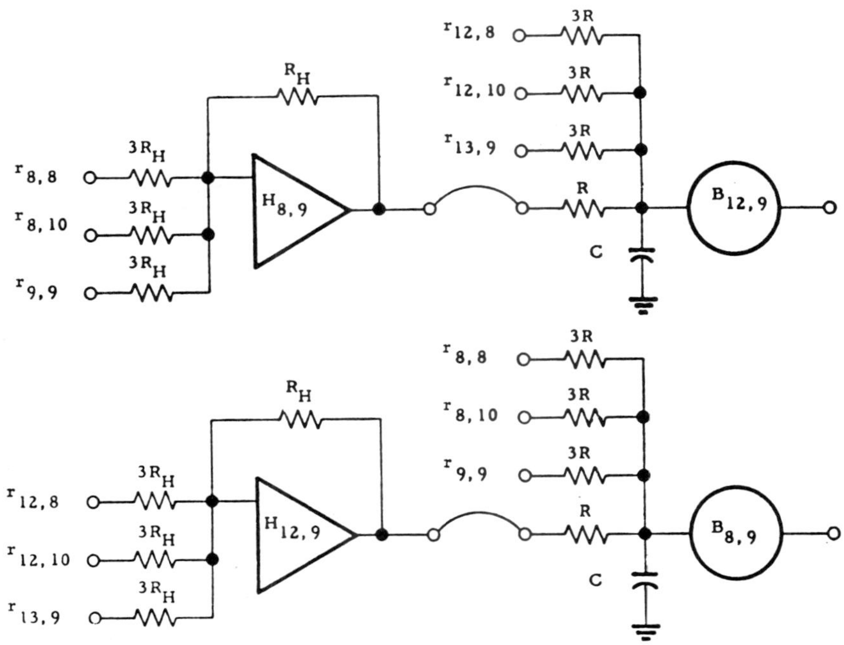

Figure 15-10. Reciprocal contrast bipolar analog. (From R. G. Runge, M. Vemora, and S. S. Viglioue, 1968.)

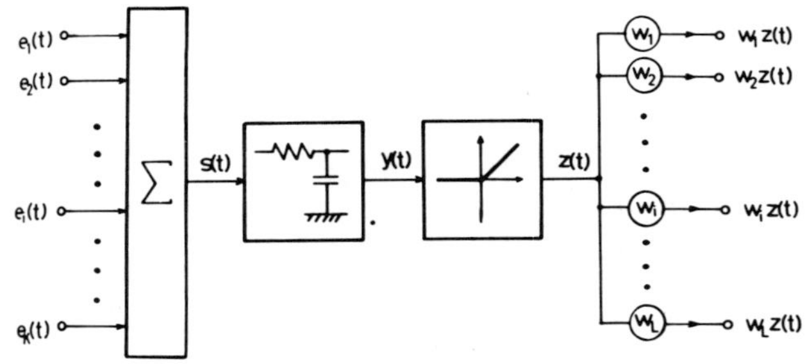

Figure 15-11. A continuous-variable model of the neuron. The adder, the low-pass filter, and the nonlinearity represent the dendrites, the nerve cell body, and the axon hillock, respectively. The weight-coefficient setting units represent the axon terminals. The inputs $e_i(t)$ represent postsynaptic potentials in the dendrites, and $y(t)$ and $z(t)$ represent the nerve-cell membrane potential and the frequency of nerve impulses that travel along the axon, respectively. (From Morishita and Yajima, 1972.)

to y_i if y_i is positive, whereas z_i is equal to 0 if y_i is negative; i.e., they are supposing that the output firing rate is approximately proportional to the level of the cell generator potential when that potential is positive. The x_i corresponds to the input to the ith cell; w_{ij} is the interconnection matrix of the network itself, while v_{ij} is the interconnection matrix between the laterally inhibiting network, on the one hand, and some other network, on the other hand, whose activity is represented by u_j. Morishita and Yajima then represent u_i, x_i, y_i, and z_i by vectors, and v_{ij} and w_{ij} by matrices, thus allowing them to rewrite this equation system as follows:

$$\begin{aligned}\tau\frac{dy(t)}{dt} &= x(t) - W(\phi[y(t)])y(t) \\ x(t) &= Vu(t) \\ z(t) &= y_i(t)\phi[y_i(t)]\end{aligned} \tag{15-2}$$

These authors then proceed to examine steady-state solutions of this equation system under certain simplifying assumptions. They show that, depending upon weight coefficients and the inputs, the steady-state solution has a unique equilibrium solution, a multiple equilibrium solution, or a periodic solution. They suggest that these three cases correspond, respectively, to three types of information processing: the sharpening of input patterns, the temporary storage of information, and the generation of periodic signals.

The concept of lateral inhibition, then, is extremely powerful, primarily in enabling one to conceptualize spatial sculpturing of input patterns.

However, one cannot yet relate the specific details of retinal circuitry to this concept. For example, lateral influences are effected in the retina by at least two classes of cells, horizontals and amacrines. Details of the operation of this circuitry are being worked out in several laboratories. Moreover, the mechanics whereby lateral inhibition relates to coloring coding remains as yet a mystery.

A particularly interesting feature of the outer plexiform layer is that the horizontal cells make outgoing connections on subsequent horizontal cells as well as on the retinal bipolar cells. It is easy to see that such an interconnection pattern provides pathways whereby any given retinal point is potentially connected with every other retinal point and, moreover, because of the finite time required to conduct potentials, provides an extremely interesting basis for spatiotemporal coding of information concerning the visual field. We have asked several questions concerning the spatial properties implied by these horizontal cell chains: What is the locus of receptive horizontal cell endings which influence a given retinal point? (This is tantamount to: What is the likely receptive field of a single midget bipolar cell?); or, turning the question about: What is the likely spatial distribution of effects when a single retinal point is stimulated?

The concern is with the locus of the receptive endings of horizontal cells whose activation eventually influences a given reference point in the retina. Such influence can be mediated after conduction through 1, 2, . . . , n horizontal cells. For simplicity, suppose that all the horizontal cells have the same length and neglect the spread of their dendritic trees and of their axonal terminations. Suppose that each axon in the chain can be oriented at any angle and that all orientations are equally likely. This problem is a two-dimensional isotropic random-walk problem, as illustrated in Figure 15-12. We wish to find the radial probability density for the nth-order horizontal cell endings, i.e., the probability $P_n(r)$ that a given nth-order ending will be localized between r and $r + dr$, where r is the radial distance from the point of reference.

We have found it convenient to solve this problem by finding first the probability density for the x component of the chain (i.e., the probability that the x component of the nth-order ending will be between x and $x + dx$), and then using an established relation to go from this density to the radial distribution.

The problem can be described by a series of random variables X_i, $i = 1, 2, \ldots, n$, denoting the projections of the n axons on the x axis. Let $f_i(x_i)dx_i$ denote the probability that X_i takes on a value between x_i and $x_i + dx_i$. It is easy to show that, when all orientations are equally likely,

$$f_i(x_i)dx_i = \begin{cases} \dfrac{1}{2\pi} \dfrac{dx_i}{(1 - x_i^2)^{1/2}}, & |x_i| < 1 \\ 0, & |x_i| > 1 \end{cases} \qquad (15\text{-}3)$$

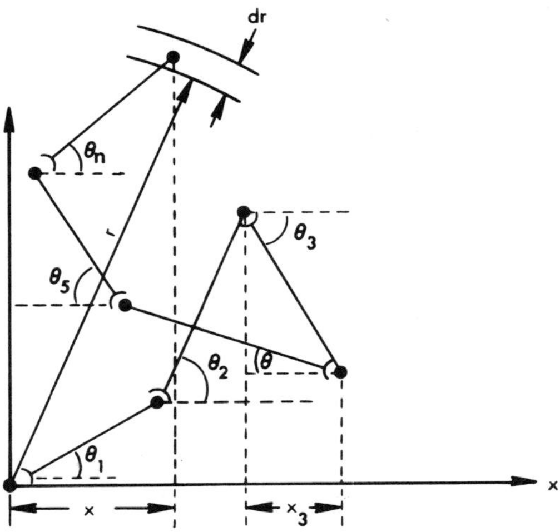

Figure 15-12. Schematic of a horizontal cell chain.

We will need below the Fourier transform of f_i, denoted by F_i and given by

$$F_i(t) = (2\pi)^{-\frac{1}{2}} J_0(t) \tag{15-4}$$

J_0 is the zeroth-order Bessel function.

We are interested in the probability $p_n(x)$ that the x component of the vector from the origin to the nth-order ending has the value x, where $x = \sum_{i=1}^{n} x_i$. Using an iteration process, one can write

$$\begin{aligned} p_n(x)dx \\ &= dx \int^{x_{n-1}} \int^{x_{n-2}} \cdots \int^{x_1} dx_1\, dx_2 \cdots dx_{n-1} f_1(x_1) f_2(x_2) \cdots f_{n-1}(x_{n-1}) \\ &\quad \times f_n(x - [x_1 + x_2 + \cdots + x_{n-1}]) \end{aligned} \tag{15-5}$$

Then, by a convenient theorem

$$p_n(x) = (2\pi)^{\frac{1}{2}n-1} \int_{-\infty}^{\infty} F_1(t) F_2(t) \cdots F_n(t) e^{-ixt}\, dt \tag{15-6}$$

where $F_i(t)$ is the Fourier transform of $f_i(x_i)$. Since in our case all the f_i are assumed to be identical, we have

$$
\begin{aligned}
p_n(x) &= (2\pi)^{\frac{1}{2}n-1} \int_{-\infty}^{\infty} F^n(t) e^{-ixt}\, dt \\
&= \frac{1}{2\pi} \int_{-\infty}^{\infty} J_0^n(t) e^{-ixt}\, dt \\
&= \frac{1}{\pi} \int_{-\infty}^{\infty} \cos(xt) J_0^n(t)\, dt
\end{aligned} \tag{15-7}
$$

A convenient relation for obtaining the radial distribution function $V_n(r)$ from the density of the x component, $p_n(x)$, is

$$
1 - V_n(r) = r \int_0^{\pi/2} p_n\left(\frac{r}{\sin\theta}\right) \frac{d\theta}{\sin^2\theta} \tag{15-8}
$$

The radial density $P_n(r)$ is related to the radial distribution function according to

$$
P_n(r) = dV_n(r)/dr \tag{15-9}
$$

Performance of the operations indicated in equations (15-7) to (15-9) leads to the following expressions for the radial density and distribution functions for the nth-order endings.

$$
\begin{aligned}
P_n(r) &= r \int_0^{\infty} t J_0^n(t) J_0(rt)\, dt \\
V_n(r) &= 1 + r \int_0^{\infty} t J_0^n(t) J_1(rt)\, dt
\end{aligned} \tag{15-10}
$$

To complete the picture, the following equations contain the limiting expressions for these functions, valid for large n:

$$
\begin{aligned}
P_n(r) &\to (2r/n) e^{-r^2/n} \\
V_n(r) &\to 1 - e^{-r^2/n}
\end{aligned} \tag{15-11}
$$

From equation (15-10) one can show that for $n = 1$, the density $P_1(r)$ is the Dirac delta function with the singularity at $r = 1$, while for $n = 2$, $P_2(r)$ is given as

$$
P_2(r) = \begin{cases} \dfrac{2}{\pi} \dfrac{1}{(4 - r^2)^{1/2}}, & r < 2 \\ 0, & r > 2 \end{cases} \tag{15-12}
$$

$P_3(r)$ can be expressed in terms of a complete elliptic integral, but for n equal to or greater than 4 the integral in equation (15-10) is intractable. We have performed numerical integration of the expression in equation (15-10) for $P_n(r)$ for several n. These results are shown in Figure 15-13.

A particularly interesting feature of these results is the peaks which occur in the curves for $n = 2$ and $n = 3$. Thus, we find that one-half of the second-order endings will occur in an annular region $\sqrt{2} < r < 2$, and one-third of the second-order endings will end in the narrow band $\sqrt{3} < r < 2$. Also, about one-fourth of the third-order endings will fall in an annulus around $r = 1$. For fourth- and higher-order endings the curves are seen to be relatively smooth.

One can offer tentative suggestions based on the results of Figure 15-13. A reasonable working hypothesis is that all outgoing horizontal cell synapses are inhibitory. It is easy to see that with respect to a given point in this case odd-ordered endings would be inhibitory while even-ordered endings would be disinhibitory, or essentially excitatory. Stimulation of a single point then (superimposed on a uniform background illumination) would result initially in a highly localized excitation, followed at one conduction time by a wave of inhibition localized in a narrow annulus around $r = 1$, followed at two conduction times by a wave of excitation spread throughout a circle with a diameter of 4 but tending to localize between $r = 1.5$ and 2, followed at three conduction times by a wave of inhibition spread throughout a circle of radius 6 with a slight tendency to localize in an annulus around $r = 1$, and followed then by alternating waves of excitation and inhibition which probably tend to decrease in effectiveness with time and exhibit little spatial individuality aside from a progressive increase in size.

Turning this around, one should predict that a wave of excitation (i.e., disinhibition) from second-order endings should influence a given retinal point and should tend to be localized in an annulus between about 1.5 and 2 horizontal cell lengths from the point. This higher-order disinhibition is

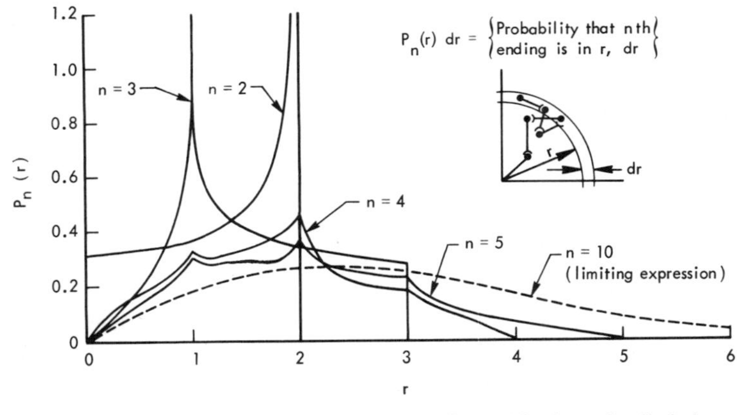

Figure 15-13. Probability distributions for endings of horizontal cell chains.

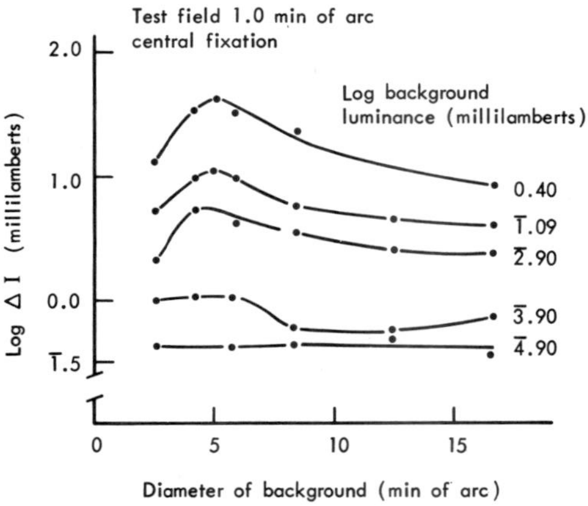

Figure 15-14. Differential intensity threshold for test field within larger background field.

indeed observed in a particularly relevant psychophysical study provided by Westheimer (1967). Westheimer measured the sensitivity in foveal vision to a small, brief test flash as influenced by a uniform background illumination of varying diameter *d*. His results are shown in Figure 15-14.

As the diameter of the background approaches 5 min, the effectiveness of the spot stimulus decreases, as is expected, according to the conventional picture of first-order laterally inhibiting endings. As the background diameter increases to, say, 7.5 or 10 min, however, the sensitivity to the spot increases as the interpretation of second-order disinhibition predicts.

PART IV

CONCLUSION

16

Neural Modeling: The State of the Art

We have seen in the pages of this book that a considerable amount of neural modeling activity exists and that many valuable accomplishments and contributions have been made. We have seen the establishment of physical theories which describe the essential physicochemical basis of neuroelectric activity. Such physical theories exist for both passive and active neuroelectric events. A general quantitative theory has been established to describe the spatial distribution and propagation of electrical potentials in neurons. Quantitative physically-based models for synaptic processes and receptor action are available. We have seen the establishment of general quantitative techniques and tools for simulating passive and active activity in neurons of arbitrary morphological configuration with a wide range of physiological characteristics. Generalized techniques for analyzing and displaying the characteristics of single and simultaneously recorded trains of action potentials exist, and a variety of workable quantitative tools are available for simulating and studying the dynamic activity of neuronal networks with optional degrees of verisimiltude. Thus, many basic neurophysiological mechanisms are understood with various degrees of certainty and are describable quantitatively, and many powerful and general quantitative tools are available for use in the study of their characteristics and ramifications.

There is, however, a need to satisfactorily integrate these activities within the rest of contemporary brain research. The purpose of neural modeling as identified in Chapter 1 is to relate function to structure on the basis of operation. We have seen in the pages of this book considerable development with respect to basic neurophysiological mechanisms underlying neural operations, and with respect to quantitative tools for studying such operations. However, the task of identifying and conceptualizing salient neural operations has been barely begun, and the significance of brain electrical operations within the functioning of the neurobehavioral systems in which they occur has been only touched upon. In view of these deficiencies, it seems to us worthwhile to provide what suggestions we can concerning the context of neural modeling and brain theory, goals, priorities, and operational guidelines for ongoing research.

16.1. The Stratification of Variables

The essential substantive content of brain modeling can be defined by a principle of stratification of variables. The main assumptions in this approach are that the variables, constructs, and knowledge which define the problem context of brain modeling arrange themselves naturally (or can be arranged) into strata such that individual strata identify highly interrelated concepts, and causal (or at least meaningful and more or less direct) relationships can be identified or sought for between adjacent strata.

A candidate stratification is indicated in Table 16-1. There are many imperfections in this particular stratification plan. First, the relative significance of structure and physiology in determining stratification is not entirely clear. On the basis of structure, one might feel inclined to relegate neurons and neural networks to different levels, whereas on the basis of electrophysiology one might be inclined to consider them in a common stratum because the electrophysiological and chemical signals which characterize their functional activity are intimately related and identical in some cases. (This latter grouping is suggested here.) Second, the clarity of the *inferred constructs* stratum indicated in Table 16-1 leaves much to be desired. For example, the concepts of ego and superego in Freudian psychology represent constructs inferred in an attempt to explain interrelationships of variables observed at the level of consciousness and behavior.

TABLE 16-1. Stratification of Variables

Variables	Examples
Behavior, consciousness	Will, mood
Inferred constructs	Ego, superego, libido (Freud), phase sequence (Hebb), circularities (Kubie)
Electrical signals, chemical secretions	EEG, EMG, spike trains, neurohormones
Cellular neurophysiology	Ionic fluxes, synaptic permeability changes, mechanisms of membrane excitation
Molecular processes	Molecular 'gating' processes, flux kinetics

The concept of "circularities" in theorizing on neural networks by Kubie, on the other hand, represents an inferred construct used to conceptualize the overall electrical activity of neural networks which can be only most imperfectly measured experimentally. The former case represents an instance of constructs inferred in a lower level from the behavior of variables of a higher level, whereas the latter case represents an instance of constructs inferred in a higher level from variables observed in a lower level. Should such constructs share the same stratum, be separated, or placed within different substrata within a given parent stratum? Nonetheless, the inferred constructs stratum is a significant feature of our stratification and, as elaborated below, represents an important meeting ground for psychologically-oriented and biologically-oriented theorists and modelers whose work approaches the stratum from opposite sides, as it were. Third, it is not entirely clear how consciousness and behavior should be stratified. Should these be in two distinct levels? If so, which should be above, and which below? If not, should there be two separate strata conceived to exist in parallel at the same level?

The main difficulty in establishing an entirely satisfactory stratification is that its validation presupposes a relatively complete development of the main sciences at hand, namely, the neurosciences and the psychological sciences, the progress of which it is precisely the purpose of such a stratification to facilitate. Thus, any particular candidate stratification will encounter objection until such time as the underlying sciences are considerably more complete than they are now. In the meantime, nonetheless, it seems to us that the idea of stratification of variables is a powerful general concept to keep in mind with respect to guiding progress in the fields, and, moreover, that the candidate stratification indicated in Table 16-1 in particular is sufficiently valid to be used to good advantage in considering priorities, goals, and operational guidelines for contemporary research in brain modeling.

One of the main values of stratification of variables is that it allows for the enunciation of several working principles which can be used to characterize certain models and to indicate directions for improvement in particular modeling efforts. The first such principle is that models which relate adjacent strata within the stratification scheme are the most powerful. There is a definite sense in which a causal relationship is implied in that the interrelations and operations of the variables at the lower level are used to explain the behavior of variables at the higher level. The model of Hodgkin and Huxley (1952*d*) for the generation of action potentials is an excellent example of a model which relates variables of adjacent strata. In their model, observed electrophysiological signals (action potentials in individual neurons) are related to variables of cellular neurophysiology. In the model, the properties of action potentials are explained in terms of transmembrane ionic fluxes. Another salient example of this kind of model is the model of cognitive psychology (Klatzky, 1975) which relates variables of the behavioral stratum to the operation of certain inferred constructs. In particular, this model explains the information-processing characteristics of people learning lists of items in terms of the inferred constructs of sensory register, short-term memory, long-term memory, pattern recognition, and attention. This principle reflects the classical position that the most satisfactory models are those that get to the heart of causal mechanisms involved in the phenomena at hand and describe their operation in essentially mechanistic fashion.

A second principle is that models which interrelate variables within one and the same stratum are not satisfying as ends in themselves and are severely limited in scope. Such an approach is often a necessary first step in modeling studies, but one should recognize clearly the limitations of the approach as well as its potential payoffs. Models which are so restricted generally do not reflect a grasp of the causal or driving apparatus at work in the phenomena. They often reflect heuristically or experimentally gained knowledge which, although often useful within its domain of validity, is generally powerless to predict behavior in ranges or conditions outside the particular experimental arrangement in which it was originally obtain. The so-called "black box" approach, often used in the very first steps of systems-theoretic investigation, is of this variety and is so limited. Models which have this characteristic may be said to exhibit the "unilevel" limitation.

A third principle is that models which skip one or more levels, i.e., which purport to interrelate variables of two distinct strata and do not consider variables of an intervening stratum or strata, have very low believability. Thus, for example, models which purport to describe or explain inferred processes or constructs relating to intelligence, cognitive processes, or other mental operations on the basis of the behavior of, say, cellular interconnections and rules for interaction between neurons must generally be treated with considerable skepticism. This is not to say that such study should not necessarily be undertaken, but rather that the theorist or modeler should recognize his responsibility to make the model testable by providing explicit predictions for the behavior of variables at the intervening level—in this example the intervening level is that of observed electrical signals in neurons or networks. Models which violate this principle may be said to exhibit the "gaping middle" syndrome.

A fourth principle states that the highest payoff is attained from theories which interrelate three or more adjacent strata. To avoid the pitfalls indicated above such theories should include or be composed of individual models which explicitly relate the variables of all pairs of adjacent strata in the scope of the theory. A salient example of this kind of model is the model for lateral inhibition in the retina (Ratliff, 1965), which interrelates variables from the level of cellular neurophysiology, the level of electrical signals and chemical secretions, and the level of inferred constructs. Thus, the model explains certain inferred prosses in visual coding (such as boundary enhancement) on the basis of certain observed properties of electrical recording from retinal neurons, and explains these in turn on the basis of neurophysiological mechanisms of the neurons in the retina. Seen in this context, the general model for lateral inhibition appears to us as the most powerful and broadly successful of all neural network models that have yet been attained. Another model which shares this property of interrelating three adjacent strata is the Melzack–Wall theory for pain (Melzack, 1973). This model relates inferred processes in pain to characteristics of electrical recordings, and these in turn to hypothetical "gating" mechanisms involving mechanistic interaction between structural elements.

These principles represent fair and rather clear implications of the concept of stratification of variables.

The stratification of variables also helps illuminate the level of neural networks as a singularly significant stratum in contemporary brain research in two contexts. First, neural networks, conceived as devices for coordinating and organizing electrical signals and chemical secretions, are extremely and fundamentally crucial to the overall development of comprehensive theory in the brain sciences (including both the psychological sciences and the neurosciences). Thus, as indicated above, although we do not yet have satisfactory concepts of even the identity of dynamic variables which most adequately characterize the operations of entire neural networks (let alone the organization of these variables), we do have considerable knowledge concerning the presumed basic neurophysiological mechanisms underlying such variables, and quantitative tools for studying them. Moreover, knowledge of these variables and operations constitutes precisely the stratum needed to substantiate those inferred constructs which serve in models purporting to deal with the basis of aspects of consciousness and behavior; i.e., the underlying premise of biological psychology is that the dynamic operations of inferred constructs underlying behavior and consciousness are to be explained in terms of the coordination of electrical signals and chemical secretions. This, in turn, is exactly the achievement we attribute to neural networks. In the very broadest sense, then, the level of neural networks conceived in this sense is seen as a vital stratum with respect to linking the physiological and psychological domains of brain research.

Second, the stratification of variables illuminates the level of neural networks as the critical dividing area between two relatively isolated domains of brain modeling activities, and in the process helps suggest how these activities can better serve each other and thereby serve the field generally. These two domains of brain modeling can be referred to as the "top-down" domain and the "neural modeling" domain, respectively.* There has been a rather clear and distinct separation in orientation between researchers in these two domains and a resulting communication void as well. The top-downers have been concerned primarily with constructing models which describe variables of the consciousness and behavior stratum in terms of variables of the inferred-constructs stratum without a great deal of concern for physiological detail. Neural modelers, on the other hand, have been generally concerned with physiological mechanisms and have generally been forced to forgo concern with variables of the consciousness and behavior stratum or to restrict their attention to more or less directly mediated functions such as respiratory and cardiac control or various reflex actions, due to the great complexity of the relevant and intricate physiological details. Top-downers emphasize the need for an overall guiding theory in studying a brain system; they have been known to consider the neural modelers as excessively concerned with minutiae to the point of losing sight of broader principles, and have on occasion thought of the latter as "neural muddlers." The neural modelers, on the other hand, feel that one theory solidly based on empirical evidence is more satisfying than many theories whose underlying premises are only loosely tied to empirical foundations; they have from time to time considered the models of the top-downers to be generally unbelievable and, on occasion, have considered the work of the latter to constitute a field better labeled as "false intelligence."

The stratification of variables suggests to us that the level of inferred constructs is the natural meeting ground for these two approaches, and that the elaboration of neural network studies is the essential ingredient for collating them. Thus, neural network studies represent, on the one hand, the highest level of achievement attainable by the "neural modeling" division of brain modeling, and, on the other hand, and according to the reasoning given above, the essential substrate for allowing physiological credence to the models of the "top-down" division of brain modeling.

16.2. Goals of Neural Network Modeling

Stratification of variables thus provides certain operational principles for guiding brain modeling and suggests that a high priority should be given to the level of neural networks as a key stratum in the brain sciences

*The present book attempts to survey the area defined on this page as "neural modeling"; the book *The Metaphorical Brain* by M. A. Arbib (1972), and other publications by Dr. Arbib and his group indicate the character and the flavor of the top-downer approach. Dr. Arbib is to be credited with coining the phrase "top-downer" and has repeatedly argued for the necessity of a theory of function to guide models of neural operation in studies of particular nervous components.

generally and in modeling particularly. It now seems possible to state, for the narrower field of neural network modeling, four main goals or charges which seem to comprehensively cover the responsibilities of the field and at the same time indicate the direction that progress should take within it.

The first charge is to attain an adequate understanding of the operational dynamics of the main neural networks of the mammalian nervous system. This is a clear and obvious goal. Priorities for various networks should be established based on the behavioral and clinical significance of the network considered, the degree of anatomical and physiological information available, the likelihood of significant achievement, the likelihood of generalization to other systems, and related factors. Certainly, we should hope to understand the operational dynamics of the various cerebral cortical regions, the hippocampus, the cerebellum, the reticular formation, hypothalamic coordinating nuclei, brain stem centers, spinal motor mechanisms, and so on. The main proof of achievement in the field generally can be best measured by the extent to which we understand the dynamics and operations of these systems.

The second charge is to attain adequate definition of and understanding of the operational dynamics of essential canonical abstract systems. Systems such as lateral inhibition, recurrent excitation, recurrent inhibition, and various combinations of these and perhaps other basic network arrangements should be understood in detail. An important element in a mature science of neural networks should be the understanding of main model systems which may be realized repeatedly with slight variations in different networks throughout nervous systems. Thus, for example, if contrast enhancement is a primary operation of lateral inhibition, we should understand what significance that operation has when it occurs within the retina, perhaps in the thalamus, perhaps in the cortex, perhaps in the spinal cord under different guises, and so on. We should have a feeling for how these model systems exhibit slightly different variations of the same main operational component when various parameter values are altered or different input patterns effected. This task is clearly a modeling task which deals theoretically with abstract model systems implying a variety of particular applications, rather than exclusively with the intricacies and vagaries of any one particular system.

The third charge is to attain realization of efficient multipurpose simulation or "working tool" models. These are primarily digital-computer programs or electronic devices to simulate the electrical activity of arbitrary configurations of neurons. A mature science of neural networks should be able to spell out in some detail dynamic particularities of a wide variety of particular configurational schemes with arbitrary physiological parameter settings. One should be able to use such devices to estimate the influence on activity of modulating certain parameter values or adjusting input configurations. Such general-purpose computational models are necessary because the complexity of detail surrounding any particular network theory or set of ideas is so high and the resulting reliability of qualitative theorizing so low that quantitative and precise elaboration of the implications of certain theoretical positions is essential to maintain reliable progress in the field. This is spelled out in some detail in Chapter 1. Such general-purpose models should be established in important brain research centers and made available to qualified researchers working on bona fide research activities within either of the two goals indicated above.

The fourth charge is to attain a concise and elegant mathematical description of the mechanics of neural networks. We have in mind a theory similar to the statistical mechanics of gases. The theory would not have to be useful as a "working tool" as discussed above, but rather serve to comprehend the essential mechanisms and relations operating in neural networks, and to identify salient variables and properties. Such an attainment is the crowning glory of any mature science and should reflect the most pristine and ethereal achievements of the field. It seems to us that such endeavors have an essential aesthetic value and as such should represent a clear goal in the ongoing development of the field. Within the context of the brain sciences, however, it seems also that these efforts should take second priority behind the three goals specified above, first because the three previous goals fully encompass many vital questions within areas of psychology, mental health, and the like, and second because an adequate mathematical formulation may quite likely be necessarily dependent on some certain degree of achievement in at least the first two of the above-mentioned goals and perhaps in all three of them.

The explicit establishment of such goals has the advantage that progress in the field can be measured by the extent to which achievement has been made toward those goals. Thus, if the science of neural network modeling is developing satisfactorily, one should see corresponding explicit achievement in the understanding of the operations of the main neural networks of the vertebrate nervous system, in the identification and understanding of main canonical systems, in the attainment of general-purpose simulation devices, and in the general mathematical formulation of the essential mechanistic basis of the area.

16.3. Guidelines for Brain Modeling

Stepping back again to the broader and more general field of brain modeling, we will conclude this discussion with some final suggestions for operational guidelines in the pursuance of research in brain modeling. First, modeling studies should be undertaken with, and where possible guided by, a clear appreciation for the functional context of the phenomena studied. The "top-downer's" point in this regard is well taken. Modelers should generally avoid attempting to project operational peculiarities of model systems in the absence of guidelines from function.

Second, the fundamental importance of the empirical basis of models should be recognized. Models and theorizing presented in the area should be accompanied by explicit and relatively detailed indications of critical experimental tests to corroborate or contradict critical model assumptions or operational trends. It is clearly time to request a higher level of scientific veracity from models than has been generally adhered to thus far. We must assume that the modeler who does not spell out how his model can be contradicted experimentally has not lived up to his responsibility for encouraging progress in the field generally. It is, after all, precisely this insistence upon active search for possible contradiction and this insistence only which maintains science as our only reliable tool for achieving rational understanding. It is no longer sufficient to construct equation or computer systems that seem reasonably to represent certain processes and then merely show the behavior of those systems, or to imagine a model system that seems a reasonable representation of some brain system and merely describe the workings of that model system. In both cases we should now ask that efforts at modeling be effected within a general context of relevance to brain science generally. Model systems should be designed for specific systems, or generalized models should be given parameter values representative of a specific real system and model results should be compared with data from that system. Modelers should not focus efforts exclusively on generalized abstracted canonical systems in the search for broad general principles that apply to many systems without recognizing the responsibility to spell out the information required by experimental refutation, and actively encourage or perform experimental test.

Third, it seems to us that experimentation and modeling work should go hand in hand. We conceive quantitative modeling as an activity on the same level with laboratory experimentation. Ideally, theoretical considerations should help determine and guide both experimentation and the accompanying quantitative modeling studies necessary to help interpret the experimental results within the theoretical context.

Fourth, it seems to us that modelers should keep in mind the principle of stratification of variables in an effort to see that they are getting maximum efficiency from their modeling efforts (see the four modeling principles suggested earlier in this chapter) and if their work can be related in any meaningful fashion to other theoretical and modeling work at adjacent strata. In this way we might hope to see a synergistic effect in the ongoing work of brain modeling which is certain to keep many of us busy for many decades.

References

Accardi, L., 1971, Rank and reverberations in neural networks, *Kybernetik* **8**, 163–164.

Adelman, W. J., 1971, *Biophysics and Physiology of Excitable Membranes*, Van Nostrand Reinhold Company, New York.

Adey, W. R., 1967, Intrinsic organization of cerebral tissue in altering, orienting, and discriminative responses, *The Neurosciences*, G. C. Quarton *et al.*, eds., pp. 615–633.

Adey, W. R., 1969, Neural information processing: Windows without and the citadel within, *Biocybernetics of the Central Nervous System*, L. D. Proctor, ed., Little, Brown and Company, Boston, pp. 1–27.

Adey, W. R., 1970, Cerebral structure and information storage, *Progress in Physiological Psychology*, Eliot Stellar and James M. Sprague, eds., Vol. 3, Academic Press, New York pp. 181–200.

Adolph, A. R., 1964, Spontaneous slow potential fluctuations in the *Limulus* photoreceptor, *Journal of General Physiology* **4B**, 297–322.

Aidley, D. J., 1971, *The Physiology of Excitable Cells*, Cambridge University Press, Cambridge, England.

Aiello, A., Burattini, E., and Caianiello, E. R., 1970, Synthesis of reverberations in neural networks, *Kybernetik* **7**, 191–195.

Aldrich, J. A., McCulloch, W. S., Lettvin, J. Y., and Pitts, W. H., 1962, The iron-wire neural nets, *Quarterly Progress Report of the Laboratory of Electronics, M.I.T.* **64**, 303–305.

Amari, S., 1974 A method of statistical neurodynamics, 1974, *Kybernetik* **14**, 201–215.

Amassian, V. E., 1961, Patterns of activity of simultaneously recorded neurons in midbrain reticular formation, *Annals of the New York Academy of Sciences* **89**, 883–895.

Amassian, V. E., and Waller, H. J., 1958, Spatiotemporal patterns of activity in individual reticular neurons, *Reticular Formation of the Brain*, H. H. Jasper, L. D. Proctor, R. S. Knighton, W. C. Noshay, and R. T. Costello eds., Little Brown and Company, Boston, Massachusetts, pp. 69–108.

Andersen, P., and Andersson, A., 1968, *Physiological Basis of the Alpha Rhythm*, Appleton-Century-Crofts, New York, 235 pages.

Andersen, P., Gillow, M., and Rudjord, T., 1966, Rhythmic activity in a simulated neuronal network, *Journal of Physiology* **185**, 418–428.

Anninos, P. A., 1972, Cyclic modes in artificial neural nets, *Kybernetik* **11**, 11–15.

Anninos, P. A., 1973, Evoked potentials in artificial neural nets, *Kybernetik* **13**, 24–29.

Anninos, P. A., and Raman, S., 1975, Derivation of a mathematical equation of the EEG and the general solution within the brain and in space, *International Journal of Theoretical Physics* **12** (1), 1–9.

Anninos, P. A., Beek, B., Csermely, T. J., Harth, E. M., and Pertile, G., 1970, Dynamics of neural structures, *Journal of Theoretical Biology* **26**, 121–148.

Arbib, M. A., 1972, *The Metaphorical Brain*, Wiley-Interscience, New York.

Barr, L., 1965, Membrane potential profiles and the Goldman equation, *Journal of Theoretical Biology* **9**, 351–356.

Baylor, D. A., and Hodgkin, A. L., 1974, Changes in time scale and sensitivity in turtle photoreceptors, *Journal of Physiology* **242**, 729–758.

Beament, J. W. L., ed., 1962, Biological receptor mechanisms, *Symposia of the Society for Experimental Biology*, Vol. 16.

Beidler, L. M., and Reichardt, W. E., 1971, Sensory transduction, *Neurosciences Research Symposium Summaries*, F. O. Schmitt *et al.*, eds., M.I.T. Press, Cambridge, Massachusetts, pp. 441–538.

Bell, C. C., and Grimm, R. J., 1969, Discharge properties of Purkinje cells recorded on single and double microelectrodes, *Journal of Neurophysiology* **32**, 1044–1055.

Bell, C. C., and Kawasaki, T., 1972, Relations among climbing fibers to responses of nearby Purkinje cells, *Journal of Neurophysiology* **35**, 35–155.

Bennet, M. V. L., ed., 1974, *Synaptic Transmission and Neuronal Interaction*, Raven Press, New York.

Bennett, M. R., 1967, An analysis of the surface fixed-charge theory of the squid giant axon membrane, *Biophysical Journal* **7**, 151–164.

Berger, H. 1969, *The Electroencephalogram of Man, Collected Papers*, Elsevier, New York.

Beurle, R. L., 1956, Properties of a mass of cells capable of regenerating pulses, *Philosophical Transactions of the Royal Society of London, Series A* **240**, pp. 55–94.

Bonhoeffer, K. F., 1948, Activation of passive iron as a model for excitation of nerve, *Journal of General Physiology* **32**, 69–91.

Borsellino, A., and Fuortes, M. G. F., 1968, Interpretation of responses of visual cells of *Limulus*, *Proceedings of the I.E.E.E.* **56**, 1024–1032.

Borsellino, A., Fuortes, M. G. F., and Smith, T. G., 1965, Visual responses in *Limulus*, *Cold Spring Harbor Symposia on Quantitative Biology* **30**, 429–443.

Boyde, I. A., and Martin, A. R., 1956, Spontaneous subthreshold activity at mammalian neuromuscular junctions, *Journal of Physiology* **132**, 74–91.

Bradley, K., and Somjen, G. G., 1961, Accommodation in motoneurons of the rat and the cat, *Journal of Physiology* (*London*) **156**, 75–92.

Brindley, G. S., 1970, *Physiology of the Retina and Visual Pathway*, Williams & Wilkins Co., Baltimore.

Brown, J. E., and Blinks, J. R., 1974, Changes in intracellular free calcium concentration during illumination of invertebrate photoreceptors, *Journal of General Physiology* **64**, 643–665.

Bryant, H. L., Marcos, A. R., and Segundo, J. P., 1973, Correlations of neuronal discharges produced by mono-synaptic connections and common inputs, *Journal of Neurophysiology* **36**, 205–225.

Bullock, T. H., 1959, Neuron doctrine and electrophysiology, *Science* **129**, 997–1002.

Bullock, T. H., 1970, The reliability of neurons, *Journal of General Physiology* **55**, 563–584.

Bullock, T. H., and Hagiwara, S., 1957, Intracellular potentials in pacemaker and integrative neurons of the lobster cardiac ganglion, *Journal of Cellular and Comparative Physiology* **50**, 25–48.

Bullock, T. H., and Horridge, G. A., 1965, *Structure and Function in the Nervous Systems of Invertebrates*, W. H. Freeman, San Francisco.

Caianiello, E. R., 1961, Outline of a theory of thought processes and thinking machines, *Journal of Theoretical Biology* **2**, 204–235.

Caianiello, E. R., ed., 1968, *Neural Networks: Proceedings of the School on Neural Networks, June, 1967, in Ravello*, Springer-Verlag, New York.

Caianiello, E. R., de Luca, A., and Ricciardi, L. M., 1968, Neural networks: Reverberations, constants of motion, general behavior, *Neural Networks: Proceedings of the School on Neural Networks, June, 1967, in Ravello*, E. R. Caianiello, ed., Springer-Verlag, New York, pp. 92–99.

Capocelli, R. M., and Ricciardi, L. M., 1971, Diffusion approximation and first passage time problem for a model neuron, *Kybernetik* **8**, 214–223.

Capocelli, R. M., and Ricciardi, L. M., 1972, On the inverse of the first passage time probability problem, *Journal of Applied Probability* **9**, 270–287.

Carnay, L. D., and Tasaki, I., 1971, Ion exchange properties and excitability of the squid giant axon, *Biophysics of Excitable Membranes*, W. J. Adelman, ed., Van Nostrand Reinhold Co., New York, pp. 379–422.

Casey, K. L., 1969, Somatic stimuli, spinal pathways, and size of cutaneous fibers influencing unit activity in the medical medullary reticular formation, *Experimental Neurology* **25**, 35–56.

Castillo, J., del, and Katz, B., 1954, Statistical factors involved in neuromuscular facilitation and depression, *Journal of Physiology* **124**, 574–585.

Charlton, J. S., 1973, Characterization of an intermediate process in the lateral eye photoreceptors of *Limulus polyphemus*, Dissertation, College of Engineering, University of California, Berkeley.

Clay, J. R., 1976, A stochastic analysis of the graded excitatory responses of nerve membrane, *Journal of Theoretical Biology* **59**, 141–158.

Cohen, E., and Cooley, J. W. 1965, The numerical solution of the time-dependent Nernst–Planck equations, *Biophysical Journal* **5**, 145–162.

Cole, K. S., 1968, *Membranes, Ions and Impulses*, University of California Press, Berkeley.

Cole, K. S., 1971, Some aspects of electrical studies of the squid giant axon membrane, *Biophysics and Physiology of Excitable Membranes*, W. J. Adelman, ed., Van Nostrand Reinhold, New York, pp. 125–142.

Cole, K. S., Antosiewicz, H. A., and Rabinowitz, P., 1955, Automatic computation of nerve excitation, *Journal of the Society for Industrial and Applied Mathematics* **3**, 3.

Cole, K. S., Guttman, R., and Bezanilla, F., 1970, Nerve membrane excitation without threshold, *Proceedings of the National Academy of Sciences of the U.S.A.* **65**, 884–891.

Connor, J. A., and Stevens, C. F., 1971, Prediction of repetitive firing behavior from voltage clamp data on an isolated neuron soma, *Journal of Physiology* (*London*) **213**, 31–54.

Cooley, J. W., and Dodge, F. A., Jr., 1966, Digital computer solutions for excitation and propagation of the nerve impulse, *Biophysical Journal* **6**, 583–599.

Coombs, J. S., Eccles, J. C., and Fatt, P., 1955, The specific ionic conductances and the ionic movements across the motoneuronal membrane that produce the inhibitory post-synaptic potential, *Journal of Physiology* **130**, 326–373.

Cowan, J. D., 1968, Statistical mechanics of nervous nets, *Neural Networks: Proceedings of the School on Neural Networks, June, 1967, in Ravello*, E. R. Caianiello, ed., Springer-Verlag, New York, pp. 182–188.

Cowan, J. D., 1972, Stochastic models of neuroelectric activity, *Proceedings International Union for Pure and Applied Probability Conference on Statistical Mechanics*, S. Rice, J. Light, and K. Freed, eds., University of Chicago Press, Chicago.

Cowan, J., and Winograd, S., 1963, *Reliable Computation in the Presence of Noise*, M.I.T. Press, Cambridge, Massachusetts.

Dalton, J. C., and FitzHugh, R., 1960, Applicability of Hodgkin–Huxley model to experimental data from the giant axon of lobster, *Science* **131**, 1533–1534.

Deutsch, S., 1967, *Models of the Nervous System*, John Wiley & Sons, Inc., New York.

Dill, J. C., Randall, D. L., and Richer, I., 1968, Plexus—An on-line system for modeling neural networks, *Proceedings of the Association for Computing Machinery* **11**, 622–629.

Dodge, F. A., 1963, A study of ionic permeability changes underlying excitation in myelinated nerve fibers of the frog, Thesis, The Rockefeller University, New York, University Microfilms No. 64-7333.

Dodge, F. A., and Frankenhaeuser, B., 1958, Membrane currents in isolated frog nerve fibre under voltage clamp conditions, *Journal of Physiology* **143**, 76–90.

Dodge, F. A., and Frankenhaeuser, B., 1959, Sodium currents in the myelinated nerve fiber of *Xenopus laevis* investigated with the voltage clamp technique, *Journal of Physiology* **148**, 188–200.

Dodge, F. A., Knight, B. W., and Toyoda, J., 1968, Voltage noise in *Limulus* visual cells, *Science* **160**, 88–90.

Eccles, J. C., 1957, *The Physiology of Nerve Cells*, Johns Hopkins Press, Baltimore.

Eccles, J. C., 1964, *The Physiology of Synapses*, Academic Press, New York.

Eccles, J. C., 1970, *Facing Reality: Philosophical Adventures by a Brain Scientist*, Springer-Verlag, New York, 210 pages.

Efremova, T. M., and Trush, V. D., 1970, *Study of the Organization of Neuronal Activity in the Cerebral Cortex*, M. N. Livanov, ed., Nauka, Moscow.

Efremova, T. M., Zhadin, M. N., Kaburneeva, L. I., Kondrat'eva, I. N., Melekhova, A. M., and Sul'gina, G. I., 1950, Correlation analysis of spike activity of adjacent cortical neurons in rabbits, *Rosinor*, Institute of Higher Nervous Activity and Neurophysiology, Academy of Sciences of the U.S.S.R., Moscow, pp. 113–118.

Eigen, M., and DeMaeyer, L., 1971, Carriers and specificity in membranes, *Neuroscience Research Program Bulletin* **9**, 300–437.

Elul, Rafael, 1972, The genesis of the EEG, *International Review of Neurobiology* **15**, 227–272.

Erdélyi, A., Magnus, W., Oberhettinger, F., and Tricomi, F. G., 1954, *Tables of Integral Transforms*, McGraw-Hill, New York.

Farley, B. G., and Clarke, W. A., 1961, Activity in networks in neuronlike elements, *Information Theory (Fourth London Symposium)*, C. Cherry, Ed., Butterworth Scientific Publications, London.

Fehmi, L. G., and Bullock, T. H., 1967, Discrimination among temporal patterns of stimulation in a computer model of a coelenterate nerve net, *Kybernetik* **3** (5), 240–249.

Feldman, J. L., and Cowan, J. D., 1975, Large-scale activity in neural nets. I. Theory with application to motoneuron pool responses, *Biological Cybernetics* **17**, 29–38.

Feller, W., 1971, *An Introduction to Probability and Its Applications*, Vol. 2, Wiley, New York, pp. 311–323.

Fernald, R., 1971, A neuron model with spatially distributed synaptic input, *Journal of Biophysics* **11**, 323.

Fienberg, S. E., 1970, A note on the diffusion approximation for single neuron firing problems, *Kybernetik* **6**, 227–229.

Fienberg, S. E., 1972, Stochastic models for single neuron firing trains: A survey, Technical Report No. 187, Department of Applied Statistics, School of Statistics, University of Minnesota, 36 pages.

Fienberg, S. E., and Hochman, H. G., 1972, Modal analysis of renewal models for spontaneous single neuron discharge, *Kybernetik* **11**, 201–207.

Finkelstein, A., and Mauro, A., 1963, Equivalent circuits as related to ionic systems, *Biophysical Journal* **3**, 215–237.

FitzHugh, R., 1955, Mathematical models of threshold phenomena in the nerve membrane, *Bulletin of Mathematical Biophysics* **17**, 257–278.

FitzHugh, R., 1960, Thresholds and plateaus in the Hodgkin–Huxley nerve equations, *Journal of General Physiology* **43**, 867–896.

FitzHugh, R., 1961, Impulses and physiological states in theoretical models of nerve membrane, *Biophysical Journal* **1**, 445–466.

FitzHugh, R., 1962, Computation of impulse initiation and saltatory conduction in a myelinated nerve fiber, *Biophysical Journal* **2**, 11–21.

FitzHugh, R., and Antosiewicz, H. A., 1959, Automatic computation of nerve excitation-detailed corrections and additions, *Journal of the Society for Industrial and Applied Mathematics* **7**, 447–458.

Franck, U. F., 1956, Models for biological excitation processes, *Progress in Biophysics* **6**, 171–206.

Frankenhaeuser, B., 1960, Sodium permeability in toad nerve and in squid nerve, *Journal of Physiology* **152**, 159–166.

Frankenhaeuser, B., 1963, A quantitative description of potassium currents in myelinated nerve fibers of *Xenopus laevis*, *Journal of Physiology* **169**, 424–430.

Frankenhaeuser, B., and Hodgkin, A. L., 1956, The aftereffects of impulses in the giant nerve fibers of *Loligo*, *Journal of Physiology* **131**, 341–376.

Frankenhaeuser, B., and Valbo, H., 1965, Accommodation in myelinated nerve fibers of *Xenopus laevis* as computed on the basis of voltage clamp data, *Acta Physiologica Scandanavica* **63**, 1–20.

Freeman, W. F., 1972, Waves, pulses, and the theory of neural masses, *Progress in Theoretical Biology*, Vol. 1., John Wiley and Sons, New York, pp. 1–34.

French, A. S., and Stein, R. B., 1970, Flexible neural analog using integrated circuits, *I.E.E.E. Transactions on Biomedical Engineering* **BME-17**, 248–253.

Fujita, Y., and Sato, T., 1964, Intracellular records from hippocampal pyramidal cells in rabbit during theta rhythm activity, *Journal of Neurophysiology* **27**, 1011–1025.

Fuortes, M. G. F., 1959, Initiation of impulses in visual cells of *Limulus*, *Journal of Physiology* **148**, 14–28.

Fuortes, M. G. F., and Hodgkin, A. L., 1964, Changes in time scale and sensitivity in the ommatidia of *Limulus*, *Journal of Physiology* **172**, 23–263.

Fuortes, M. G. F., and Yeandle, S., 1964, Probability of occurrence of discrete potential waves in the eye of *Limulus*, *Journal of General Physiology* **47**, 443–463.

Furshpan, E. J., and Potter, D. D., 1959, Slow post-synaptic potentials recorded from the giant motor fiber of the crayfish, *Journal of Physiology* **145**, 326–335.

Gawronski, R. R., 1971, *Bionics*: *The Nervous System as a Control System*, Elsevier Publishing Company, New York.

Gerstein, G., and Kiang, N.-S., 1960, An approach to quantitative analysis of electrophysiological data, *Biophysical Journal* **1**, 15–28.

Gerstein, G. L., and Mandelbrot, B., 1964, Random walk models for the spike activity of a single neuron, *Biophysical Journal* **4**, 41–68.

Gerstein, G. L., and Perkel, D. H., 1969, Simultaneously recorded trains of action potentials: Analysis and functional interpretation, *Science* **164**, 828–830.

Gilbert, D. L., 1971, Fixed surface charges, *Biophysics and Physiology of Excitable Membranes*, W. J. Adelman, Jr., ed., Van Nostrand Reinhold, New York, pp. 359–378.

Gluss, B., 1967, A model for neuron firing with exponential decay of potential resulting in diffusion equations for probability density, *Bulletin of Mathematical Biophysics* **29**, 233–243.

Griffith, J. S., 1963, A field theory of neural nets. I. *Bulletin of Mathematical Biophysics* **25**, 111–120.

Griffith, J. S., 1965, A field theory of neural nets. II. *Bulletin of Mathematical Biophysics* **27**, 187–195.

Griffith, J. S., 1971, *Mathematical Neurobiology*: *An Introduction to the Mathematics of the Nervous System*, Academic Press, New York, Sec. 4-3.

Grossberg, S., 1969*a*, On learning, information, lateral inhibition, and transmitters, *Mathematical Biosciences*, **4**, 255–310.

Grossberg, S., 1969*b*, Some networks that can learn, remember, and reproduce any number of complicated space-time patterns. I, *Journal of Mathematics and Mechanics* **19** (1), 53–91.

Grossberg, S., 1970, Neural pattern discrimination, *Journal of Theoretical Biology* **27**, 291–337.

Hagiwara, S., 1954, Analysis of interval fluctuation of the sensory nerve impulse, *Japanese Journal of Physiology* **4**, 23–240.

Hagiwara, S., and Tasaki, I., 1958, A study on the mechanism of impulse transmission across the giant synapse of the squid, *Journal of Physiology* **143**, 114–137.

Harmon, L. D., 1961*a*, Artificial neuron, *Science* **129**, 962–963.

Harmon, L. D., 1961*b*, Studies with artificial neurons. I. Properties and functions of an artificial neuron, *Kybernetik* **1**, 89–101.

Harmon, L. D., 1964, Neuromimes: Action of a reciprocally inhibitory pair, *Science* **146**, 1323–1325.

Harmon, L. D., and Lewis, E. R., 1966, Neural modeling, *Physiological Reviews* **46**, 113–192.

Harris, T. R., and Weaver, R. E. C., 1962, The transmission of nerve signals along the axon: A mathematical simulation, *Chemical Engineering in Medicine* **62**, (66), 109–122.

Harth, E. M., Csermely, T. J., Beek, B., and Lindsay, R. D., 1970, Brain functions and neural dynamics, *Journal of Theoretical Biology* **26**, 93–120.

Heathcote, H. L., 1907, The passivifying, passivity, and activifying of iron, *Journal of the Society of Chemical Industry* (*London*) **26**, 899–917.

Hebb, D. O., 1949, *The Organization of Behavior*, Wiley, New York.

Hellerstein, D., 1968, Passive membrane potentials. A generalization of the theory of electrotonus, *Biophysical Journal* **8**, 358–379.

Herschel, J. F., 1833, Lettre fur la maniere d'agir de l'acide nitrique fur le fer, *Annales de Chimie et de Physique* **54**, 87–94.

Hill, A. V., 1936, Excitation and accommodation in nerve, *Proceedings of the Royal Society* (*London*), *Series B* **119**, 305–355.

Hille, B., 1967, A pharmacological analysis of the ionic channels of nerve, Thesis, The Rockefeller University, New York, University Microfilms No. 68-9584.

Hille, B., 1971, Voltage clamp studies on myelinated nerve fibers, *Biophysics and Physiology of Excitable Membranes*, W. J. Adelman, ed., Van Nostrand Reinhold, New York, pp. 230–246.

Hodgkin, A. L., and Huxley, A. F., 1952*a*, Currents carried by sodium and potassium ions through the membrane of the giant axon of *Loligo*, *Journal of Physiology* **116**, 449–472.

Hodgkin, A. L., and Huxley, A. F., 1952*b*, The components of membrane conductance in the giant axon of *Loligo*, *Journal of Physiology* **116**, 473–496.

Hodgkin, A. L., and Huxley, A. F., 1952*c*, The dual effect of membrane potential on sodium conductance in the giant axon of *Loligo*, *Journal of Physiology* **116**, 497–506.

Hodgkin, A. L., and Huxley, A. F., 1952*d*, A quantitative description of membrane current and its application to conduction and excitation in nerve, *Journal of Physiology* **117**, 500–544.

Hodgkin, A. L., Huxley, A. F., and Katz, B., 1952, Measurement of current–voltage relations in the membrane of the giant axon of *Loligo*, *Journal of Physiology* **116**, 424–448.

Holmes, O., and Houchin, J., 1966, Units in the cerebral cortex of the anesthetized rat and correlations between their discharge, *Journal of Physiology* (*London*) **187**, 651–671.

Holmes, O., and Houchin, J., 1967, Analysis of the activity of one type of spontaniously discharging unit in the cortex cerebri of the anaesthetized rat, *Journal of Physiology* **193**, 173–186.

Horridge, G. A., 1968, *Interneurons*, W. H. Freeman, San Francisco.

Hubbard, J. I., Llinas, R., and Quastel, D. M. J., 1969, *Electrophysiological Analysis of Synaptic*

Transmission, Physiological Society Monograph No. 19, Williams and Wilkins, Baltimore.

Hubbard, R., and Wald, G., 1960, Visual pigment of the horseshoe crab, *Limulus polyphemus*, *Nature* **186**, 212–215.

Huxley, A. F., 1959, Can a nerve propagate a subthreshold disturbance? *Journal of Physiology* **148**, 80P–81P.

Iansek, R., and Redman, S. J., 1973, An analysis of the cable properties of spinal motoneurones using a brief intracellular current pulse, *Journal of Physiology* **234**, 613–636.

Jack, J. J. B., and Redman, S. J., 1971*a* The propagation of transient potentials in some linear cable structures, *Journal of Physiology* **215**, 283–320.

Jack, J. J. B., and Redman, S. J., 1971*b*, An electrical description of the motoneurone, and its application to the analysis of synaptic potentials, *Journal of Physiology* **215**, 321–352.

Jack, J. J. B., Noble, D., and Tsien R. W., 1975, *Electric Current Flow in Excitable Cells*, Oxford University Press, Oxford, England.

Jenik, F., 1965 An outline of the development of a neuron model, *Progress in Brain Research* **17**, 356–368.

John, E. R., 1967, *Mechanisms of Memory*, Academic Press, New York, 468 pages.

Josephson, R. K., Reiss, R. F., and Worthy, R. M., 1961, A simulation of a diffuse conducting system based on coelenterate nerve nets, *Journal of Theoretical Biology* **1**, 460–487.

Julian, F. J., Moore, J. W., and Goldman, D. E., 1962, Current–voltage relations in the lobster giant axon membrane under voltage clamp conditions, *Journal of General Physiology* **45**, 1217.

Kado, R. T., 1973, Aplysia giant cell: Soma-axon voltage-clamp current differences, *Science* **182**, 843–845.

Katchalsky, A. K., Rowland, V., and Blumental, R., 1974, *Dynamic Patterns of Brain Cell Assemblies*, MIT Press, Cambridge, Massachusetts, 187 pages.

Kater, S. B., 1974, Feeding in *Helisoma trivolvis*: The morphological and physiological bases of a fixed action pattern, *American Zoologist* **14**, 1017–1036.

Katz, B., 1966, *Nerve, Muscle, and Synapse*, McGraw-Hill, New York.

Katz, B., 1969, *The Release of Neural Transmitter Substances*, C. C. Thomas, Springfield, Illinois.

Katz, B., and Miledi, R., 1965, The effect of temperature on the synaptic delay at the neuromuscular junction, *Journal of Physiology* **181**, 656–670.

Keir, J., 1790, Experiments and observations on the dissolution of metals in acids and their precipitation, with an account of a new compound acid menstrum, useful in some technical operation of parting metals, *Philosophical Transactions of the Royal Society of London* **80**, 359–384.

Kernell, D., 1968, The repetitive impulse discharge of a simple neuron model compared to that of spinal motorneurones, *Brain Research* **11**, 685–687.

Kilmer, W., and Olinski, M., 1974, Model of a plausible learning scheme for CA3 hippocampus, *Kybernetik* **16**, 133–144.

Kilmer, W. L., McCulloch, W. S., and Blum, J., 1968, An embodiment of some vertebrate command and control principles, *Currents in Modern Biology* **2**, 81–97.

Kittel, C., 1958, *Elementary Statistical Physics*, Wiley, New York.

Klatzky, R. L., 1975, *Human Memory*: *Structure and Processes*, W. H. Freeman, San Francisco.

Kling, I. V., and Szekely, G., 1968, Simulation of rhythmic neurons activity. I. Function of nets with cyclic inhibitions, *Kybernetik* **5**, 89–103.

Lashley, K. S., 1960, *The Neuropsychology of Lashley*, F. A. Beach, D. O. Hebb, C. T. Morgan, and H. W. Nissen, eds., McGraw-Hill Book Company, Inc., New York, 564 pages.

Leibovic, K. N., 1972, *Nervous System Theory*, 1972, Academic Press, New York, 294 pages.

Levinson, J., and Harmon, L. D., 1961, Studies with artificial neurons. III. Mechanisms of flicker fusion, *Kybernetik* **1**, 107–117.

Lewis, E. R., 1965, Neuroelectric potentials derived from an extended version of the Hodgkin–Huxley model, *Journal of Theoretical Biology* **10**, 125–158.

Lewis, E. R., 1968*a*, Using electronic circuits to model simple neuroelectric interactions, *Proceedings of the I.E.E.E.* **56**, 931–949.

Lewis, E. R., 1968*b*, The iron wire model of the neuron: A review, *Cybernetic Problems in Bionics*, H. L. Oestreicher and D. R. Moore, eds., Gordon & Breach, New York, pp. 247–273.

Lewis, E. R., 1968*c*, An electronic model of neuroelectric point processes, *Kybernetik* **5**, 30–46.

Lewis, E. R., 1974, A note on transfer and driving-point functions of iterated ladder networks, *I.E.E.E. Transactions on Circuits and Systems* **CAS-211**, 334–338.

Lewis, E. R., and J. S. Charlton, 1972, Systems models of visual transduction, *Proceedings of the International Conference on Systems Science* **5**, 146–148.

Lewis, E. R., and Li, C. W., 1975, Hair cell types and distributions in the otolithic and auditory organs of the bullfrog, *Brain Research* **83**, 35–50.

Lewis, E. R., Zeevi, Y. Y., and Werblin, F. S., 1969, Scanning electron microscopy of vertebrate visual receptors, *Brain Research* **15**, 559–562.

Lillie, R. S., 1909, On the connection between changes of permeability and stimulation and the significance of changes in permeability to carbon dioxide, *American Journal of Physiology* **24**, 14–44.

Lillie, R. S., 1914, The conditions determining the rate of conduction in irritable tissues and especially in nerve, *American Journal of Physiology* **34**, 414–445.

Lillie, R. S., 1915, The conditions of conduction of excitation in irritable cells and tissues and especially in nerve, *American Journal of Physiology* **37**, 348–370.

Lillie, R. S., 1916 The conditions of physiological conduction in irritable tissues, *American Journal of Physiology* **41**, 126–136.

Lillie, R. S., 1922, Transmission of physiological influence in protoplasmic systems, especially nerve, *Physiological Reviews* **2**, 1–37.

Lillie, R. S., 1928, Analogies between physiological rhythms and the rhythmical reactions in inorganic systems, *Science* **15**, 593–598.

Lillie, R. S., 1936, The passive iron wire model of protoplasmic and nervous transmission and its physiological analogues, *Biological Reviews of the Cambridge Philosophical Society* **11**, 181–209.

Liu, S. F., 1966, *Digital computer simulation of a simplified nerve net*, Defense Documentation Center, AD-634, p. 289.

Llinas, R., and Nicholson, C., 1969, Electrophysiological analysis of alligator cerebellar cortex: A study on dendritic spikes, *Neurobiology of Cerebellar Evolution and Development*, R. Llinas, ed., American Medical Association, Chicago, pp. 431–465.

Loewenstein, W. R., ed., 1966, Biological membranes: Recent progress, *Annals of the New York Academy of Science* **137**, 403–1044.

McCulloch, W. S., and Pitts, W., 1943, A logical calculus of the ideas immanent in nervous activity, *Bulletin of Mathematical Biophysics* **5**, 115–137.

MacGregor, R. J., 1966*a* A digital-computer model of spike elicitation by postsynaptic potentials in single nerve cells, RM-4877-ARPA, The Rand Corporation, Santa Monica, California.

MacGregor, R. J., 1966*b*, Pulse trains in lateral geniculate and retinal ganglion nerve cells, RM-4870-ARPA, The Rand Corporation, Santa Monica, California.

MacGregor, R. J., 1967*a*, Neural organization in the primate retina, RM-4912-ARPA, The Rand Corporation, Santa Monica, California.

MacGregor, R. J., 1967*b*, Input–output relations for axosomatic activation in a neuron model, P-3672, The Rand Corporation, Santa Monica, California.

MacGregor, R. J., 1968*a*, A model for continuous neuroelectric activity: The encoding of stimulus intensity, P-3747, The Rand Corporation, Santa Monica, California.

MacGregor, R. J., 1968*b*, A model for responses to activation by axodendritic synapses, *Biophysical Journal* **8** (3), 305–318.

MacGregor, R. J., 1969*a*, Spatial organization implied by horizontal cell chains in the vertebrate retina, P-42118, The Rand Corporation, Santa Monica, California.

MacGregor, R. J., 1969*b*, A mechanism for coincidence detection in dendrites, RM-5598-RC, The Rand Corporation, Santa Monica, California.

MacGregor, R. J., 1970, A model to relate flicker fusion and intensity coding to graded potentials in single units of the outer plexiform layer of the retina, WN-7100-RC, The Rand Corporation, Santa Monica, California.

MacGregor, R. J., 1971, Intrinsic oscillations of neural networks: A linear model for the *n*th-order loop, *Mathematical Biosciences* **11**, 311–336.

MacGregor, R. J., 1972, A model for reticular-like networks: Ladder nets, recruitment fuses, and sustained responses, *Brain Research* **41**, 345–363.

MacGregor, R. J., 1973, Intrinsic oscillation of neural networks: A linear model for parallel, single-unit pathways, *Mathematical Biosciences* **17**, 121–135.

MacGregor, R. J., and Oliver, R. M., 1973, A general-purpose electronic model for arbitrary configuration of neurons, *Journal of Theoretical Biology* **38**, 527–538.

MacGregor, R. J., and Oliver, R. M., 1974, Model for repetitive firing in neurons, *Kybernetik* **16**, 53–64.

MacGregor, R. J., and Palasak, R., 1974, Computer simulation of rhythmic oscillations in a neuron pool, *Kybernetik* **16**, 79–86.

MacGregor, R. J., and Radcliffe, C., 1973, A random walk model for reverberatory pathways in reticular-like networks, *Brain Research* **63**, 374–379.

MacGregor, R. J., and Sharpless, S. K., 1973, Repetitive discharge rate of a simple neuron model with accumulation of after-hyperpolarization conductance, *Brain Research* **64**, 387–390.

MacGregor, R. J., Prieto-Diaz, R., Miller, S. W., and Groves, P. M., 1973, Statistical properties of neurons in the rat mesencephalic reticular formation, *Brain Research* **64**, 167–187.

MacGregor, R. J., Miller, S. W., and Groves, P. M., 1975*a*, Slow rhythms and correlations in spike trains from midbrain neurons, *Experimental Neurology* **47**, 581–598.

MacGregor, R. J., Sharpless, S. K., and Luttges, M. W., 1975*b*, A pressure vessel model for nerve compression, *Journal of Neurological Sciences* **24**, 299–304.

Marimont, R., 1965, Numerical studies of the Fuortes–Hodgkin *Limulus* model, *Journal of Physiology* **179**, 489–497.

Marr, D., 1969, A theory of cerebellar cortex, *Journal of Physiology* **202**, 437–470.

Marr, D. O., 1970, A theory for cerebral neocortex, *Proceedings of the Royal Society* (*London*), *Series B* **176**, 161–234.

Martin, A. R., 1966, Quantal nature of synaptic transmission, *Physiological Reviews* **46**, 51–66.

Matsuyama, Y., Shirai, K., and Akizuki, K., 1974, On some properties of stochastic information processes in neurons and neuron populations, *Kybernetik* **15**, 127–145.

Matumoto, M., 1955, A new type of nerve conduction model, *Gunma Journal of Medical Sciences* **4**, 37–40.

Mauro, A., 1961, Anomalous impedance, a phenomenological property of time-variant resistance, *Biophysical Journal* **1**, 353–372.

Mayeri, E., 1969, Integration in the lobster cardiac ganglion, Ph.D. Thesis, University of California, Berkeley.

Mayeri, E., 1973*a*, Functional organization of the cardiac ganglion of the lobster, *Homarus americanus*, *Journal of General Physiology* **62**, 448–472.

Mayeri, E., 1973*b*, A relaxation oscillator description of the burst-generating mechanism in the cardiac ganglion of the lobster, *Homarus americanus*, *Journal of General Physiology* **62**, 473–488.

Maynard, D. M., 1955, Activity in a crustacean ganglion. II. Pattern and interaction in burst formation, *Biological Bulletin* (*Woods Hole, Massachussetts*) **109**, 420.

Melzack, R., 1973, *The Puzzle of Pain*, Basic Books, New York.

Møller, A. R., ed., 1973, *Basic Mechanisms in Hearing*, Academic Press, New York.

Monnier, A., 1934, *L'Excitation Électrique des Tissus*, Hermann, Paris.

Moore, J. W., 1971, Voltage clamp methods, *Biophysics and Physiology of Excitable Membranes*, W. J. Adelman, ed., Van Nostrand Reinhold, New York, pp. 143–167.

Moore, G. P., Perkel, D. H., and Segundo, J. P., 1966, Statistical analysis and functional interpretation of neuronal spike data, *Annual Review of Physiology* **28**, 493–522.

Moore, G. P., Segundo, J. P., and Perkel, D. H., 1970, Statistical signs of synaptic interaction in neurons, *Biophysical Journal* **10** (9), 876–900.

Morishita, I., and Yajima, A., 1972, Analysis and simulation of networks of mutually inhibiting neurons, *Kybernetik* **11**, 154–165.

Mountcastle, V. B., 1953, Response properties of neurons of cat somatic cortex to peripheral stimuli, *Journal of Neurophysiology* **20**, 374–407.

Mousson, A., 1843, Versuch einer Erklärung des Verhaltens der Salpetersäure zu den Oxydierbaren Metallen, *Poggendorf Annalen* **39**, 330–342.

Murray, M. J., and Lewis, E. R., 1964, Sensory control of prey capture in *Navanax inermis*, *Veliger* **17**, 156–158.

Nachmansohn, D., and Neumann, E., 1976, *Chemical and Molecular Basis of Nerve Activity*, Academic Press, New York.

Nakahama, H., Ishii, N., and Yamamoto, M., 1972, Markov process of maintained impulse activity in central single neurons, *Kybernetik* **11**, 61–72.

Nakahama, H., Ishii, N., Yamamoto, M., and Fujii, H., 1974, Statistical inference on Markov process of neuronal impulse sequences, *Kybernetik* **15**, 47–64.

Narahashi, T., 1971, Neurophysiological basis for drug action: Ionic mechanism, site of action and active form in nerve fibers, *Biophysics of Excitable Membranes*, W. J. Adelman, ed., Van Nostrand Reinhold Company, New York, pp. 423–462.

Nicholson, C., 1973, Theoretical analysis of field potentials in anisotropic ensembles of neuronal elements, *I.E.E.E. Transactions on Biomedical Engineering* **BME-20** (4), 278–288.

Noble, D., 1962, A modification of the Hodgkin–Huxley equations applicable to Purkinje fiber action and pace-maker potentials, *Journal of Physiology* **160**, 317–352.

Noble, D., 1966, Applications of Hodgkin–Huxley equations to excitable tissues, *Physiological Reviews* **46**, 1–50.

Noble, D., 1972, Conductance mechanisms in excitable cells, *Biomembranes, III: Passive Permeability of Cell Membranes*, F. Kreuger and J. F. G. Slegers, eds., Plenum Press, New York, pp. 427–447.

Noble, D., and Tsien, R. W., 1969, Outward membrane currents activated in the plateau range of potentials in cardiac Purkinje fibers, *Journal of Physiology* **200**, 205–231.

Noda, H., and Adey, W. R., 1970*a*, Firing variability in cat association cortex during sleep and wakefulness, *Brain Research* **18**, 513–526.

Noda, H., and Adey, W. R., 1970*b*, Neuronal firing in sleep and wakefulness, *Brain Research* **19**, 263–275.

Noda, H., and Adey, W. R., 1970*c*, Paired neuron discharges in cortex during sleep and wakefulness, *Journal of Neurophysiology* **33**, 672.

Noda, H., and Adey, W. R., 1972, Behavior of mesencephalic reticular formation neurons in sleep and wakefulness, *Experimental Neurology* **34**, 140–157.

Noda, H., and Adey, W. R., 1973, Neuronal activity in association cortex of cat during sleep, wakefulness, and anesthesia, *Brain Research* **54**, 243–259.

Noda, H., Monahar, S., and Adey, W. R., 1969, Correlated firing of hippocampal neuron pairs in sleep and wakefulness, *Experimental Neurology* **24**, 232–247.

Nunez, P. L., 1974*a*, The brain wave equation: A model for the EEG, *Mathematical Biosciences* **21**, 279–297.

Nunez, P. L., 1974*b*, Wavelike properties of the alpha rhythm, *I.E.E.E. Transactions on Biomedical Engineering* **BME-21** (6), 473–482.

Okuda, M., Yoshida, A., and Takahashi, K., 1974, A dynamical behaviour of active regions in randomly connected neural networks, *Journal of Theoretical Biology* **48**, 51–73.

Perkel, D. H., 1964*a*, A digital computer model of nerve-cell functioning, RM-4132-NIH, The Rand Corporation, Santa Monica, California.

Perkel, D. H., 1964*b*, Neurophysiological models: Methods and applications, RM-4247-NIH, The Rand Corporation, Santa Monica, California.

Perkel, D. H., 1965, Applications of a digital computer simulation of a neural network, *Biophysics and Cybernetic Systems*, M. Maxfield, A. Callahan, and L. J. Fogel, eds., Spartan Books, Washington, D.C.

Perkel, D. H., Schulman, J., Bullock, T., Moore, G., and Segundo, J., 1964, Pacemaker neurons: Effects of regularly spaced synaptic input, *Science* **145** (3627), 61–63.

Perkel, D. H., Gerstein, G. L., and Moore, G. P., 1967, Neuronal spike trains and stochastic point processes. I. The single spike train, *Biophysical Journal* **7** (4), 419–440.

Petsche, H., Stumpf, C., and Gogolak, G., 1962, The significance of the rabbit's septum as a relay station between the midbrain and the hippocampus. I. The control of hippocampus arousal activity by the septal cells, *Electro-encephalography and Clinical Neurophysiology* **14**, 202–211.

Pfaffmann, C., ed., 1969, *Olfaction and Taste*, Rockefeller University Press, New York.

Pinter, R. B., 1966, Sinusoidal and delta function responses of visual cells of the *Limulus* eye, *Journal of General Physiology* **49**, 565–593.

Pitts, W., and McCulloch, W. S., 1947, How we know universals. The perception of auditory and visual forms, *Bulletin of Mathematical Biophysics* **9**, 127–147.

Plant, R. E., and Kim, M., 1975, On the mechanism underlying bursting in the *Aplysia* abdominal ganglion R-15 cell, submitted to *Mathematical Biosciences*.

Plonsey, R., 1969, *Bioelectric Phenomena*, McGraw-Hill, New York.

Poggio, C. F., and Vernstein, L. J., 1964, Time-series analysis of impulse sequences of thalamic sensory neurons, *Journal of Neurophysiology* **27**, 517–545.

Posch, T. E., 1968, Models of the generation and processing of signals by nerve cells: A categorically indexed abridged bibliography, USCEE Report, Electronic Sciences Laboratory, University of Southern California, 65 pages.

Rall, W., 1957, Membrane time constant of motoneurons, *Science* **126**, 454.

Rall, W., 1960, Branching dendritic trees and motoneuron membrane resistivity, *Experimental Neurology* **2**, 503–532.

Rall, W., 1962*a*, Theory of physiological properties of dendrites, *Annals of the New York Academy of Sciences* **96**, 1071–1092.

Rall, W., 1962*b*, Electrophysiology of a dendritic neuron model, *Biophysical Journal* **2**, 145–167.

Rall, W., 1964, Theoretical significance of dendritic trees for neuronal input–output relations, *Neural Theory and Modeling*, R. F. Reiss, ed., Stanford University Press, Stanford, California.

Rall, W., 1967, Distinguishing theoretical synaptic potentials computed for different soma-dendritic distributions of synaptic input, *Journal of Neurophysiology* **30**, 1138–1168.

Rall, W., 1969, Time constants and electrotonic length of membrane cylinders and neurons, *Biophysical Journal* **9**, 1483–1508.

Rall, W., 1970, Cable properties of dendrites and effects of synaptic location, *Excitatory Synaptic Mechanisms*, P. Ander and J. K. S. Jansen, eds., Universitetsforlaget, Oslo, Norway.

Rall, W., and Rinzel, J., 1973, Branch input resistance and steady attenuation for input to one branch of a dendritic neuron model, *Biophysical Journal* **13**, 648–688.

Rashevsky, N., 1933, Outline of a physico-mathematical theory of excitation and inhibition, *Protoplasma* **20**, 42–56.

Ratliff, F. 1965, *Mach Bands*, Holden-Day, New York.

Redman, S. J., 1973, The attenuation of passively propagating dendritic potentials in a motoneurone cable model, *Journal of Physiology* **234**, 637–664.

Reichardt, W. E., 1970, The insect eye as a model for analysis of uptake, transduction, and processing of optical data in the nervous system, *The Neurosciences: Second Study Program*, F. O. Schmitt *et al.*, eds., Rockefeller University Press, New York, pp. 494–511.

Reiss, R. F., 1962, A theory and simulation of rhythmic behavior due to reciprocal inhibition in small nerve nets, *Joint Computer Conference* **21**, 171–194.

Reiss, R. F., ed., 1964, *Neural Theory and Modeling*, Stanford University Press, Stanford, California.

Ricciardi, L. M., and Esposito, F., 1966, On some distribution functions for nonlinear switching elements with finite dead time, *Kybernetik* **3**, 148–152.

Roberts, G. E., and Kaufman, H., 1966, *Table of Laplace Transforms*, Saunders, Philadelphia.

Rochester, N., Holland, J. H., Haibt, L. H., and Duda, W. L., 1956, Tests on a cell assembly theory of the action of the brain, using a large digital computer, *IRE Transactions on Information Theory* **IT-2**, 80–93.

Rodieck, R. W., 1967, Maintained activity of cat retinal ganglion cells, *Journal of Neurophysiology* **30**, 1043.

Rosenblueth, A., and Wiener, N., 1945, The role of models in science, *Philosophy of Science* **12**, 312–321.

Runge, R. G., Uemura, M., and Viglione, S. S., 1968, Electronic synthesis of the avian retina, *I.E.E.E. Transactions on Biomedical Engineering* **BME-3**, 138–151.

Sabah, N. H., and Spangler, R. A., 1970, Repetitive response of the Hodgkin–Huxley model for the squid giant axon, *Journal of Theoretical Biology* **29**, 155–172.

Schneider, D., 1970, Olfactory receptors for the sexual attractant (Bombykol) of the silk moth, *The Neurosciences: Second Study Program*, F. O. Schmitt *et al.*, eds., Rockefeller University Press, New York, pp. 511–518.

Schwartz, T. L., 1971, The thermodynamic foundations of membrane physiology, *Biophysics and Physiology of Excitable Membranes*, W. J. Adelman, ed., Van Nostrand Reinhold, New York, pp. 47–95.

Sensory Receptors, 1965, special issue, *Cold Spring Harbor Symposia on Quantitative Biology* **30**, 1–649.

Shannon, C., and Weaver, W., 1949, *The Mathematical Theory of Communication*, University of Illinois Press, Urbana, Illinois.

Sherrington, C., 1906, *The Integrative Activity of the Nervous System*, Yale University Press, New Haven, Connecticut.

Sherry, C. J., and Marczynski, T. J., 1973, Higher-order patterns of neuronal interspike intervals (tigrams and quadgrams) analyzed as self-adjusting sets of ratios, *International Journal of Neuroscience* **5**, 171–182.

Smith, D. R., and Davidson, C. H., 1969, Maintained activity in neural nets, *Journal of the Association for Computing Machinery* **9**, 268–279.

Spencer, W. A., and Kandel, E. R., 1969, in *Basic Mechanisms of the Epilepsis* (H. H. Jasper *et al.*, eds.), Little, Brown and Company, Boston, Massachusetts.

Stein, R. B., Leung, K. V., Mangeron, D., and Oguztoreli, M. N., 1974*a*, Improved neuronal models for studying neural networks, *Kybernetik* **15**, 1–9.

Stein, R. B., Leung, K. V., Oguztoreli, M. N., and Williams, D. W., 1974*b*, Properties of small neural networks, *Kybernetik* **14**, 223–230.

Stevens, C. F., 1968, Synaptic physiology, *Proceedings of the I.E.E.E.* **56**, 916–930.

Stewart, R. M., 1962, Theory of structurally homogeneous logic nets, *Biological Prototypes and Synthetic Systems*, H. F. Bernard and A. S. Kare, eds., Plenum Press, New York, p. 370.

Takeuchi, A., and Takeuchi, N., 1960, On the permeability of the endplate membrane during the action of transmitter, *Journal of Physiology* **154**, 52–67.

Tasaki, I., and Bak, A. F., 1959, Voltage clamp behavior of iron–nitric acid system as compared with that of nerve membrane, *Journal of General Physiology* **42**, 899–915.

Tasaki, I., and Spyropoulos, C. S., 1958, Nonuniform response in the squid axon membrane under voltage clamp, *American Journal of Physiology* **193**, 309–317.

Taylor, R. E., 1963, Cable theory, *Physical Techniques in Biological Research*, W. L. Nastuk, ed., Academic Press, New York.

Van der Pol, B., 1926, On relaxation oscillations, *Philosophical Magazine* **2**, 978.

Varjü, D. 1965, On the theory of lateral inhibition, *Cybernetics of Neural Processes*, E. R. Caianiello, ed., CNR, Rome.

Wallach, D. F. H., 1972, *The Plasma Membrane*, Springer-Verlag, New York.

Weber, H., 1873, Ueber die Stationären Strömungen der Elektricität in Cylindern, *Borchardt's Journal of Mathematics* **76**, 1–20.

Weinberg, A. M., 1941, Weber's theory of the Kernleiter, *Bulletin of Mathematical Biophysics* **3**, 39–55.

Werner, G., and Mountcastle, V. B., 1965, Neural activity in mechano-receptive cutaneous afferents—Stimulus-response relationships, *Journal of Neurophysiology* **28**, 359–397.

Westheimer, G., 1967, Spatial interaction in human cone vision, *Journal of Physiology* **190**, 139–154.

Wetzlar, G., 1827, Ueber die Reduction der Metalle Durcheinander auf Nassem Wege, *Schweiggers Journal* **50**, 88–109.

Wilson, D. M., and Waldron, I., 1968, Models for the generation of the motor output pattern in flying locusts, *Proceedings of the I.E.E.E.* **56**, 1058–1064.

Wilson, H. R., and Cowan, J. D., 1973, A mathematical theory of the functional dynamics of cortical and thalamic nervous tissue, *Kybernetik* **13**, 55–80.

Wilson, W. A., and Wachtel, H., 1974, Negative resistance characteristic essential for the maintenance of slow oscillations in bursting neurons, *Science* **186**, 932–934.

Winfree, A. T., 1967, Biological rhythms and the behavior of populations of coupled oscillators, *Journal of Theoretical Biology* **16**, 15–42.

Wittie, L. D., 1976, A digital computer system for large brain cortex models: Creation and simulation of anatomically correct models of regular neuronal structures such as the cerebellar cortex, submitted to *Kybernetik*.

Wong, R., and Harth, E., 1973, Stationary states and transients in neural populations, *Journal of Theoretical Biology* **40**, pp. 77–106.

Yamagiwa, K., 1951, The electro-saltatory transmission of nervous impulse (model experiments and theoretical considerations), *Japanese Journal of Physiology* **2**, 79–92.

Zeevi, Y. Y., 1972, Structural functional relationships in single neurons scanning electron microscopy and theoretical studies, Thesis, University of California, Berkely (Biophysics Group).

Index

The numbers in italics refer to text pages on which a definition appears.